The Microsoft Windows 2000 Professional Handbook

The Microsoft Windows 2000 Professional Handbook

Louis Columbus

CHARLES RIVER MEDIA, INC.
Hingham, Massachusetts

Publisher: David Pallai
Production: Publishers' Design and Production Services
Cover Design: The Printed Image
Printer: InterCity Press, Rockland, MA.

CHARLES RIVER MEDIA, INC.
20 Downer Avenue
Suite 3
Hingham, MA 02043
781-740-0400
781-740-8816 (FAX)
www.charlesriver.com

This book is printed on acid-free paper.

ISBN: 1-58450-009-3

Printed in the United States of America
00 01 7 6 5 4 3 2 First Edition

CHARLES RIVER MEDIA titles are available for site license or bulk purchase by
institutions, user groups, corporations, etc. For additional information,
please contact the Special Sales Department at 781-740-0400.

Requests for replacement of a defective CD must be accompanied by the original disc, your mailing
address, telephone number, date of purchase and purchase price. Please state the nature of the
problem, and send the information to CHARLES RIVER MEDIA, INC., 20 Downer Avenue, Suite 3,
Hingham, MA 02043. CRM's sole obligation to the purchaser is to replace the disc, based on defective
materials or faulty workmanship, but not on the operation or functionality of the product.

Table of Contents

1 Introducing Windows 2000 Professional

In 1992 a fundamental shift occurred in how operating systems were created. Instead of focusing on the internals of a computer system, then working outward to a series of user commands that at times were difficult to use and understand, Microsoft began developing Windows NT by listening to its customers and building this operating system from the outside going in. By putting the customers and their needs first, the design goals of Windows NT quickly followed. These design goals drove the development of an operating system that has proven to be bulletproof in the management of memory and seamless with regard to interoperability with other systems. Windows NT continues to evolve as an operating system due in large part to the willingness of the product design team to listen to what you, the customer, are interested in seeing. As you read this book you'll find examples of how Windows NT has matured as development has continually focused on how to meet and exceed the expectations of customers interested in an enterprise-level operating system.

What is Windows 2000 Professional? And what about the versions of Windows 2000 Server? How do these operating systems complement one another? These questions and more are answered in this chapter. Specifically, the role of compatibility in the Windows NT product strategy is also explored, as well as the enhancements made in both Windows NT Workstation and Windows 2000 Server. There are also key points on how to navigate the Windows NT desktop and a case study examining how one company is planning to migrate from Windows NT Workstation 4.0 and NetWare to Windows 2000.

One of the most significant developments in the Windows 2000 Professional and Server is the integration of Plug-and-Play compatibility. While many industry observers originally thought that this functionality would be integrated into Windows NT by removing the Hardware Abstraction Layer, Microsoft has, in fact, integrated a subsystem into the Windows NT kernel to accomplish Plug-and-Play compatibility. This chapter also discusses Windows 2000's improved networking capabilities and compares Windows 2000 to UNIX. You'll also be interested in Windows 2000's ability to coexist with UNIX and the transition path for moving from UNIX to Windows 2000.

WHAT IS WINDOWS 2000 PROFESSIONAL?

Now in its third generation, the Windows NT operating system continues to evolve. An operating system is actually the "wake-up call" for a computer system in that its role is to coordinate all the activities in a computer system. The focus on operating system design has changed dramatically in the last 15 years, from designers focusing exclusively on the needs of a hardware platform to streamlining code to maximize system performance. For DOS designers needed to provide the core features required to make a computer work. The lesson learned in the DOS years continues to influence the development of compatibility-oriented features in Windows NT, among other operating systems being developed today.

From designers focusing on the needs of the hardware to product managers and design leads turning to focus on the needs of the corporate customer operating systems are now evolving into highly manageable, strong foundations for information systems. The transition from hardware-centric to customer-centric is exemplified in the new features profiled in this chapter. The design and creation of an operating system is now more complex than ever because the design objectives are more challenging—they are focused outside to the hardware, on the customer.

Windows NT is actually a multitasking operating system that uses pre-emptive multitasking memory management to ensure reliability and consistency of performance across applications. Windows NT is truly a 32-bit, multitasking, preemptive operating system that has as its networking foundation TCP/IP. The goal of making Windows NT more accessible than previous operating systems was accomplished by using a Windows-like graphical user interface for the desktop.

Let's now tour the new features of Windows 2000 Professional. Keep in mind that if you're an experienced user of Windows NT 3.51 and 4.0, you'll find Windows 2000 easy to navigate. Keep in mind as well that many of the changes to NT 4.0 were apparent; the changes in Windows 2000 were designed to streamline connectivity and compatibility with other operating systems located throughout a network. Envisioning Windows NT as one of many operating systems in a networked environment, Microsoft focused on interoperability and making the architecture of Windows NT match the needs of those users who have adopted other networking operating systems such as Novell NetWare, UNIX, SCO XENIX, or any other mainframe-based operating system.

New features in Windows 2000 Professional include the following:

On-now, advanced Configuration and Power Interface (ACPI).

Support for the Plug-and-Play IEEE 1394 standard previously supported in Windows95. Many industry analysts and members of the press believed that Microsoft would do away with the Hardware Abstraction Layer (HAL for short) that provides cross-platform compatibility, ensuring that NT can run on the Intel, Motorola PowerPC (up to NT 4.0), MIPS, and other processors. Instead of deleting the HAL, Microsoft chose to fully integrate the IEEE 1394 subsystem into the modular structure of the operating system itself. This provided for a more robust Plug-and-Play solution than has ever been possible before due to the complementary role dedicated memory partitions play in the NT architecture. This book will discuss in detail the Plug-and-Play features customizable through Windows 2000 Professional. Figure 1.1 shows an example of the Plug-and-Play subsystem that has been integrated into Windows 2000 Professional and Server's kernel architecture.

Power management for laptops. You'll recall that in Windows NT Workstation 4.0, power management was lacking, especially as more and more laptops were shipped for use as corporate desktops that could also travel. Toshiba estimated docking station sales at just over 45 percent of its total laptop sales—which meant there was a strong market need for laptops that could be configured for use on corporate desktops and for travel with their owners. Up until Windows 2000 Professional, loading a laptop with NT was an interesting proposition: Only those laptops configured with utilities that allowed for power management at the BIOS level could be counted on for solid performance and reliable feedback to their users. Microsoft Windows 2000 has made great strides in the area of power management. Chapter 2,

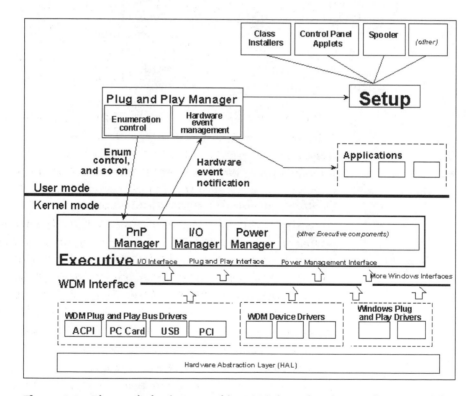

Figure 1.1 Plug-and-play integrated into NT's kernel makes application stability possible.

"Touring the Active Desktop," covers how to use the Power Management applet found in the Control Panel.

Accelerated Graphics Port (AGP) and multimonitor support for graphics adapters. With the increasing trend of system manufacturers to differentiate themselves with graphics displays, the AGP standard promises to deliver video performance previously seen only in high-resolution display market segments, which include CAD/CAM, medical imaging, and graphic arts.

Support for the Intelligent I/O (I^2O) architecture. This focuses on specialized board support for storage cards and extensive support for redundant arrays of inexpensive disk (RAID) cards. Support for the Infrared (IR) interface has been augmented in Windows 2000. Using the Network Driver Interface Specification (NDIS), 4.0 is used as the basis for an expanded support for the IR interface. Microsoft has developed the IR Link Access Protocol (IrLAP), IR Link Management Protocol (IrLMP), TinyTP, and the IAS subset as Windows NT protocols. IrDA is exposed through Windows Socket 23.0 (WinSock) using the transport driver interface (TDI). Printing through the IR interface is handled by the IrLMP protocol, which is exposed through WinSock.

Intellimirror management capabilities. These capabilities synchronize the data on local clients with servers.

Task Scheduler. One of the key advantages UNIX had over NT 3.51 and 4.0 was its ability to schedule batch-oriented tasks for completion at a later time. This is a strong feature in UNIX, where many applications, which are client/server-based, schedule their communications to other systems at a time when network traffic is low. Task Scheduler is found in My Computer, which is located on the Active Desktop. Task Scheduler can be a very useful tool for you as a system administrator. You can use this tool to queue up tasks to occur during the evening or weekends when you are out of the office. The Task Scheduler uses a wizard to configure tasks. Once a task has been defined using this process, Task Scheduler saves the tasks and in effect creates a queue of them. This is very useful for completing recurring tasks.

Computer Management. This is a snap-in software program used in the Microsoft Management Console. Included in Windows 2000 Professional and Server, the Computer Management snap-in is used for configuring both local and remote systems.

An entire new set of Administrative Tools. In addition to the Windows NT 4.0 Administrative Tools of Backup, Performance Monitor, Remote Access Services, and Windows NT Diagnostics, there are now new tools never before released in NT. These are under the Administrative Tools selection in the Start menu.

Enhanced printing support. Major improvements have been made in Internet printing, a simplified user interface, Active Directory integration, and image color management APIs (Application Programmer Interfaces). Chapter 5, "Learning to Use Printers in Windows 2000 Professional," discusses printing in detail. To summarize, the major enhancements to Windows 2000 printing include printer sharing over the Internet or an intranet and the ability to install HTTP printers using point-to-point protocol, install printer drivers from a URL, and even change printer status using HTML-based forms. User interface changes include a Web view of the Printers folder and the queues on each printer, plus an enhanced Add Printer wizard that allows users to search for printers in the directory as well as browse a network. There are also more expanded printer properties associated with printers. Last, the Image Color Management series of APIs provides for a mechanism to send high-quality color documents from a Windows 2000 Workstation to your printer or to another system more efficiently and with greater image quality.

EXPLORING THE WINDOWS 2000 ARCHITECTURE

When you hear the word "architecture" you no doubt think of the skylines of your favorite city or a familiar landmark. In the world of operating systems, the architecture refers to the way in which the operating system's components function and interact. Windows NT has an architecture based on its need to ensure system reliability and compatibility across a variety of applications, and its need to communicate with a wide variety of other operating systems at the client or even the server level. Windows 2000's architecture is significantly different from that of MS-DOS, Windows 3.1, or even Windows95.

At the root of its differences from other network operating systems is the modularity of the kernel within Windows NT. Starting with the first release of Windows NT in 1992, Microsoft has continually pursued a modular approach to the operating system to ensure extensibility of key features including support for subsystems that ensured compatibility with a variety of application types. From the beginning, there has also been support for Windows 16-bit-based applications, Win32-based, POSIX and OS/2 –character-based application.

Microsoft has also worked to extend the functionality of the kernel by working with companies that have subsystems that would complement the overall breadth of support NT could provide. Examples of companies that continue to work with Microsoft to extend the functionality delivered through the operating systems' kernel include Softway Systems. Figure 1.2 shows an example of the enhancements made by Softway Systems to the Windows 2000 kernel to ensure compatibility for specific government markets.

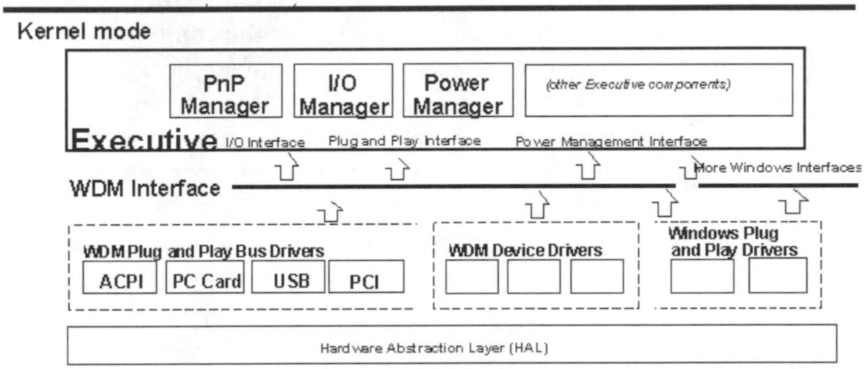

Figure 1.2 The Windows 2000 kernel is modular in structure and modifiable through the use of APIs.

WHAT IS WINDOWS 2000 SERVER?

There are four versions of Windows 2000 with Workstation and Server being the most pervasively used. Both versions share the same architecture, have the same networking features built in, and have the same Plug-and-Play capabilities, which are briefly described in this chapter. Both have the same level of functionality, with NT Server having additional tools for managing an entire group or enterprise of systems. NT Server's role in a network is specifically to add value as a communications gateway in the case of two or more diverse systems that need to be connected to one another. Windows 2000 Server also includes tools for making the task of managing a diverse network environment easier. With the inclusion of Active Directory in Windows 2000 Server, the task of managing interdomain trusts—and establishing, then maintaining intertrust relationships throughout a domain—is no longer required. Active Directory makes the task of configuring and viewing servers on multiple LANs easier than it has been in previous versions of NT Server. With Windows 2000 Server there are also enhancements to the Remote Access Services, which provide dial-in capabilities for those companies or universities with a highly mobile population so that they can stay in contact with their fellow team members in another location.

Let's take a closer look at the new features of Windows 2000 Server:

Active Directory. This is a new directory service that stores information about all the objects on the computer network and makes this information accessible in a standard format to administrators. Active Directory is a significant new development in the Windows NT product line because it alleviates the need to create domains and to manage and maintain interdomain trust relationships; both of these tasks consume the administrator's time. Active Directory provides the functionality of domains at a more global level. Additional features of Active Directory include a single network logon for anyone who is a member of the Active Directory; a single point of administration for all network objects or resources; full query capability on any attribute of any object; and a directory structure that is fully replicable, partitionable, hierarchical in appearance, and extensible in structure.

Microsoft Directory Service Migration Tool. Novell's focused product strategy of being one of the best file and print services-based network operating systems has served the growth of that company well. Within the arena of application servers, both Microsoft and Novell continue to compete for customers. Many market research companies report that many of the customers once loyal to Novell are migrating to Windows NT at a rate higher than ever seen before. One research company, International Data Corporation, has said that one in every three Novell customers in a given study had either considered moving to Windows NT or had already made the move. Given this market dynamic, providing Directory Service Migration Tool in Windows 2000 Server is a prudent product management decision.

This tool gives Novell users the opportunity to first model the architectural differences between their Novell servers and Windows NT's Active Directory. Users can model NetWare servers in an Active Directory environment. Users can define NetWare user and group properties for both binderies of existing network interface card resources and connections, and they can define properties for migrating the NetWare Directory Services (NDS) directories to Active Directory. There is also a utility that can be used for migrating files through an export Active Directory option. The graphical interfaces used for accomplishing these tasks are navigated through a series of dialog boxes that ensure the user is specifically interested in testing his or her migration from NDS to Active Directory before completing a full migration.

Enhanced Networking Features in Windows 2000 Server

One of the most competitive areas in operating systems development is the integration of networking functionality into the core product or system. While UNIX was the first operating system to have what many call "native IP" capability due to networking being built into the core of that operating system, today both Microsoft and Novell are proactive in their efforts to position their capabilities relative to one another. This is evident in the additions to networking functionality found in Windows 2000 Server. Presented in this section are the five major networking features added to Windows 2000 Server in response to customers' evolving demands and requirements for increasing the connectivity of their enterprises using Windows 2000 Server.

Distributed File System. The Distributed File System (Dfs) debuts for the first time in Windows 2000 Server, and it implements a single namespace for file system resources located at a site. These file system resources can have a multitude of namespace variations—in short, this feature provides for an organized and hierarchical structure of logical volumes, independent of a given disk's physical location. The Dfs is then actually a "logical mapping" of physical drives that makes the process of making resources available easier than it has been before. To users on a network, a Dfs volume provides a unified and transparent access to network resources. In addition to all the functionality found in Dfs v4.1 (now shipping on NT Server 4.0), the Windows 2000 Server release provides for fault-tolerant Dfs roots, support for DNS naming, integration of Dfs with Active Directory, and support for nested junctions.

Dynamic DNS. The latest implementation of DNS for Windows 2000 Server supports dynamic DNS name server capabilities. This feature is compliant with open and approved Internet standards for DNS. How does Dynamic DNS work? Updates to distributed DNS records data are made and replicated automatically to all affected DNS name servers throughout a network. This is particularly useful for those administrators who manage multiple DNS gateways and need to make sure the correct files are present on the server for naming recognition. In addition, Microsoft DNS Server (included in Windows 2000 Server) integrates the capability to pass dynamic updates through DNS with other network services such as Active Directory Services, Dynamic Host Configuration Protocol (DHCP), and the Windows Internet Naming Service (WINS).

IP Security Management. With the increasing focus on the need for security on both Internet and general IP-based network traffic, there is continuing focus on the part of all operating system vendors on this issue. Microsoft's approach is embodied in the Microsoft IP Security Management feature, seen for the first time to this extent in Windows 2000 Server. What is included in this Security Management capability? First, a series of safeguards ensure that sensitive data sent across a network is protected from unauthorized access. Once an administrator has implemented IP security using IP Security Management, communications are secured transparent to the end user. IP Security Management is completely configurable by the administrator; no interaction is required from the user. IP security also protects communication between hosts using any protocol in the TCP/IP command set. Finally, IP Security Management eliminates the need for separate packages for each protocol.

Admission Control Services. These services provide for management of subnetwork resources by preventing applications from consuming more traffic than a subnet can handle. Applications that implement Quality of Service (QoS) standards can reserve bandwidth and establish priority for transmission of data. QoS is a set of service requirements that the network must meet while transmitting a flow or stream of data. QoS-based services and protocols provide an end-to-end express delivery system for Internet Protocol (IP) traffic. The Admission Control Service (ACS) enables QoS-aware applications to reserve bandwidth and establish priority for transmission of critical data without overcommitting network resources.

Multiprotocol routing. Since version 3.5 Windows NT has supported Remote Access Services, which provide for dial-in connection to an NT server from a remote client. Windows 2000 Server 5.0 for the first time now includes Routing and Remote Access Service (RAS) Admin, a utility that makes it possible to route connections over IP and IPX networks on LANs and WANs.

Internet Integration Is New in Windows 2000 Server

Both intranet- and Internet-based tools and features within operating systems are increasingly being provided in response to the needs of customers who are interested in leveraging the many communications- and commerce-oriented aspects of Intranets and the Internet. What's new in Windows 2000 Server in this area?

Microsoft Internet Information Server. Microsoft includes Microsoft Internet Information Server (IIS) version 4.0 with Windows 2000 Server. IIS is a comprehensive application used for creating and managing intranet and Internet Web sites.

User-defined Internet Special Groups. This feature gives the administrator the option to apply and grant access to resources at the group level, such as printers and file servers. This feature is designed for administrators who have responsibility for supporting a large population of networked computer users who are also sharing information on an intranet. Security Enhancements to Windows 2000 Server

Making sure you and your audience are viewing only the messages you send and receive is one of the biggest FUD (Fear, Uncertainty, and Doubt) factors that people think of when they consider large-scale migration from a network to an intranet or even the Internet. Security of data and transmission of messages are key concerns for many companies and with good reason—as the structure and strength of tools have grown, so has the demand for computer security consultants who can safeguard data warehouses that hold information worth millions of dollars. Where once the machinery, buildings, and physical assets made a company what it was in terms of net worth, today information and the knowledge of market dynamics and customer behavior are the true assets. What does Windows 2000 Server have to alleviate these issues? Here are the new features that ensure greater security when using Windows 2000 Server:

Kerberos, public key, and DPA support. In creating enhanced security for Windows 2000 Server, the development team has continually focused on the integration of security on Active Directory with security features. Specifically, Windows 2000 Server now has a scalable policy for account management for domains, the ability to selectively delegate security administration across domains. As has been mentioned in this chapter, Active Directory has the ability to integrate resources seamlessly across domains. This ability to create transitive domain trusts and authentication using the Kerberos approach provides users with a single network logon, regardless of where they physically log on to a network. Due to the Kerberos authentication logic, a single account can have multiple security credentials, depending on the server and applications being used. Public key authentication through SSL/TLS connections is also supported through domain accounts.

Security Configuration Editor. This tool is a one-stop security configuration and analysis tool for Windows NT. This tool allows configuration of various security-sensitive registry settings, access controls on files and registry keys, and security configuration of system services. All versions of NT Server 5.0 from the first beta have had this tool included for review. Graphically it is very similar to User Manager for Domains in NT Server 4.0.

New Administration Tools for Windows 2000 Server

With the fourth generation of NT Server, Microsoft has steadily been listening to its customers—many of NT Server's customers are administrators, managers of networks, directors, even CIOs who are responsible for keeping an entire enterprise up and running. The latest versions to Windows 2000 Server include the Microsoft Management Console, which is also found in Windows 2000 Workstation.

Microsoft Management Console. This feature provides for a graphical overview of the resources on a network, including support for snap-ins or applets that administer network functions. The MMC actually acts as a base component that provides snap-ins to administer a network. The different snap-ins in a console are organized in a tree structure and provide all the tools and information an administrator may need to complete a task. The MMC specifically has a series of separate windows in the console that can display different views of the console as tasks are completed. The MMC makes it possible for a system administrator to include Web pages on a console, so others can get the status of a network as it is performing daily tasks.

Application installation management with Microsoft Installer Technology. This feature is a transaction-based application installer for managing not only the installation process but also the history of applications installed over time. Microsoft Installer Technology plays a key role in the Zero Administration Windows (ZAW) initiative by providing benefits for standardizing the installation and application process.

Group Policy Editor. A snap-in for the MMC, Group Policy Editor is responsible for managing the settings for Group Policy as it is applied to a given site, domain, or organizational unit. This editor also acts an anchor point for third-party applications to build snap-in extensions.

New Storage Management Features in Windows 2000 Server

Microsoft is in the process of turning all of its businesses into ones which can leverage the Internet. The storage strategy is one in particular where the growth of Storage Area Networks is being planned for today with storage management features in Windows 2000.

Disk Management. The Disk Management MMC snap-in is a graphical tool for managing disk storage. It replaces the Disk Administrator and offers new capabilities including online disk management and simplified disk management commands through a graphical user interface. This plug-in is provided with the first releases of Windows 2000 Server.

Windows NT Backup. Unlike other versions of NT Server, Windows 2000 Server now is media-centric instead of tape-centric. Using the Windows 2000 Server version of Backup, you can copy data to a variety of magnetic and optical storage devices as well as tape. Management tasks such as mounting and dismounting media or drive functions are now done by a utility called Windows NT Media Services (NTMS). NTMS presents a common interface to media libraries and robotic changes, and it also enables multiple applications to share local libraries and tape or disk drives and to control removable media within a single-server system. The NTMS snap-in gives administrators the ability to add NTMS objects, view and modify properties of NTMS objects, inject and eject media, perform inventories on media, mount and dismount media, and check status information.

WHAT'S NEW IN WINDOWS 2000 PROFESSIONAL?

With the significant number of additions to Windows 2000 Professional for ease of manageability and administration, there is a definite bias in the latest generation of the NT operating system toward ease of integration with heterogeneous environments, including integration with the Internet. New features included in Windows 2000 Professional are profiled here. Where Windows 2000 Server includes many administration-level features, Windows 2000 Professional is built to empower individual users to the maximum of their computing needs. Let's take a look at the new features of Windows 2000 Professional.

Seamless Internet Access in Windows 2000 Professional

From the now familiar Internet Explorer icon on the Desktop to the new Active Desktop, resources on the Internet or on an intranet are just a click or URL away. The integration of HTTP and ActiveX technology with the Windows 2000 Professional Desktop is as thorough as any company building an intranet site for hundreds of employees or any group of students needing access to the Internet for research from their dorm rooms would need.

> **Introducing the Active Desktop.** Remember that in previous versions of Windows NT Workstation, Server, and even in Windows95 and Windows98, the option of configuring the Desktop was provided using a series of graphics files, all static. Windows 2000 Professional has changed all that. Now you can configure the Desktop of your workstation for your favorite Internet or intranet site—using the same series of dialog boxes used before for creating static displays. Steps to configure the Active Desktop to support a URL as wallpaper are provided in later sections of this chapter.
>
> **Offline web browsing.** In conjunction with Internet Explorer 4.0 the latest release of Windows NT includes the option of queuing up a series of Web sites to be viewed and then quickly getting to each location with minimal loss of time.

Ergonomic Desktop Is New in Windows 2000 Professional

Looking to their customers for the direction of the Desktop's appearance and navigation, Microsoft extensively tests the usability of their new products, especially their newest operating system, Windows 2000. Many of the features explained in this section are the result of ongoing ergonomics research where the ease of use and operability aspects of features are tested and refined.

Quick Launch Toolbar. In streamlining the ways users work with Windows 2000 Professional, Microsoft learned from speaking with customers that adding in access points to the Toolbar would increase productivity. The result of ongoing customer conversations was the development of the Quick Launch Toolbar and the option of creating a customized Toolbar. This combination makes possible a single button click to any of the profiled applications.

Control Panel contents simplified. Applets included in the Control Panel have had their feature sets increased, making each capable of completing many administrative and user preference-oriented tasks on their own. In addition, the Plug-and-Play wizard is also located in the Control Panel.

Increased use of wizards. Quite a few more wizards are used to streamline repetitive tasks. The wizards for configuring Plug-and-Play peripherals and the one for queuing up tasks are also welcome tools for those accustomed to using UNIX for their batch-oriented requests.

New Communication Tools in Windows 2000 Professional

With more than 60 percent of all personal computers networked and a higher percentage of Windows NT Workstations networked, the need for streamlining communications in the latest release of NT has been given priority. Presented here are the new communications features in Windows 2000 Professional:

Internet Explorer 4.0. This browser includes support for sending e-mail, entering and participating in chat rooms, and viewing and participating in newsgroups.

Outlook Express client. This is the full Internet standards-based e-mail client that is included with standard or full installation of Internet Explorer 4.0. Outlook Express provides flexibility for sending either HTML or standard ASCII messages, or both.

Microsoft Personal Fax for Windows NT Version 5.0. This provides users with the ability to send, receive, monitor, and administer faxes directly from the desktop. Several utilities are available from the Start menu in Windows 2000 Professional. These include a Cover Page Editor, Fax Document Viewer, Fax Send Utility, Fax Configuration (in the Control Panel), and Online Help.

Microsoft Telephony API (TAPI) support. This provides both PSTN telephony and telephony over IP networks. IP telephony enables voice, data, and video transmission over LANs, WANs, and the Internet. Support for major telephony service providers is provided in Windows 2000 Professional, providing the translation between hardware and software to enable multimedia computers to act as telephony devices. Supported service providers include Microsoft H.323 TAPI Service Provider and Microsoft IP Conference Service Provider.

Windows ATM services. Increasing the level of support for hardware and software that uses the Asynchronous Transfer Mode (ATM), Windows 2000 Professional includes new user and kernel-mode APIs to enable applications and drivers to create and manage ATM virtual circuits (VCs). These APIs can be used to determine Quality of Service (QoS) and stream multiple information types (data, voice, video) through each VC. There is also support for an ATM LAN Emulation client module for enabling existing applications and protocols that use Ethernet or Token Ring to run over an ATM network. There is also support for TCP/IP over ATM.

Searching Tools in Windows 2000 Professional

In staying consistent with its focus on the Internet, Windows 2000 Professional offers enhanced searching capabilities from the Active Desktop. Here are the new search tools:

Textual Find File support. Built-in indexing support is accessible from the Find command accessible from the Start button.

Search button on Internet Explorer 5.0. This button provides for the first time the option of being able to see both the search engine configured for use with IE 5.0 and the results. In previous versions of IE 4.0 you would need to press the Back button to see the search engine's syntax; today there are windows for both the search engine and the results.

New Tools in Windows 2000 Professional

Consolidating and grouping the most commonly used tools into new ones, the product management and development teams who work on Windows 2000 think about how to make functions the everday user would need more accessible and valuable. Many of the tools described in this section are from previous versions of Windows NT, while others are for the first time appearing in a Microsoft operating system.

OnNow power management for laptops. This feature is specifically aimed at laptops or even desktops where the user wants to have the workstation go into a state of "hibernation." After a period of computer inactivity, OnNow puts the workstation on an off-but-ready mode that in effect makes the workstation look as if it has been turned off. The OnNow API prevents systems from drive wear and reduces drive noise. This support is compatible with ASCPI-based workstations and laptops.

Increased focus on Accessibility applet. Included in Windows 2000 Professional are three new accessibility tools. The Accessibility Settings wizard helps you adapt Windows options to the needs and preferences of those users wishing to customize their system for easier access. The Microsoft Magnifier enlarges a portion of the screen for easier viewing. The Microsoft Screen Reader uses text-to-speech to read the contents of the screen.

Introduction of the Windows Scripting Host. This is a language-independent scripting host for 32-bit Windows platforms that includes Visual Basic Scripting Edition (VBScript) and Jscript scripting engines.

Task Scheduler. This feature provides a graphically based tool for scheduling tasks and applications. This interface is the same on Windows95 and Windows NT and Windows 2000, with the exception of added security features in Windows NT. The user interface is fully integrated into the operating system and is accessible from the My Computer icon. This new service replaces the System Agent that was included in the Windows Plus! Pack on Windows95 and the AT service on Windows NT and now Windows 2000. It also offers a Component Object Model (COM) programming interface for developers.

Computer Management snap-in. Integrated into the Microsoft Management Console (MMC), the Computer Management snap-in is an administrator's primary computer configuration tool. It is designed to work with a single computer, and all of its features can be used from a remote computer, allowing an administrator to troubleshoot and configure a computer from any other computer on the same network.

Device Manager snap-in. This resides in the MMC and provides for the configuring of devices on your workstation.

Hardware wizard. This wizard is a well-done series of screens that make it possible to configure Plug-and-Play devices to your workstation. It allows you to add, remove, repair, upgrade, and customize hardware. Throughout this book this wizard will be extensively explored.

New Storage, Security, and Windows9X Compatibility Features

When asked what their primary criteria are for new operating systems, the leading companies in the world cite platform and application compatibility. The need for having migration paths available is critical to having an operating system which can span multiple needs within a company.

Windows95 upgrade support. Windows 2000 Professional includes an Upgrade wizard for seamlessly moving from Windows95 to Windows 2000 Professional.

NTBackup. User interface enhancements include Backup and Restore wizards, along with a Windows Explorer-like look and feel. NTBackup is compatible with all file systems and includes support for the Windows NT Media Services (NTMS).

FAT32 support. Support for this file system originally appeared first in Windows95 OSR2 and later. Support for this file system gives Windows 2000 Professional users the flexibility of using FAT-based file systems to their fullest performance extent. FAT32 is typically used because it supports disk volumes larger than 2 GB.

NTFS enhancements. This new version of NTFS offers many performance enhancements and a host of new features, including per-disk quotas, file encryption, distributed link tracking, and the ability to add disk space to an NTFS volume without rebooting.

UDF. Support for the first time is included for the Universal Disk Format, a new file system designed for interchanging data on DVD and CD media.

Encrypting File System. Using Public Key technology, this file system provides transparent on-disk file encryption for NTFS files.

Disk quotas. For the first time administrators can set disk quotas and monitor and limit disk space use.

Disk defragmentation utility. A utility is now provided for defragmenting FAT, FAT32, and NTFS disk volumes.

Enhanced Printing in Windows 2000 Professional

The continual refinement of printing in Windows 2000 is the culmination of years of commitment to strenghtening Windows printing technologies and making the actual printing process more flexible yet also more complete in terms of features. One area of improvement is the formatting of data being printed from the Internet.

Advances in Internet printing. Now users will be able to gain access to networked printers and send a URL or Web site location to a networked printer, browse printer status using HTML to change printer or job characteristics, install HTTP printer for use as a point and print function, and install drivers from a URL.

Simplified Print Manager interface. Support for an HTML view of printers includes Web views of the Printers folder for each printer. There is also an enhanced Add Printer wizard that gives users the flexibility to users to search for printers in the in addition to browsing the network.

Active Directory integration. Windows 2000 Professional makes all shared printers in a domain available in the directory. Publishing printers in the Active Directory allows users to quickly locate the most convenient printing resources.

Image Color Management 2.0 API. Using Image Color Management 2.0 you can send high-quality color documents from your computer to your printer or to another computer faster, more easily, and with greater consistency than ever before.

Hardware Innovations in Windows 2000 Professional

Think of Windows95 as the proving ground where Microsoft learned how to get Plug and Play right. The implementation in Windows 2000 has the reliability Microsoft must have been hoping for in Windows95. After having used both extensively, it's very clear that Windows 2000 is architected and therefore performs at a much higher level than the Windows95 counterpart. The features here also show the progression of Windows NT from an operating system that had a high-end clientele to one suitable for many mainstream tasks.

Plug and Play. Windows NT Workstation and Server now supports Plug and Play, making it easy to install and troubleshoot new hardware. Plug and Play support includes a new Hardware wizard, the Device Manager, and improved support for laptops.

USB support. This feature automatically detects the new device and installs the appropriate device driver.

Win32 Driver Model. This is designed to provide a common architecture of I/O services and binary-compatible device drivers for both Windows and Windows NT operating systems. Included in the Win32 Driver Model is the IEEE1394 standard for cameras and storage, plus support for still image and scanner support for parallel, SCSI, and USB interfaces. Input support includes DirectX, HID layer, and support for joysticks.

Additional support from Windows 2000 Professional hardware includes the following:

Graphics and multimedia support.

I^{2}O Fibre Channel and Smart Card support.

Large memory support (64-bit) for Alpha reduced instruction set computing-based (RISC-based) systems.

Media changer support for CD-ROM, tape, and optical disc changers.

PnP and Power Management through ACPI. ACPI is a specification that defines a new interface to the system board that enables the operating system to implement operating system-directed power management and system configuration.

Multiple joystick model support, with WDM support for HID-class devices. HID (Human Interface Devices) is a uniform way to access input devices.

CUSTOMIZING THE WINDOWS 2000 ACTIVE DESKTOP

If you have ever used Windows95 or Windows NT 4.0 you will find your way around Windows 2000 Professional quickly. The actual appearance of Windows 2000 Professional and Server is very similar to both Windows95 and NT 4.0. You'll find the same taskbar and Start button and many of the same applets.

One of the key differences is the role of the Active Desktop, in that it can be customized to a high degree with Internet content. You can have the wallpaper on your Windows 2000 Professional be your favorite Web site. How do you configure the Active Desktop? Let's take a look at a quick series of steps for getting this done.

1. From the main Desktop of your Windows 2000 Professional, right-click once on the Desktop. Notice the series of selections in the abbreviated menu. The first selection is for the Active Desktop. Figure 1.3 shows the abbreviated menu with the Active Desktop entries shown.
2. You can see from Figure 1.3 that a series of selections pertain to the Active Desktop. The first option, View Desktop as A Web Page,

FIGURE 1.3   Selecting the Active Desktop entries from the Shortcut Menu on the Desktop.

needs to be toggled for the Web Properties as defined in the Display Properties dialog box.

3. Right-click on the Desktop again, and select Active Desktop. From the Active Desktop submenu, select Customize My Desktop…. The Display Properties dialog box then appears, as shown in Figure 1.4.

4. Click once on the Properties page titled Web. For the first time, Windows 2000 Workstation provides for the integration of Web sites as the background to your Desktop. Using the options in this dialog box you can also get a Web page enabled at any time. You'll

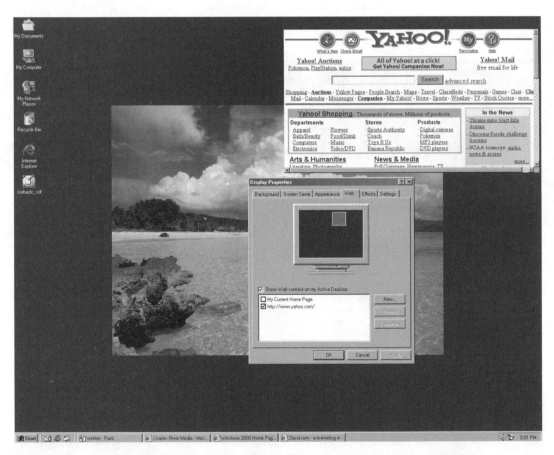

FIGURE 1.4 Using the Display Properties dialog box to customize the Active Desktop.

need to have reasonably good Internet access to provide the band-width to have the wallpaper, as a Web site, not obstruct the re-mainder of your activities. From experience, if you are running a modem line it is recommended that you use one of the predefined wallpapers or settings delivered with Windows 2000 Professional, leaving the Web-paged background to the systems that have high-speed Internet connections.

5. Click once on New… on the Web Properties page. The New Active Desktop Item wizard is then invoked, which guides you through the definition of a URL to be used as your system's Desktop. You can specifically type in the URL you want; for example, if you spend quite a bit of time at your company intranet site, you can type its URL here.

6. Click once on Browse… to open up the Favorites folder, where additional URLs you may have saved are listed.

7. For purposes of example, for the URL: type in any Web site you use the majority of your time online.

8. Click once on OK in the New Active Desktop Wizard dialog box. A secondary dialog box appears titled Subscribe, which is shown in Figure 1.5.

9. Click once on the Customize… button in the Subscribe dialog box. The New Active Desktop wizard again appears. The following series of screens will guide you through three questions.

   ■ Click once on the option, Yes, download the channel content. This selection is intuitively the better choice for Web-based content. While the entries on this screen of the wizard call it a channel, the majority of users will be using this feature for Web sites.

   ■ Click once on Next. The second page of the wizard is then shown, asking when the Active Desktop should be updated. By default the option selected is the first one, AutoSchedule. Leave the value set at this selection.

   ■ Click once on Next. The last of three pages is shown, with a username and password required for access to the Active Desktop when Internet-based content is being provided.

10. Type in a username and password, then click OK. The Subscribe dialog box is again shown. Click OK. The Subscribe dialog box closes, leaving the Display Properties dialog box, with the URL

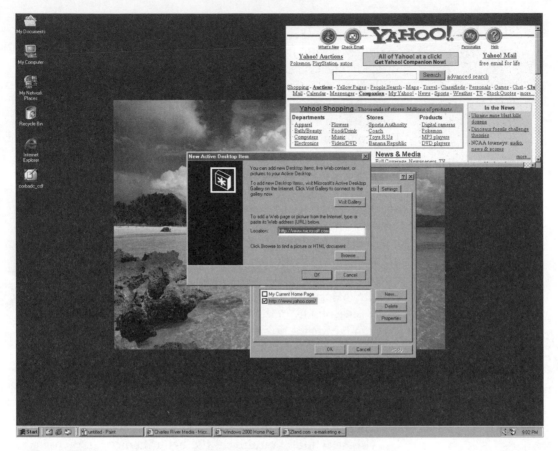

FIGURE 1.5 Using the subscribe dialog box to customize URL updates.

listed in the open space titled Items on the Active Desktop. Figure 1.6 shows an example of this dialog box with the URL listed.

11. Once the Subscribe dialog box closes, Windows 2000 immediately begins the process of contacting the Web site selected (or any other Web site you have entered using this process). If you have a modem dial-up, Windows NT will continue to try to connect. In your system uses a modem you'll notice a small dialog box indicating that dialing is beginning. There is an option in the dialog box to either stop the dial-out or let it continue.

12. Click once on Apply. This takes the URL defined and sends it to Internet Explorer for display as wallpaper. If your system has an

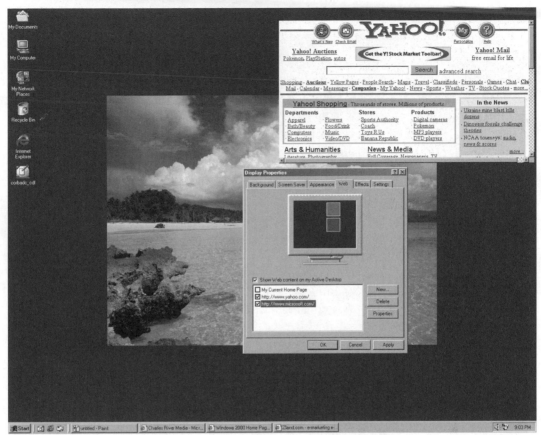

FIGURE 1.6   Using the Display Properties dialog box to enable a Web site for the Active Desktop.

active Internet connection, the wallpaper will show the Web site selected.

The term Active Desktop refers to the pace of change that occurs on many of the Web sites you'll find out on the Internet today. It's commonly known by Web page designers that a Web page is really like a magazine; it is fresh for two months, four months maximum. That's why the Active Desktop, in many respects, is like visiting your favorite news stand or bookstore; as you log into Windows 2000 Professional and see your system's wallpaper it will be like visiting a favorite news stand and seeing what's new. You could even use www.yahoo.com to create your own personalized

Web page, which could include news elements you find of interest, including stock prices, sport scores, and industry-specific news.

WHAT'S NEW IN THE TASKBAR?

Much of what is new in Windows 2000 makes the task of finding information and launching applications easier. In effect, Windows 2000 streamlines many day-to-day tasks at the Active Desktop. Enhancements to the Toolbar also provide jump points to applications, files, even the Internet, and they also give you the option to customize the selections that appear in the Toolbar. The Toolbar is the rectangular gray box that by default is located at the bottom of the screen in Windows95, Windows NT 4.0, and now in Windows 2000.

Quick tip: *It's possible to reposition the Taskbar by clicking once on it and dragging it to the left or right in 45 degree angles. Try it. Click once on the Taskbar, drag it to the left, then drag it to the top of your screen. This works in Windows95, NT 4.0 Workstation and Server, and Windows 2000 Professional and Server. This could come in handy if you want to move the Taskbar to another location so that you can better view more of the screen.*

Let's explore the Windows 2000 Professional Toolbar options, seen for the first time in this release of the operating system. For purposes of example follow through the steps provided here.

1. Right-click on the Taskbar. An abbreviated menu appears with Toolbars... as the first selection at the top of the menu.
2. Select Toolbars... from the menu. Figure 1.7 shows an example of the abbreviated menu with the Toolbars selection chosen.
3. From the extended menu for Toolbars, select Desktop. The Toolbar then displays all the elements of the Desktop as you have defined them. Each of the items on your Desktop has its icon shown in the Taskbar. You'll also see along the Taskbar a small arrow pointing to the right. This lets you scroll through the remaining icons for the Desktop as they appear in the Toolbar. Why would someone want this? Because many times there will be several, if not a dozen, applications open, and closing or minimizing each one can take time. Instead, the Taskbar can show the contents of the

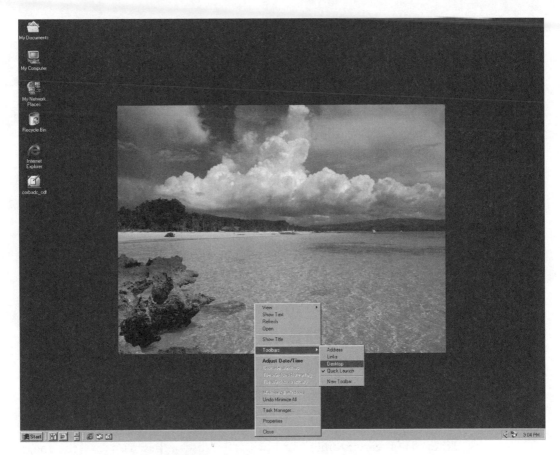

FIGURE 1.7 Customizing the Taskbar.

Active Desktop, relieving you from having to minimize each application to again view the tools you have there.

4. Right-click again on the Toolbar. The abbreviated menu again appears. This time select Toolbars... , then select Quick Launch. A series of four icons appears in the Toolbar. These icons are provided as launch points to Internet Explorer, Outlook Express, Desktop View, and View Channels. The last icon looks like a satellite dish transmitting messages. What are channels? In IE 5.5, there is the option of configuring channels created by companies including Walt Disney, Wall Street Journal, and various other content providers.

You can also create a customized Toolbar as well. Right-click on a gray area of the Toolbar once more, and select New Toolbar. The New Toolbar dialog box appears, giving you the option of selecting either a specific sub-directory or folder location on your workstation to use as the customized Toolbar or entering a URL. If you have Internet access up and running, type in a URL to see how this works. For example, you could enter www.microsoft.com, then click once on OK or press Enter.

All three selections are now visible in the Toolbar. If you have a 17-inch or 21-inch monitor, taking the Toolbar and placing it along either the left or right sides of your screen gives you a great view of all active applications and provides for quick access to many of the most commonly used tools included with Windows 2000 Professional.

How do you change the Toolbar back in case you don't want all these items shown later? Right-click on the Toolbar, select Toolbars... , and notice that all three now have a check mark next to them. To deselect them, click once on each of the items.

HOW MICROSOFT MADE PLUG-AND-PLAY WORK IN WINDOWS 2000

On any given day the product management team at Microsoft will receive thousands of product requests from customers. Compounding this flow of information are the trip reports from the top managers at Microsoft as they visit their customers worldwide. Because of these visits and the voices of customers, the Plug-and-Play standard was included in Windows 2000 Professional and Server. Microsoft had also said during the Windows NT 4.0 beta testing that there would eventually be Plug-and-Play in Windows NT, which increased functionality and ease of use compared to the first efforts on Windows95.

Just how did Microsoft integrate Plug-and-Play into Windows 2000? How could it include this standard and still preserve the hardware independence of NT? Many members of the press speculated that the Hardware Abstraction Layer (HAL) would eventually be done away with and the Plug-and-Play component would be included instead. This would have caused some serious incompatibilities for Microsoft with previous-generation customers on NT 3.51 and NT 4.0.

Instead of sacrificing hardware compatibility by deleting the Hardware Abstraction Layer, Microsoft chose to integrate Plug-and-Play directly in

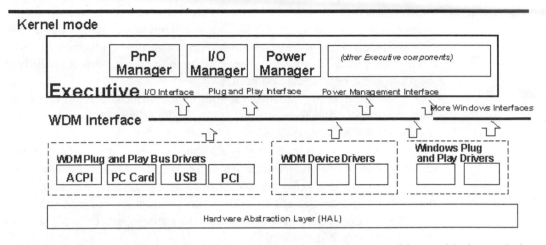

FIGURE 1.8 The modularity of the Windows 2000 kernel made it possible to add Plug-and-Play compatibility.

the Windows NT kernel. Figure 1.8 shows the Windows 2000 Professional kernel.

How is the NT operating system being modified to support the Plug-and-Play standard? Through a series of modifications to how the Hardware Abstraction Layer works in conjunction with the newly introduced Plug-and-Play subsystem in Windows 2000. Here are the key issues that Microsoft has focused on in bringing the Plug-and-Play standard to the Windows NT operating system.

First, device drivers that formerly were part of the Hardware Abstraction Layer have been moved to a location within the operating system where other essential elements of the operating system can use the device drivers. The Windows NT Executive, other device drivers, and especially the bus drivers must be able to communicate independently of the Hardware Abstraction Layer. Due to the expanded role of device and bus drivers in a Plug-and-Play environment, placing them outside the HAL made sense. Bus and device drivers in Windows 2000 play a more centralized role due to the need to ensure consistent and reliable identification and communication with Plug-and-Play devices.

Second, new methods and capabilities are available to support Plug-and-Play device installation and configuration. This new design specifically includes changes and extensions to existing user-mode components, such as the Windows 32-bit Spooler, class installers, certain Control Panel applications that rely on the Plug-and-Play device drivers and sensing logic being built into the operating system components, and the significant changes to the Setup routines, both within the graphical interface of Windows 2000 and the more pervasive changes to the device drivers and their location in the Windows 2000 kernel.

Third, for the Plug-and-Play capabilities to work correctly in Windows 2000 there also needs to be allowances in the Windows 2000 kernel architecture for communicating with and updating the registry. The registry structure itself in Windows 2000 is also being upgraded to support Plug-and-Play extensions and features. As has been and continues to be the case in architectural decisions on Windows NT, the registry structure and methods of updates as defined in previous product generations will continue to be supported in Windows 2000. Registry updates for NT 4.0 are compatible with Windows 2000.

Fourth, device drivers compatible with previous versions of Windows NT are also compatible with Windows 2000. The device drivers for Plug-and-Play compatibility, however, are specific to Windows 2000 due to the redefinition of the role of these drivers in the HAL and their new location within the kernel as a centralized resource.

From the user's standpoint, how then does Windows 2000 support plug-and-play? By ensuring that the following key features are integrated into the Windows NT implementation of this standard:

Support for dynamic and automatic recognition of installed hardware. This is the crux of the Plug-and-Play support challenge for Microsoft in implementing this standard within Windows 2000. Specifically, the ability of Windows 2000 to reliably sense compatible peripherals during system installation, recognition of Plug-and-Play hardware changes during system installation, and a response to run-time hardware events such as dock/undock and device insertion and removal. Getting this area of the Plug-and-Play standard right is the most challenging but also the core of what customers will see day to day.

Hardware resource allocation. Drivers for Plug-and-Play devices do not assign their own resources. Instead the resources needed for a given device are identified when the device is recognized by the operating system. The Plug-and-Play Manager retrieves the requirements for each device during resource allocation. Based on the resource requests made, various hardware resources are provided. These resources could include I/O ports, IRQs, DMA channels, and memory locations. The Plug-and-Play Manager works to assign resources on an as-needed basis to ensure there are sufficient resources for the device to function fully.

Device driver installation. In conjunction with the support for device drivers within the system kernel, there is now a Hardware Installation wizard that guides the user through the steps of installing device drivers and associated peripherals. The Plug-and-Play device drivers in Windows95 and Windows98 will not work with Windows 2000 due to the many differences in the functionality of the device driver, its location of Dynamic Linked Libraries, and the role of the Windows 2000 based Plug-and-Play drivers relative to their Windows95 and Windows98 counterparts.

COMPARING WINDOWS 2000 AND UNIX

Many companies, universities, and institutions of all types worldwide have standardized on UNIX as their operating system of choice. Why does UNIX continue to have a strong following? What are its strengths and weaknesses relative to Windows 2000 and vice versa? How do you migrate from UNIX to Windows NT, and what are the decisions to be made? These questions and more are defined in detail throughout this section. Let's begin by looking at the primary differences between UNIX and Windows NT, and then progress through the considerations companies make when transitioning from UNIX to Windows 2000.

How Do UNIX and Windows 2000 Compare?

Of all operating systems being used by corporations today, UNIX is most often considered the one that most closely resembles Windows NT in terms of device support, security, and networking capabilities. While many see UNIX and Windows NT as having very comparable architectures, there are fundamental differences between them. For example, the Windows NT kernel is modular and extensible by nature, while the many variants of UNIX have kernels compiled and predefined according to one of the many target audiences they serve. Windows NT's kernel is consistent across hardware platforms due to the role of the Hardware Abstraction Layer (HAL). What truly differentiates NT from UNIX is that the former also has a consistency in terms of learning time spent. From personal experience, I can tell you that with every version of UNIX you work with, there is a learning curve you must climb to understand the specific command syntax and structure for a given version of UNIX. NT's consistency of interface across the Intel and in previous versions of PowerPC and MIPS processors was the learning curve accelerator. With UNIX and its many variations there is a learning curve for each type due to the extent of the variations. To net it out: UNIX is just as powerful from a command standpoint, yet there is so much variation between versions that many administrators drill down in their specific version and know it very well. Supporting NT is easier because the user interface is consistent, freeing up the administrator for more challenging tasks such as enabling DHCP capability in a Novell environment using a Windows 2000 Workstation in a heterogeneous environment, for example.

Both Windows NT and UNIX are 32-bit, with some versions of UNIX supporting the evolving 64-bit standard. Hewlett-Packard is working with Intel to ensure that the Merced processor will be compatible with the latest generation of HP's version of UNIX, which is called HP-UX. Likewise, Microsoft and Intel continue work on the development of a 64-bit version of NT, also built on the same Merced processor. Both operating systems also have security inherent in their architecture, with UNIX having the advantage of being multiuser from the start.

In the area of specialized imaging applications such as CAD/CAM, design and drafting, and 3D animation, UNIX had built a strong reputation for years. Companies such as Intergraph Corporation had built their strongest differentiators relative to HP and Sun Microsystems on their imaging technology, called the Intergraph Graphics Design Systems, or IGDS for short. With the large-scale migration from UNIX to Windows NT for many of the applications in use on the technical desktop, the migration of imaging technologies has followed. This has included a surprising move by Intergraph to partner with Compaq on the sale of the Intergraph graphics subsystem, and the development of high-end 3D protocols including Open GL and Raydream point to Windows NT becoming the platform of choice for design and animation professionals. The role of UNIX has diminished in the area of graphics, once its stronghold, due to a wide variety of market factors that are, in turn, making the market for Windows NT grow at a rapid pace, projected to continue through the remainder of the decade. One of the primary drivers or initiators of change in the migration of users from UNIX to Windows NT is the cost of ownership of UNIX-based systems and the costs of maintenance. UNIX manufacturers, seeing an opportunity for high-margin sales from the support of their platforms, have seen Intel-based systems running Windows NT penetrate and capture market share in their larger accounts with increasing frequency. Many customers transition from UNIX to Windows NT due to the maintenance costs they save; others transition due to the increased level of performance for the price. Still, more and more software companies are basing their development decisions on the strength Windows NT is showing as a viable platform for development.

In summary, UNIX and Windows NT share many comparable features and traits as operating systems. The truly differentiating aspect of each is NT's uniformity and standards-based approach, even on multiple platforms. UNIX has many variations, each with a specific set of needs and target customers in mind. NT has already begun to gain acceptance as both an imaging platform on the workstation side and a viable server on the enterprise side.

A BRIEF HISTORY OF WINDOWS NT

The origins of Windows NT emanate from a joint development agreement between Microsoft and IBM on a version of OS/2 originally developed in the late 1980s. Microsoft and IBM failed to complete the project due to differences in architectural direction, and the companies parted before the operating system could be finalized. IBM took over OS/2 for its own development, and Microsoft decided to redesign the operating system to create what would become Windows NT. At first, Windows NT struggled to gain market acceptance, and with the release of Windows NT 3.51, Microsoft targeted its 40 million Windows 3.X users worldwide. The result is a highly differentiated operating system relative to OS/2.

The first releases of Windows NT 3.51 were met with mixed reviews, and as the majority of PCs became networked, Microsoft continued to refine NT with the goal of capturing market share from Novell, UNIX, and DOS users. In 1995, Windows NT began to steadily gain momentum, started competing with NetWare, and gained market share in the very competitive network operating systems marketplace.

Windows NT is now rapidly becoming a market standard for businesses worldwide due to its reliability, multithreaded capabilities, and proven networking abilities. The success of Windows NT has occurred partly through aggressive marketing, partly through market leverage of the large Windows and DOS installed base, but mostly because the very discriminating buyers—the CIO, system administrator, or client/server manager—have all tested NT according to their requirements, and it has passed with flying colors.

ADDITIONAL RESOURCES AND WWW SITES PERTAINING TO THIS CHAPTER

To stay current on Windows NT, you should monitor several Web sites periodically for new content. Each of these Web sites has a readership in the Windows NT marketplace, so as a consequence their content is kept current to serve the readers who rely on them.

www.bhs.com. The Web site for Beverly Hills Software, this is perhaps one of the best-supported third-party Web sites dedicated to Windows NT topics. Resources on this Web site include utilities for getting key tasks done; there is also a listing of jobs for those interested in working in the NT field. One of the most valuable areas of the Web site is the listing of user groups. These listings at times include copies of the handouts provided at the group meetings. This is a very useful site overall for staying current with Windows NT.

www.win2000mag.com. This site was formerly the Windows NT Magazine Web site. You'll find this an invaluable aid in making sure you stay current with the latest and greatest technologies, including hands-on articles that define how best to navigate the intricacies of the operating system and also how to manage a Windows NT network if you are a system administrator.

www.microsoft.com/windows 2000. This is the home page for Windows 2000 on the Microsoft Web site. This is the epicenter of all Windows 2000 information.

ntbeta.microsoft.com. This is the beta site for Windows 2000 Professional; it has a series of white papers and online documentation. This Web site requires a password and login.

CHAPTER SUMMARY

With the latest generation of this powerful new operating system, Windows 2000 Server and Workstation 5.0 have embraced the World Wide Web and its applications both inside and outside companies. Windows 2000 Professional has revolutionary features that include support for plug-and-play, extensive printing and file system enhancements, and new hardware support for the ACPI standard. This chapter has focused on the key features of Windows NT Workstation and Server 5.0; the following chapters will include key hands-on exercises and insights that will maximize the value of Windows NT in your school, business, university, or organization.

2 Touring the Windows 2000 Interface

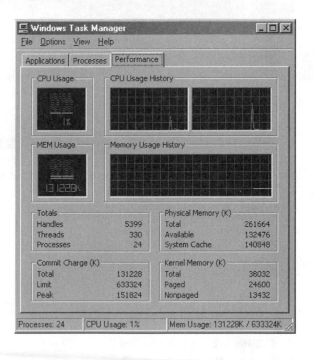

One of the truly revolutionary aspects of Windows 2000 Professional is the Active Desktop, the control over properties for many of the essential tools and utilities included in Windows 2000 Professional, and an entirely new approach to managing customization and network configuration functions. You'll also notice when you first start Windows 2000 Professional that the Active Desktop looks a lot like a Web page. You'll also find that any references to properties also have references to Internet resources as well, making the steps to researching any aspect of Windows 2000 Professional much more efficient than having to switch over to a browser and look up information on topics of interest. The intent of this chapter is to give you a thorough overview of the properties customizable directly from the Windows 2000 graphical interface. Figure 2.1 shows the Active Desktop.

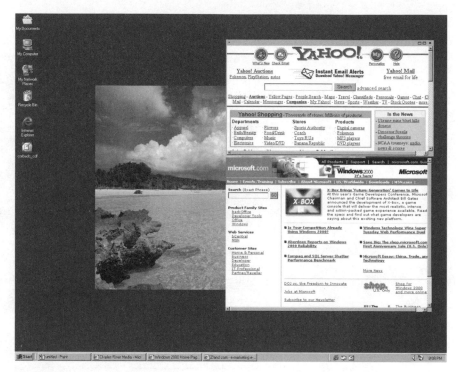

FIGURE 2.1   Introducing Windows 2000 Professional's Active Desktop.

WHAT'S NEW ON THE ACTIVE DESKTOP?

The approach Microsoft continues to take on all aspects of its next generation of operating systems, including Windows 2000 Professional, is to make Internet-based resources transparent to the user. This gives users who have T1 or ISDN lines to the Internet quick access to resources that are as up to date as possible. The Active Desktop, for the remote user or even the stand-alone workstation user, can be cumbersome. Customizing the Desktop is easy to do using the series of steps shown in this section. If you use Windows 2000 Professional on a laptop or if you support a group of systems that dial out frequently to the Internet, you'll probably want to change the Active Desktop from being Internet-centric to being more stand-alone in nature.

Let's look in detail at how the Active Desktop is becoming more and more focused on being Web-centric in its approach to handling resources, and how properties on the Active Desktop are easily managed through a series of dialog boxes.

The focus on how to streamline communications between Windows 2000 Professional and other operating systems is also defined to the property level in this chapter. You'll also see that you have the option of completely or partially enabling a Windows 2000 installation for Internet access. You can also use the options defined in this chapter to disable Internet access in the event your company provides a central area for Web surfing instead of from workstations.

Exploring the Taskbar

Instead of using file groups and groupings of applications like Windows 3.1 or even 3.51, 4.0, Windows95, and Windows 2000 Professional all have moved toward a Desktop approach to presenting applications and operating system resources to users. An essential element of this approach to organizing the Desktop is the development of navigational assistants such as the Toolbar, the shortcut, and the pervasive use of properties on many aspects of the operating system.

What is the Taskbar? And how customizable is it? As a system administrator you'll need to be able to customize this tool for users. Let's take a look at the Taskbar and the properties you can use to customize it for yourself and others you support. The Taskbar is by default located along the bottom of the screen; as applications are started, they are added to the Taskbar. Notice in Figure 2.2 the Active Desktop with the Taskbar across the lower section of the screen.

The Taskbar also includes the Start button, time, and icons representing utilities and tools as they are installed on an NT workstation. Many applications compatible with NT 4.0 and Windows 2000 add their icons along the bottom right portion of the Taskbar during the installation process. As an administrator you'll be able to see which utilities are initialized during set-up by looking at the lower right section of the Taskbar.

The Taskbar originally was created as a method for quickly changing between applications from the Desktop. In previous versions of Windows, you would have used the Alt+Tab keystroke sequence to toggle between

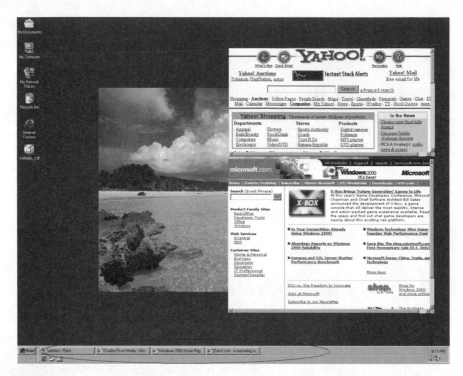

FIGURE 2.2   Introducing the Taskbar.

open applications. Windows 2000 Professional supports this keystroke sequence as well and makes it possible to choose either approach to shift between applications.

Hands-On with the Taskbar

The best approach to learning about how to manage the Taskbar is to complete a series of hands-on tutorials presented here. These tutorials are structured to assist you with learning how to customizing the Taskbar and they provide shortcuts on how to complete the more common customization tasks using the various properties associated with the Taskbar.

TIP

Quick tip: The Taskbar can be rotated in 45 degree increments, making it possible to align it along the right, upper, left, or bottom parts of the screen. Click once on the Taskbar and drag it to the right. Notice how the Taskbar pivots to the right. You can now size the Taskbar to any width you want. Click on the Taskbar again, and rotate it up to the top of the screen. Notice the Taskbar pivots and is placed along the top of the screen. You can continue to rotate the Taskbar around the screen by simply clicking once and holding down the button on your mouse, moving the Taskbar around the screen.

Exploring Taskbar Properties

As a system administrator you'll sooner or later encounter the need to understand how to customize the Taskbar for a specific user, group of users, or even for yourself. How do you do it? Options are available for the Taskbar options and Start Menu programs, and to explore these capabilities, follow the steps profiled here:

1. Right-click on the Taskbar. A pop-up menu appears, showing selections for Toolbars, Cascade Windows, Tile Windows Horizontally, Tile Windows Vertically, Minimize All Windows, Task Manager, and Properties. Figure 2.3 shows the pop-menu with Properties selected.

FIGURE 2.3 Taskbar Options Menu selected by right-clicking on the Taskbar.

2. Select the Properties selection from the menu. The Taskbar Properties dialog box appears, as shown in Figure 2.4.

Figure 2.4 Introducing the Taskbar Properties dialog box.

3. By default the Taskbar Options page of this Taskbar Properties dialog box is shown. Many Windows 2000-equipped workstations need all the display area they can get, so toggling the Auto hide feature is a good idea. Click once on this option. This hides the Taskbar, as the applications are active in the main part of the Desktop. The option within this dialog box is the option of having the Taskbar always in the foreground, or on top of other applications. The option Always on top is selected by default to ensure that the Taskbar is always visible from the Desktop. As the users you support become more comfortable with the Taskbar they can change this on their own setting if they wish.

4. Click once on the Start Menu Programs page. This page is dedicated to the customization of the Start menu. You can add, remove, or manage the Taskbar's contents from these selections. Figure 2.5 shows the Start Menu Program page of the Taskbar Properties dialog box. This is the basis for customizing the contents of the Start menu's contents.

FIGURE 2.5 The Start Menu Program page in Taskbar Properties makes it possible to customize the Start menu.

FIGURE 2.4 The Taskbar Properties Dialog Box.

This can be accomplished through the use of drag-and-drop as well.

5. Click once on Advanced… . An Explorer-like interface titled the Exploring—Start Menu appears on screen and lists the contents of the Start menu. Using this Explorer-like interface you can add, change the location of, and delete application links within the Start menu. Figure 2.6 shows an example of the Explorer-like interface with the Start menu contents shown.

6. How would you use this interface? Let's say, for example, that you want to have Internet Explorer launched when you first log on to Windows 2000 Professional. To accomplish this, you'd use the Exploring–Internet Explorer interface to move the Internet Explorer from its existing location to the Startup group. Microsoft's Internet Explorer by default under Office95 and Office97 are installed into the subdirectory called Internet Explorer. To move the

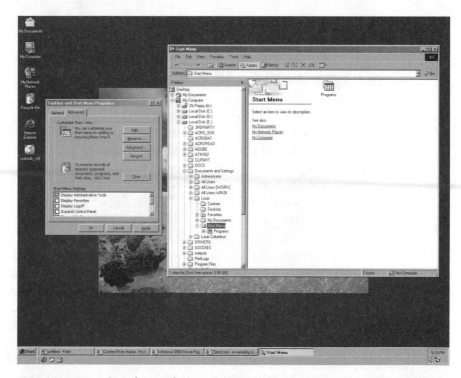

FIGURE 2.6 Using the Exploring—Programs interface for managing the Start Menu Program contents.

Internet Explorer to the Startup group, simply select the file and drag and drop it into the Startup folder. It's that simple.

7. Select Close from the Exploring–Internet Explorer window. The Taskbar Properties dialog box appears on the Desktop again. Click OK. The Taskbar Properties dialog box closes, and the Active Desktop is again shown.

8. Want to check to see if Internet Explorer is actually part of the Startup group of the Start menu? Go to the Start button, holding down your mouse button and navigating to the Startup Program group. Select this and you'll find Internet Explorer there. Next time you reboot your system you'll have Internet Explorer launched.

9. Continuing to explore (no pun intended!) the Start menu is a good idea, as you will most likely need to move applications around, provide shortcuts in the menu to applications that may have lost their shortcuts, or realign the location of applications on the Start menu.

GETTING A HANDLE ON SYSTEM PERFORMANCE— THE WINDOWS 2000 TASK MANAGER

How does performance monitoring or estimating the system resources being used relate to the Taskbar? What can the Taskbar tell you about the performance of your system? If you've used NT 3.51 or 4.0 you're probably using the Performance Monitor today to check on system performance; specifically, you're most likely monitoring object/counter relationships that show you how the performances of your server, network, and individual workstations are faring. Microsoft has listened to customers like you and me who wanted to get a quick snapshot of system performance without having to start up the Performance Monitor and check object/counter parameters and options just to monitor the local workstation's performance. Enter the Windows 2000 Task Manager. This is a very handy utility for checking system performance and the tasks running on a local workstation. Unlike Performance Monitor, Task Manager doesn't provide NetBEUI or TCP/IP connectivity across a network so as to measure performance of other systems; it is stand-alone but very convenient for measuring the performance of your individual workstation. This utility also presents a small icon in the lower right of the Desktop in the Taskbar, near the clock, to illustrate relative system resource usage at any point. When Task Manager is active, a small green bar appears on the Taskbar, next to the clock. This is the CPU usage meter. This meter changes to reflect the resource load on your system as applications use system resources.

Performance Monitor can monitor a wide variety of system performance variables by using object/counter relationships that serve to quantify system performance. Chapter 13, "Learning to Use Windows 2000's Administrative Tools," covers the Performance Monitor in depth. For now, let's take a look at how you as a system administrator can troubleshoot systems' performance using the Windows 2000 Task Manager.

The Task Manager is a tool for administrators and power users for quickly troubleshooting performance issues on individual systems. This utility is very useful especially for checking the status of each running application. Figure 2.7 shows the Performance page of the Windows 2000 Task Manager utility.

You can access the Task Manager from any of these three approaches:

- By right-clicking on the Taskbar and selecting Task Manager... from the pop-up menu.

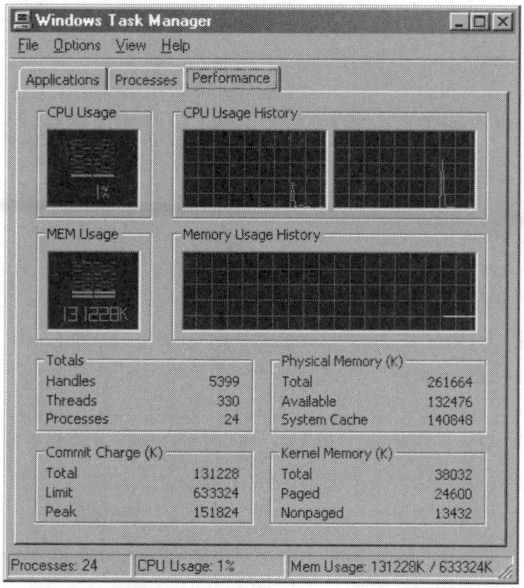

FIGURE 2.7 Windows 2000's Task Manager includes enhancements for checking on application performance.

- By typing **taskmgr** from the Start menu's Run dialog box.
- By pressing Ctrl+Alt+Del, which opens the Windows 2000 Security dialog box. From this dialog box you can also select Task Manager.

Task Manager is divided into three tabs: Applications, Processes, and Performance. Processes, CPU Usage, and Memory Usage are all listed across the bottom of the Windows 2000 Task Manager, regardless of the page selected in this dialog box. These statistics report back the virtual memory usage as first created during NT installation through the use of disk space in pagefile.sys (more on this file in the discussion of the System applet later in this chapter).

Let's take a quick tour of the Windows 2000 Task Manager and see how you can use this tool to analyze system performance.

1. Right-click on the Taskbar and select Task Manager. The Windows 2000 Task Manager opens, showing the three tabs for accessing applications, processes, and performance.

2. Click once on the Application tab. This lists all running applications and their current status. An application's status will be either Running or Not Responding. In the case of an application in a state of not responding, you can first select it from the list on this page and then click once on End Task. This eliminates the need to reboot your workstation just to stop an errant application from potentially bringing your entire system down. Thanks to the memory architecture of Windows 2000, each application has its own dedicated memory partition, so a single errant application cannot bring down the entire operating system. Virtual Device Machines, described later in this book, provide you with "insurance" against errant applications causing the entire operating system to crash.

3. Click once on New Task… . You can also launch applications from the Applications page in the Windows 2000 Task Manager as well by using a series of dialog boxes provided with the following series of steps. Figure 2.8 shows the New Task dialog box, which gives you with the option of browsing for executables. For this example the Registry Editor has been selected.

   Following the options in the Browse options for New Tasks launches the Registry Editor. Notice that in the Applications view in the Windows 2000 Task Manager Registry has now been added.

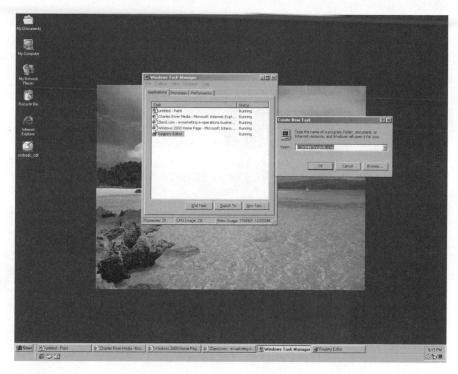

FIGURE 2.8   Using the New Task options in the Task Manager to launch the Registry Editor.

Figure 2.9 shows the Applications page with Registry running after completing these steps.

You can also close applications currently running by first highlighting the application you want to close, then selecting End Task. This will stop the application from running, providing an interim dialog box that prompts you to make sure this is an action you want to take. If an application is correctly running and you want to exit, it is highly recommended that you switch back to the original application and close it, saving your work first. Exiting a healthy-running application from the Task Manager without first saving your work can easily lead to quitting applications with work still left unsaved. Be forewarned that taking these steps will result in all unsaved work completed within an application will not be saved, when the application is exited using this approach.

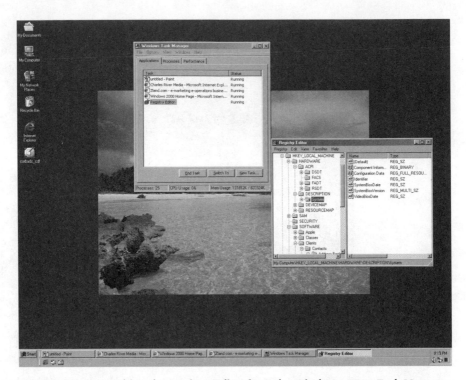

FIGURE 2.9   Launching the Registry Editor from the Windows 2000 Task Manager.

4. Click once on the Process tab of the Task Manager next. Figure
   2.10 shows the current processes running. You can see that WIN-
   WORD.EXE (Microsoft Word) and POWERPNT.EXE (Power-
   Point) are both running, using memory resources and CPU time
   to complete tasks.

Processes are actually separate programs that Windows 2000 runs concurrently. They include any applications you are currently running and all
the background programs and applications that Windows 2000 runs automatically, including various services configured during set-up and for network connecting. Each running process, in the event it becomes errant or
no longer needed, can be manually stopped by first highlighting the process
and then clicking the End Process button.

Processes can be sorted using any of the headings as defined along the
top row including Process Name, CPU Usage, CPU time (a measurement
showing how much real time, in seconds, a process has a processor's at-

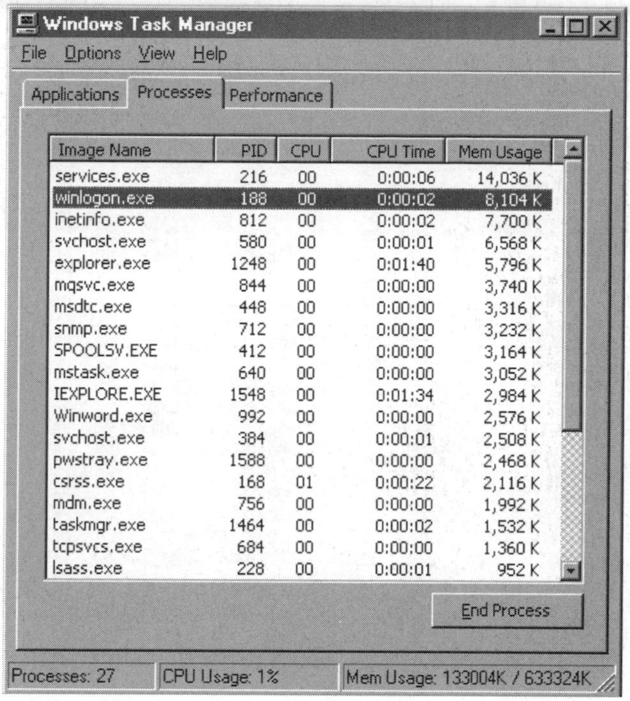

FIGURE 2.10 Task Manager lists processes currently running and the resources being used by each.

tention), and Memory Usage. By sorting the applications and processes running by CPU Time and Memory Usage (MEM Usage for short), you can learn about how various applications and processes affect the overall performance of your system. Processes and tasks are fundamentally different in Windows 2000. A process is a constant system activity that supports both applications and functions, while a task is a singular event initiated either through an application or scripting tool.

5. Click once on the Performance tab next. This page of the Task Manager records the performance of your system with regard to memory, processor time, threads in service, and physical and kernel memory use. Figure 2.11 shows the Performance page of the Windows 2000 Task Manager utility.

This specific page of the Task Manager provides a graphic for each CPU being used, along with a profile of memory usage. This page includes

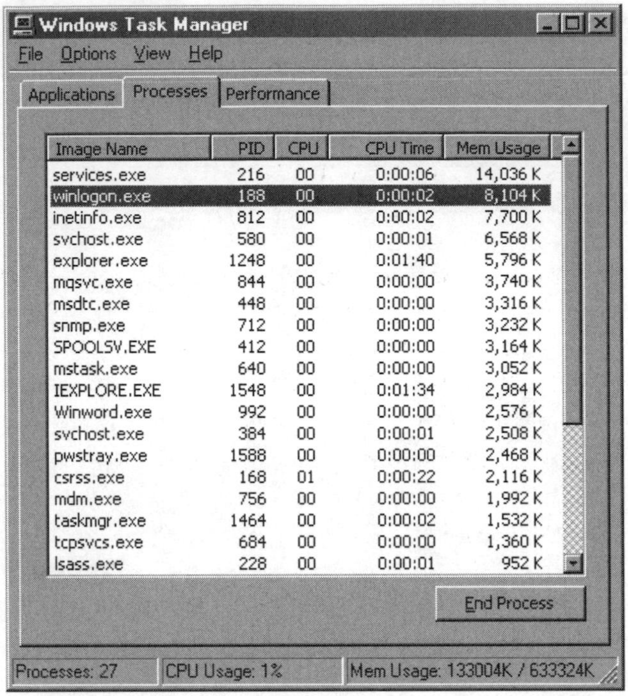

FIGURE 2.11 Task Manager also reports the CPU, memory usage, and physical and kernel memory performance metrics in real time.

information on how much physical memory is being used and how much memory the Windows 2000 kernel is using.

REVAMPED CONTROL PANEL

With the additional of new applets in the Control Panel in Windows 2000 Professional, there is also an entirely new look, as is shown in Figure 2.12.

Notice that along the top part of the Control Panel there is a series of buttons that resemble Internet Explorer. This resemblance is in keeping with the Microsoft Management Console approach to enveloping resources and information into a common graphical interface. Apart from the Microsoft Management Console, there is also the pervasive use of Internet Explorer-like functionality throughout Windows 2000 Professional. You'll also see that the hyperlinks across the left side of the screen are hy-

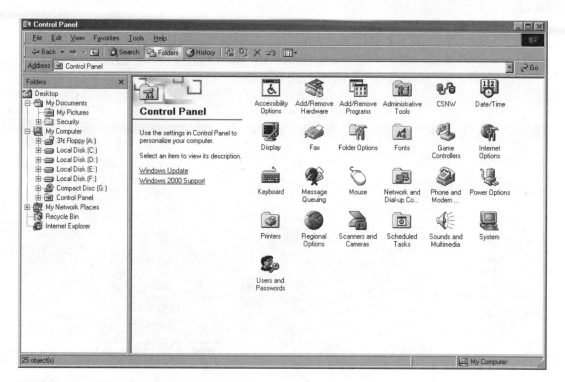

FIGURE 2.12   Windows 2000 Professional's Control Panel.

perlinks to Web sites that provide the latest content for configuration and support options.

This is very much a new graphical interface, one attuned to the needs of system administrators who must teach others how best to navigate through the files on a specific workstation and network or over the Internet. The graphical environment around the Control Panel is very browser-centric, but it is also very easy to split the screen and see the contents of various other resources on a network at any given time.

What are the benefits of this approach to arranging the Control Panel? What does this give you, the system administrator? Or the power user in your organization? Or someone called on to teach others about computers in general and NT specifically? The short answer is that this approach graphically displays not only the applets in the Control Panel, but also the selected items on a PC or a network as well. And what about the benefits of navigation in this graphical approach to presenting the Control Panel? You can quickly see, through the use of the Address: option at the top of the page, that this approach provides a global approach (from a systems perspective) to presenting files, printers, applets, and data that exists networkwide. Figure 2.13 shows an example of the Address: option in the Explorer-like graphical user interface to the Control Panel.

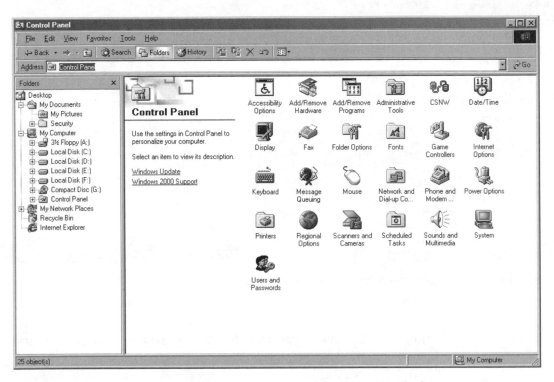

FIGURE 2.13   Using the Address Option in the Control Panel for navigating between resources.

THE ROLE OF SYSTEM-LEVEL APPLETS IN WINDOWS 2000

The applets in the Control Panel are there to guide you in the tailoring of a Windows 2000 Professional Desktop specifically for the users you support inside your company, or the customer(s) you support via system configuration, or the users at a university you support with ongoing system maintenance and troubleshooting. The system-level properties serve both to inform you about the status of the system's properties and performance variables and to provide you with a tool for specifying your minimum requirements and customization requirements as a system administrator or power user of Windows 2000. Figure 2.14 shows the Control Panel's contents when its address is defined in the Address: option field of the Windows 2000 graphical interface.

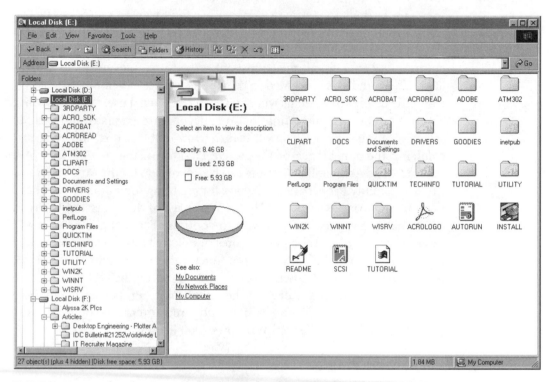

FIGURE 2.14   Using the Control Panel's graphical interface for navigating between resources.

What follows in this section is a tutorial of each of the applets found in the Control Panel, complete with a description of their functions and when you would use them to customize your Windows 2000 Professional workstation or those of others. You'll find that throughout these tutorials there is pervasive use of a browser-like user interface. The same conventions for navigating within browsers also apply to the graphical interface used for presenting the contents of the Control Panel. The following is a quick tour of the applets included in Windows 2000's Control Panel.

Accessibility Options

Represented as an icon with the international symbol for wheelchair access, this applet focuses on setting properties for the keyboard, sound, display, mouse, and general system properties that make interaction with a system for someone with handicaps possible. On the keyboard properties page, there are options for setting sticky keys (which translate a Ctrl+Alt+Del into a single keystroke sequence), filter keys, and toggle keys. There are also options for toggling on SoundSentry and ShowSounds capability, which communicate via visuals the equivalent system warnings normally associated with sound. Also included in this applet are useful options for configuring the display on systems where visually challenged users need to work. The options behind the Settings option selection are extensive and include selections for having black on white text, white on black text, and even customizing the appearance of the Active Desktop using predefined color schemes. The Accessibility Option also has an option on the Mouse page of the Accessibility Properties dialog box for enabling the numeric keyboard on your keyboard as a mouse. The General page of the Accessibility Properties dialog box provides you with the option of toggling the automatic reset on or off (it's set for five minutes of inactivity as default). SerialKey options are also included, where the functionality of serial-based devices provides convenience to users through emulations on the keyboard. Figure 2.15 shows the Keyboard page of the Accessibility Properties dialog box. The features of this applet are easily configured and can be of great assistance to physically challenged Windows 2000 Professional users.

Add/Remove Programs

If you've been involved with Windows 2000 as an administrator or even as a power user, you've spent your share of time with this applet. You'll be glad to hear that in Windows 2000 Professional, this applet is very similar to NT 4.0 Workstation and Server. The Install/Uninstall page of the Add/Remove Program Properties dialog box by default is first shown when this dialog box opens. The Install... button on the Install/Uninstall page launches an easy-to-follow Install Program Wizard that guides you through the process of finding files for launching application installs. If you're a system administrator supporting a group of users, you'll find that this wizard is a useful learning tool. There's also a batch-oriented series of tools for automating the installation of applications, covered later in this book. Also included in the Add/Remove Programs Properties dialog box is

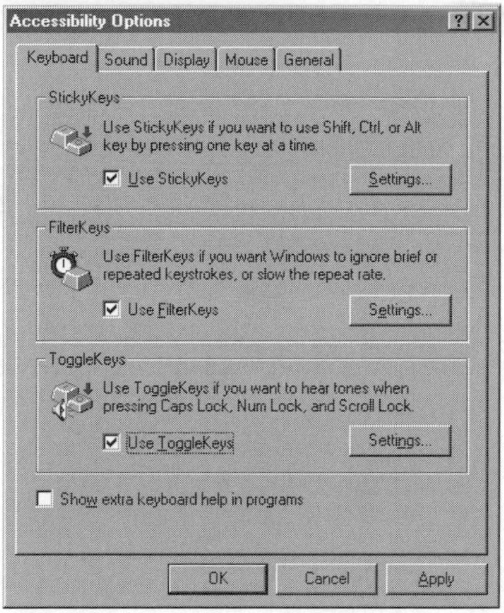

FIGURE 2.15   Keyboard page of the Accessibility Properties dialog box.

a page dedicated to set-up of the utilities shipped with Windows 2000 Professional. This page of the Windows 2000 Setup dialog box is one that system administrators and those serving others' questions on configuring applications will find very useful. You can, for example, select install components for the Accessibility Options, Accessories, Communications, Games, Multimedia, and Networking Options selections in the Control Panel. Figure 2.16 shows an example of the Add/Remove Programs dialog box with the Windows 2000 Setup properties page selected.

Double-clicking on the Networking Options selection on the previous page of the Add/Remove Programs Properties dialog box provides for installation of network components. In this specific instance, the Microsoft Network Monitor Tools, Simple TCP/IP Services, SNMP Services, and TCP/IP Print Servers are all selected. This is one location in Windows 2000 Professional that will register when a given networking protocol or service has been installed. Figure 2.17 shows an example of this dialog box with the four services previously mentioned listed.

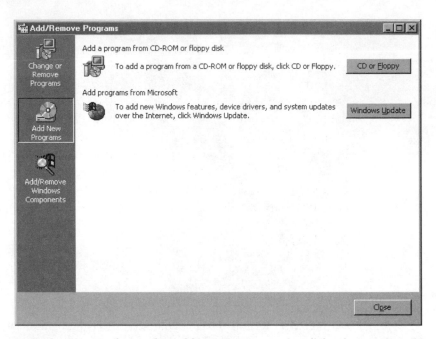

FIGURE 2.16 Exploring the Add/Remove Properties dialog box options for selectively adding components.

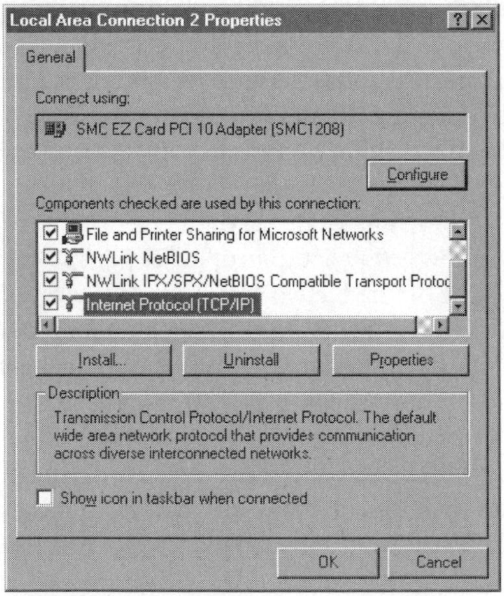

FIGURE 2.17 Using the Networking Options dialog box for adding TCP/IP and SNMP-based services.

Add/Remove Programs

Selecting this applet launches the Program wizard, which has selections for adding new applications, upgrading existing ones, repairing an existing program, or removing or modifying an existing program. This Program wizard is the first one in the NT product family to give users the option of loading software from the Internet as part of the loading process. In addition to the Internet, both CDs and corporate networks are also supported. Figure 2.18 shows an example of the second screen of the Programs wizard.

The Programs wizard continues with a series of selections for choosing the location of the set-up files, guiding you through the process of getting an application installed, modified, or removed from an application.

Date/Time

Many Windows 2000 Professional-based workstations and servers, when first shipped, have the date and time already set to the manufacturer's time zone. As an administrator, in most case you will need to reset the date and

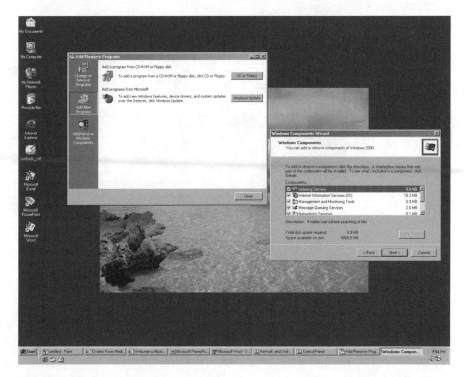

FIGURE 2.18   Using the Programs wizard to add, modify, or delete applications.

time, specifically the time zone. The Date/Time applet in the Control Panel provides pages for setting Date & Time and Time Zone. The Date/Time Properties dialog box also includes a toggle for adjusting time for daylight savings time. The support for time zones will give you the flexibility to configure a laptop running Windows 2000 for use anywhere in the world. Figure 2.19 shows an example of the Date/Time Properties dialog box. The settings included in this dialog box are reflected throughout Windows 2000 Professional's other features, including date and time stamping on files and event and application processes being logged in Event Viewer. Because the rest of the operating system keys off these values, it is not necessary to reset the date and time and then reboot. Windows 2000 Professional picks up the date and time change, then adopts the revised settings into processes requested.

Dial-up Networking Monitor

This applet provides status, summary, and preferences for modems and their device statistics. This series of properties pages also include connec-

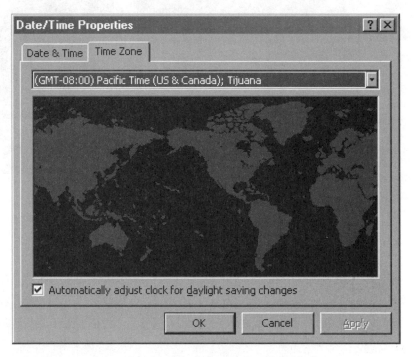

FIGURE 2.19 Using the Date/Time Properties dialog box.

tion statistics and device errors, including the number of CRC (Cyclic Redundancy Checks) that have occurred while being connected. This applet is most effective as a diagnostic tool while a Windows 2000 workstation is being used online. Three pages of properties are included in this applet, Status, Summary, and Preferences. The Summary page is useful for troubleshooting a multilink connection. This is useful as an analytical tool for seeing what links are available to the modem during its communication processes while online. The Preferences page gives you options to configure the Console to emulate the hardware-based functions of a modem. There is also the option of adding the Dial-Up Networking Monitor button to the Task List. This last feature can make the customization of systems that require dial-out capability essential for streamlining their desktops. Figure 2.20 shows an example of the contents in the Dial-Up Networking Monitor dialog box.

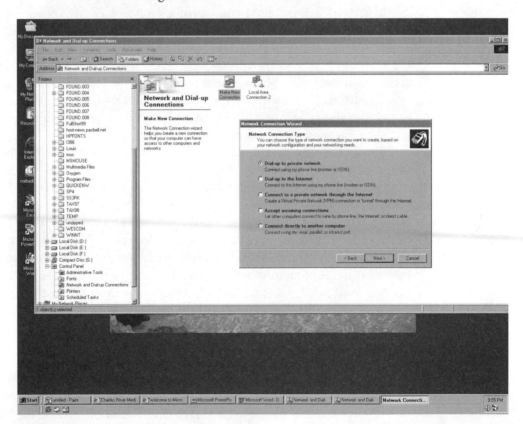

FIGURE 2.20   The Dial-Up Networking Monitor is useful for troubleshooting modem connections.

Display

One of the most commonly used applets or utilities in Windows 2000, the Display applet includes many useful options for customizing the following:

- The appearance of the Desktop
- The color scheme for the various dialog boxes
- The definition of the screen saver options
- The definition of Web pages that are linked directly to the Desktop
- The definition of device driver settings for the video card and monitor settings
- The customization of icons
- Color management options for display systems supporting color standards.

Microsoft has included color profiles for five of the more popular monitors. Profiles are included for Diamond Compatible 9300K G2.2, Hitachi Compatible 9300K G2.2, NEC Compatible 9300K G2.2., sRGB 9300K G2.2, and Trinitron Compatible 9300K G2.2. These files are accessible directly from the Color Management page of the Display Properties dialog box. Figure 2.21 shows the Settings page of the Display Properties dialog box. From this page you'll be able to set the parameters for monitors and test your selections before saving them.

One of the new developments in the Display Properties dialog box is the emphasis on customizing the Active Desktop for constant viewing of Web sites while using resources locally. The Web page of the Display Properties dialog box includes options for embedding your most frequently visited Web sites directly to the Active Desktop. You can also use HTML-based address books of e-mail addresses as a document placed directly onto the Active Desktop. You'll find this very useful if you're a system administrator in charge of an intranet site. You'll be able to have an HTML editor open, a Web site for testing your postings open, and then the actual intranet site in a third window to test the posts made. The Active Desktop provides flexibility for handling the process of integrating Web site references and HTML address books directly to the Desktop. Figure 2.22 shows the Web page of the Display Properties dialog box. You can see from this dialog box that you can preview the location of Web sites and even HTML-based address listings on the Active Desktop.

What if you don't want to have Web sites placed directly on the Desktop? Your company may have a policy of limiting Web surfing to only certain research workstations in a library or classroom, instead of from every

FIGURE 2.21 Settings in the Display Properties dialog box give you plenty of options to customize your monitor's appearance.

FIGURE 2.22 Using the Web Properties page in the Display Properties dialog box.

desk. Many companies have adopted this approach. With the pervasive availability of Internet access from Windows 2000, how do you restrict access to the Internet? By following the series of steps profiled here:

1. Log onto the workstation you want to configure as Administrator.
2. Select the Display from the Control Panel or right-click directly on the Active Desktop and select Properties.
3. Select the Web page of the Display Properties dialog box.
4. At the bottom of the Web page click once on the box, disabling all Web-related content from the Desktop.
5. Choose Apply.

The settings are now saved. Users will not be able to add Web-based content to their Active Desktop.

Fonts

This is not actually an applet; it is a subdirectory of font files, and it is presented in the context of a Web browser. Files with the TT on them are TrueType fonts, while the ones with an A are Adobe fonts. The approach Microsoft continues to use in Windows 2000 Professional is to use a very similar Add New Fonts sequence of dialog boxes as it has in previous versions of Windows NT 3.51 and 4.0. Selecting Install New Fonts… from the File menu presents a dialog box for browsing, then installing font files. The new features in this area specific to Windows 2000 include new options in the View Menu. The View Details and option to hide or display font file variations are useful when managing a larger set of fonts. In previous versions of NT, the font file would actually have to be opened, while in NT 4.0 you need to merely place your cursor over the font file and a balloon-like window pops up showing the appearance of the font. Figure 2.23 shows the Fonts folder.

Hardware Wizard

Instead of just using the same approach it used in Windows95, Microsoft has in effect rewritten its approach to how it handles plug-and-play compatibility by including the IEEE1394 standard directly in the Windows 2000 system kernel. This translates into a wizard that senses the status of devices attached to the workstation or server. The first step with the Hardware wizard is to sense which devices are either unrecognized or need device driver support. Do you remember the long sequence of scanning an

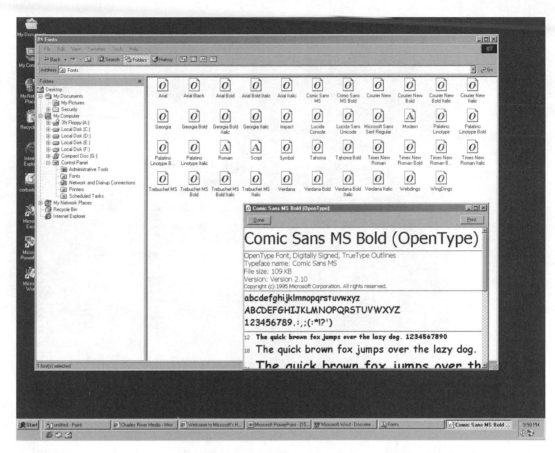

FIGURE 2.23   Using the Fonts folder to manage fonts on Windows 2000 Professional.

entire system where Windows95 would dominate all system resources looking for devices? Windows 2000 now handles this directly at the system level, eliminating the long and resource-intensive process of searching for peripherals. Figure 2.24 shows the first screen of the Hardware Wizard. By default, the option of repair devices that aren't working is highlighted. This is in keeping with the more efficient approach to managing the entire plug-and-play process. You can also see that the Hardware wizard has two paths you can take from the initial screen: one for troubleshooting and one for adding new software.

Later in this book we'll look in depth at the Hardware wizard and how you can use it to solve hardware dilemmas that in previous versions of Windows 2000 would have taken days to troubleshoot along with your

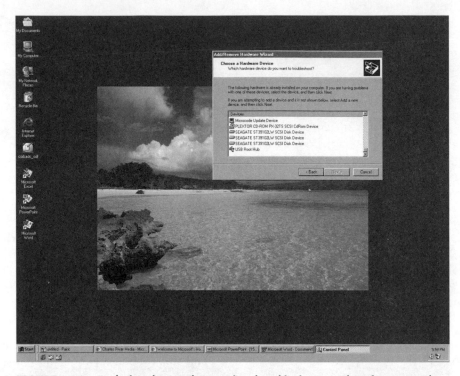

FIGURE 2.24 Exploring the Hardware wizard and its improved performance characteristics.

other tasks. Now you can isolate down to the device and get resolution to perplexing compatibility issues in a few minutes instead of wasting hours tracking them down.

Internet

If there is a single applet or utility in Windows 2000 that exemplifies the extent of Microsoft's commitment to integrating Internet capabilities in Windows 2000 Professional, the Internet utility or applet is that item. Even if you aren't doing Internet or intranet-based work today, take a few minutes and tour this entire utility, which is nearly a complete application for managing Internet/intranet connections. There are six specific areas of the Internet Properties utility, each with customization options pertaining to connections to an intranet or the Internet, storage and presentation of Web pages, definition of connections and their security aspects, even controls as to the level and type of content usable to the workstation being configured.

It is, in short, a very comprehensive series of tools for getting a workstation configured for using intranets and the Internet. Figure 2.25 shows the Internet Properties utility with the General page selected.

Clicking once on the Security tab shows the series of different security settings for different "zones" of Web content. Four zones are included in this dialog box by default. These four zones include one dedicated to local intranets, one for trusted sites, an Internet zone, and a restricted site zone. Each of these can, in turn, be configured for High (most secure) content, Medium (more secure), Low, or Custom. Figure 2.26 shows the Security page with options selected for the Internet zone access profile being created. This profile will be used when the system logs into Internet sites located outside the firewall of a company.

Clicking once on the Custom entry, then the Settings... entry shows the extent of support for defining security parameters for Java-based information delivery of information, file downloads from the Internet, and ActiveX security-related parameters. There are also options for defining scripting security options as the workstation being configured confronts these options on the Internet.

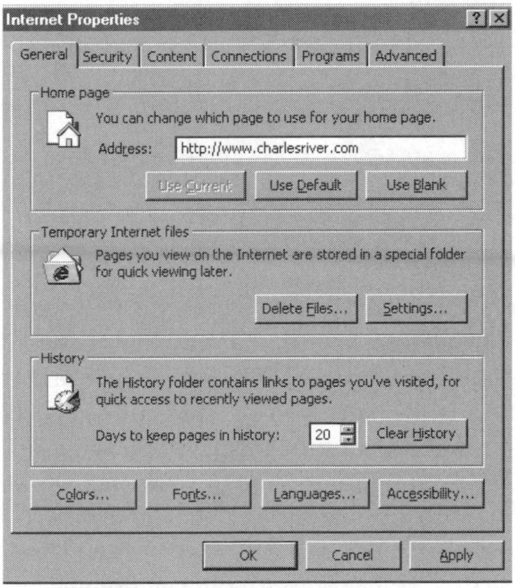

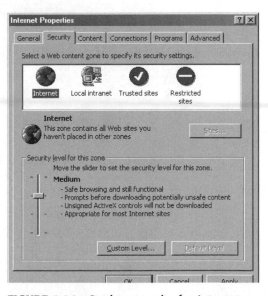

FIGURE 2.25 Setting home page parameters and managing Internet files are easy using the Internet Properties utility.

FIGURE 2.26 Setting security for Internet access.

Click once on the Content tab across the top of the Internet Properties dialog box; this specific page is shown in Figure 2.27. The Ratings options on the top of this page provides you with the flexibility of controlling the Internet content viewable by the workstation being configured.

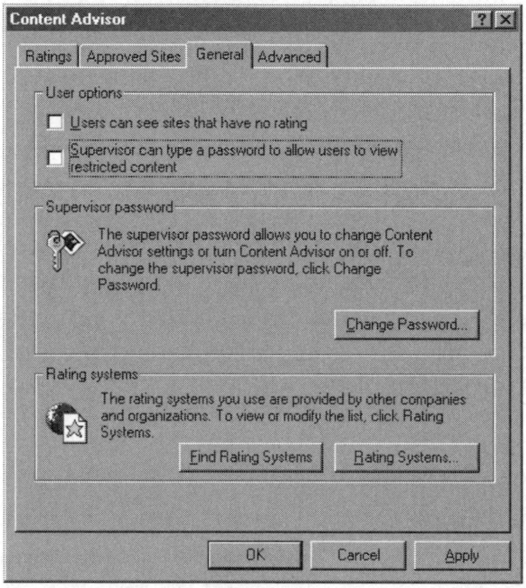

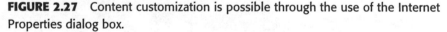

FIGURE 2.27   Content customization is possible through the use of the Internet Properties dialog box.

From within the Ratings area of the Content dialog box it's also possible to specify relative access to Web sites using the Recreational Software Advisory Council rating service for the Internet. It is based on the work of Dr. Donald F. Roberts of Stanford University, who has studied the effects of media for nearly 20 years. Microsoft has adopted this standard and given system administrators the ability to configure these standards on a workstation-by-workstation level.

Also included on the Content page are options for viewing Certificates for individuals, sites, or publishers. Certificates verify the identity of sites; are they who they say they are on the Internet? Certificate Server, as part of Microsoft's latest edition of Site Server, Enterprise Edition, supplants Microsoft's BackOffice products.

Additional pages exist for setting Connection, Programs, and Advanced properties for Internet connections. Chapter 9, "Connecting with Windows 2000 Professional," will cover these areas of the Internet Properties dialog box in the context of creating Web sites and home pages and will discuss how to enable intranets within your company or organization.

Keyboard

This applet provides customization options for character repeat delay and repeat rate for the keyboard. There's an option for specifying the cursor blink rate as well. There are two pages in the Keyboard Properties dialog box. The first is dedicated to speed parameters, and the second to Input Locales. On this second page, you can select device drivers for every language from Afrikaans to Urdu (more than 50 languages are supported). On the Keyboard Layout/Input Layout options there are more than 20 options for customizing keyboard selections for specific countries.

Licensing

The applet that looks like a certificate has been designed appropriately. Microsoft's series of initiates in its distribution channels continues to be on the integration of Electronic Multiple Onsite Licensing Program (EMOLP) and the entire issue of managing Certificates of Authenticity (COAs) and their use for customizing computer systems throughout the distribution channels Microsoft uses to sell its products today. The Licensing applet keys off the registration for applications that have been installed using COAs and licensing agreements. You'll find this useful for managing SELECT-based product purchases and the entire series of applications you have that have been purchased under license from Microsoft.

Modems

This utility is used to install a new modem or change properties for an existing modem already installed. If you've used this applet under Windows95 or even Windows 4.0 you'll find this very familiar. The two pages that make up this utility include the General page and Diagnostics. Figure 2.28 shows the Modems Properties dialog box with the General page visible.

One of your most common tasks will be to change the modem parameters of a given modem. Let's take a look at the Motorola ModemSURFR External 28.8 modem attached to COM2, the system being profiled. Substitute the name of your modem for the Motorola system.

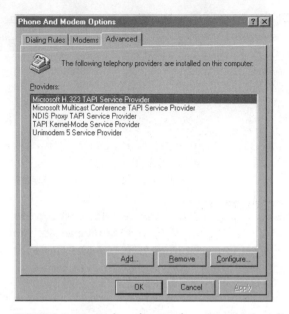

FIGURE 2.28 Using the Modems Properties dialog box to manage modem connections.

1. From the General page of the Modems Properties dialog box, select the modem for which you want to change the communication parameters.
2. Click the Properties button. The properties for the specific modem selected are shown. The properties for the Motorola Modem-SURFR External 28.8 are shown in Figure 2.29.
3. Click once on the PC to Modem Port Speed entry. Notice that there are a series of context-sensitive selections for modem speeds. It's a good idea to select the one that is one step above the one you are running today.
4. Click once on the Connection page. This is where the essence of how a modem communicates with the outside world is defined. It's a good idea to make note of the existing settings for the data bits, parity, and stop bits before changing them. These are the parameters you'll most likely change as you decide to log into different Internet service providers (ISPs), or even to bulletin boards with varying communication needs.
5. Click OK. The Modem Properties dialog box is again shown.

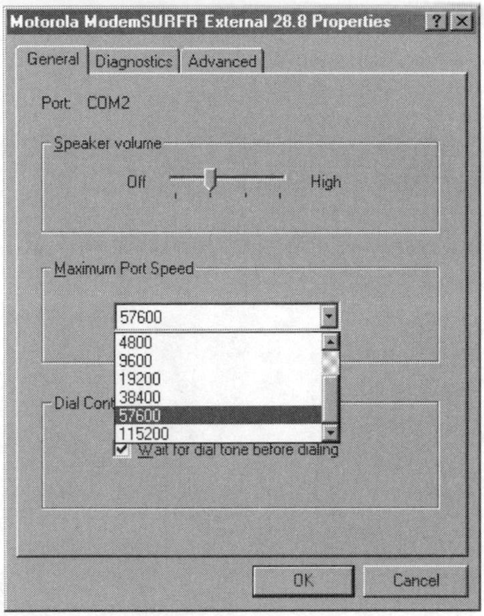

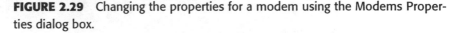

FIGURE 2.29 Changing the properties for a modem using the Modems Properties dialog box.

6. Click OK again, and the Modem Properties dialog box closes. You've just completed the change essential to modify the speed of your modem.

If you're in an administrator's role that requires extensive use of modems, consider using the Hardware wizard to have NT recognize modems during installation. Be sure to check out the Windows 2000 Hardware Compatibility List to make sure you have modems compatible with Windows 2000 Professional. The Hardware wizard expects to see an IEEE 1394-compliant device, and it is easy to use for configuring new hardware.

Monitoring Agent

This applet is used for viewing the statistics and graphics generated from the Network Monitor. This dialog box sets the passwords for the utility found in the Programs selection/Network Administration selection from the Start menu. The Microsoft Network Monitor Tool captures network utilization and frames per second used for transactions. This is a very robust application included in Windows 2000 Professional Workstation and is explored in detail in Chapter 13, "Learning to Use Windows 2000's Administrative Tools." Figure 2.30 shows the Microsoft Network Monitor window.

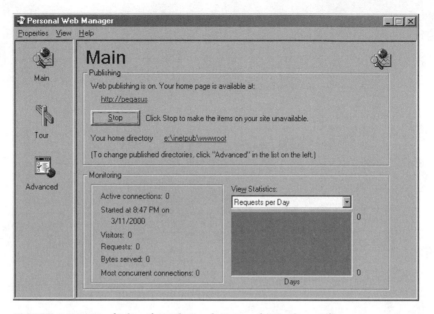

FIGURE 2.30   Exploring the Microsoft Network Monitor utility.

Mouse

This applet is familiar ground for any of you who have set up mice before on Windows95, Windows NT 4.0 Workstation, or Windows 2000 Server. There are three pages within the Mouse Properties dialog box. By default, the first one is for configuring buttons on the mouse. This page specifically focuses on configuring the mouse for left- or right-handed use and the definition of the double-click speed for the mouse's buttons. The Pointers page is used for customizing the specific scheme you want to use for your cursors. Because cursors can be in multiple states, depending on the activity of the operating system at any point in time, this page of the Mouse Properties dialog box has an entire series of cursors available for customization. Figure 2.31 shows the Pointers page of the Mouse Properties dialog box with the Dinosaur scheme selected. If any of the schemes do not fit your tastes, there are plenty available on various Internet sites dedicated to NT. There is a third page in the Mouse applet focused on motion of the mouse. The options included here include those for setting the speed of the mouse movement.

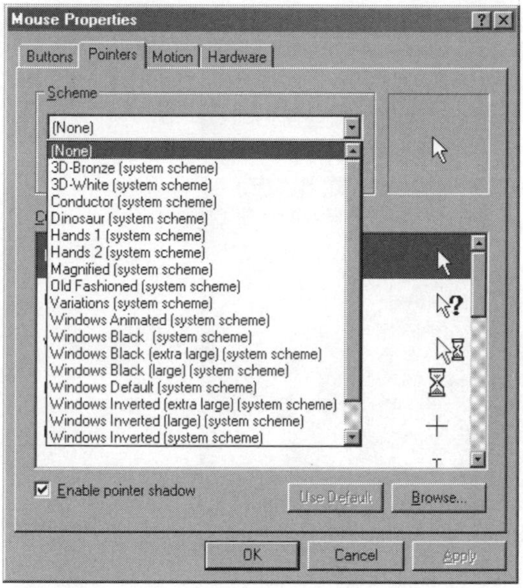

FIGURE 2.31 Configuring pointers using the options in the Mouse Properties applet.

Multimedia

This applet includes a series of options for configuring audio, video, MIDI, CD music, and multimedia devices for use in either stand-alone or networked multimedia environments. This is the applet you'll use for the following:

- Configuring a series of device drivers for sound mixers so you can play musical CDs on your workstation
- Setting the the options for playback of video files
- Configuring MIDI-compatible peripherals, specifically MIDI-compatible output and MIDI schemes for specific peripherals
- Designating a CD drive as the CD music source for your Windows 2000 workstation
- Defining properties for each of the multimedia-specific devices compatible with Windows NT.

Figure 2.32 shows the Devices page of the Multimedia Properties applet.

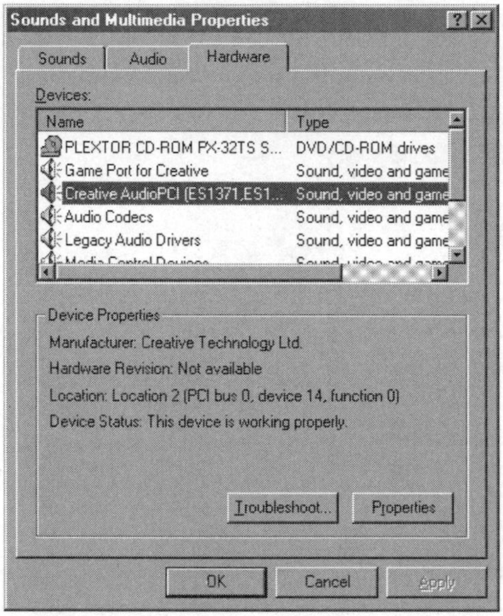

FIGURE 2.32 Configuring multimedia devices with the Devices options in the Multimedia Properties applet.

Network

Just like the Display applet, this applet is another place where administrators spend a fair amount of time. Throughout this book we'll spend time in this area of the Control Panel because of its essential nature to connectivity between Windows 2000 and other systems based on many other operating systems. Let's take a quick tour of the Network applet.

There are five separate pages in the in the Network applet. The Identification page includes the Computer Name and Workgroup entries for the specific workstation being configured. It is easy to change either of these attributes using the options available on the Identification page. You'll find this useful as an administrator who is given the task of moving systems around a network. The ability to quickly change the identity of a system is something you'll use with increasing frequency as you move systems from one department to another. You will be able to teach others how to handle this process, which is essentially a two-step process of getting into the Control Panel, then the Network applet.

The next page is the Clients page. Just as the name suggests, this is where you'll get an opportunity to configure the client-based software utilities corresponding to specific network operating systems to which you want ongoing connections. The main part of the Clients page lists the network clients that are currently installed on a system. Once a client device driver is installed on a system viewing the properties associated with a given client device driver set is painless. Let's take, for example, the Client for Microsoft Networks, which is by default installed when Windows 2000 Professional networking is installed. Figure 2.33 shows the Clients page with the Client for Microsoft Networks option selected. Notice that when it is selected on your own system the Properties... button is activated, making it possible to further delineate the characteristics of this specific client software for Microsoft Networks.

Selecting Properties from the Clients page with the Client for Microsoft Networking also selected yields the following dialog box, titled

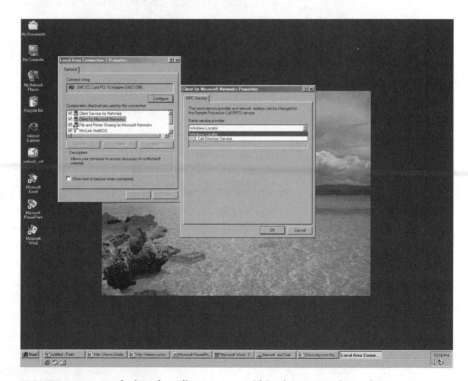

FIGURE 2.33 Exploring the Clients page within the Network applet.

Client for Microsoft Networks Properties. Notice that this dialog box, as shown in Figure 2.34, includes directions for setting client, service, and Protocol options.

The NetBIOS is critical to the overall functioning of any Microsoft networking and its interoperability with other networks, as NetBIOS is the "glue" within the OSI Model that makes it possible for NT-based implementations of TCP/IP to correspond with other operating systems from a networking standpoint. On the NetBIOS page of the Client for Microsoft Networks Properties dialog box, you can see the specific routing by LAN number for either TCP/IP protocol for WAN assignment of values on the network or, in the case of the implementation on the system being used for this book, the third LAN ID being used for NWLink (Novell NetWare) compatibility.

You can see that, from a system administration standpoint, the opportunities for customizing a network configuration are almost limitless given the context of how NetBIOS is being used in the format of the Client For Microsoft Networks implementation for connectivity. We'll be spending more time in a few later chapters on networking and connectivity. It can be argued that the strongest aspects of NT are its ability to be multilingual and speak with multiple platforms at the same time using a common series of network connectivity tools. We'll explore those throughout later chapters.

Next, let's take a tour of the Services page of the Network applet. By default, the File and Print Services and Dial-Up Client are included in the initial installation of Windows 2000. You can also install Admission Control

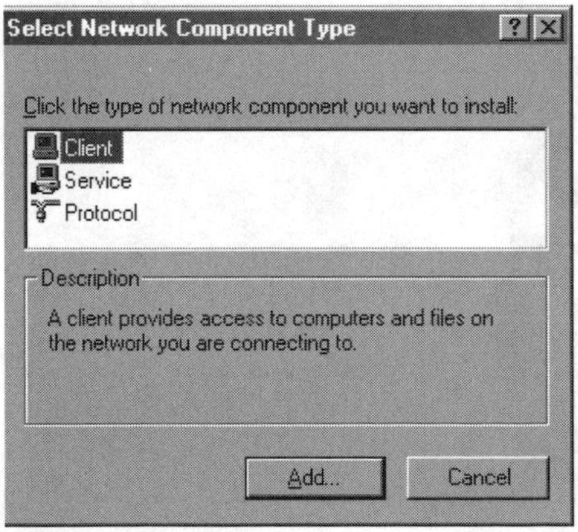

FIGURE 2.34 Using the Client for Microsoft Network Properties dialog box to set RPC values.

Service, ATM ARP/MARS Service, Dial-Up Server, and S.A.P. Agent. These are the series of dialog boxes you'll use for configuring client software to enable the workstations you're responsible for supporting to speak with each other. You'll also find that the ongoing services from Microsoft planned after the introduction of Windows 2000 Professional will provide connectivity at the client level to a wider variety of local area networks than has been possible in the past. Figure 2.35 shows the Service page of the Network utility.

As a system administrator you will undoubtedly spend most of your time on the next page of the Network utility, the Protocols page. Notice that Figure 2.36 shows how protocols are listed in the Network protocols: portion the Protocols page.

This specific page of the Network utility will be pervasively used throughout the chapters that follow to show how to configure the TCP/IP protocol. A key point to keep in mind is that a Windows 2000 Professional-based workstation can have several protocols installed and even running at the same time; Windows 2000 supports running up to two protocols at the same time on the same network interface card, assuming the card has BIOS capabilities for supporting each.

The last page to the right within the Network utility is titled Bindings. What are bindings? They are, simply put, the relation between network cards, the protocol device drivers assigned to them, and the services driving

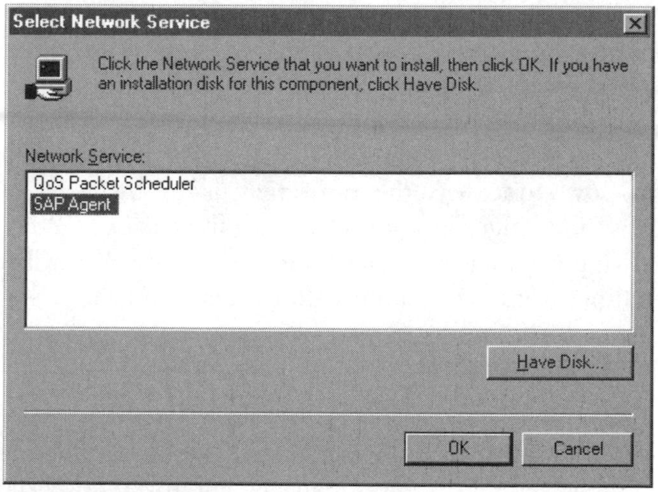

FIGURE 2.35 Configuring Services is possible using the Network utility accessible from the Control Panel.

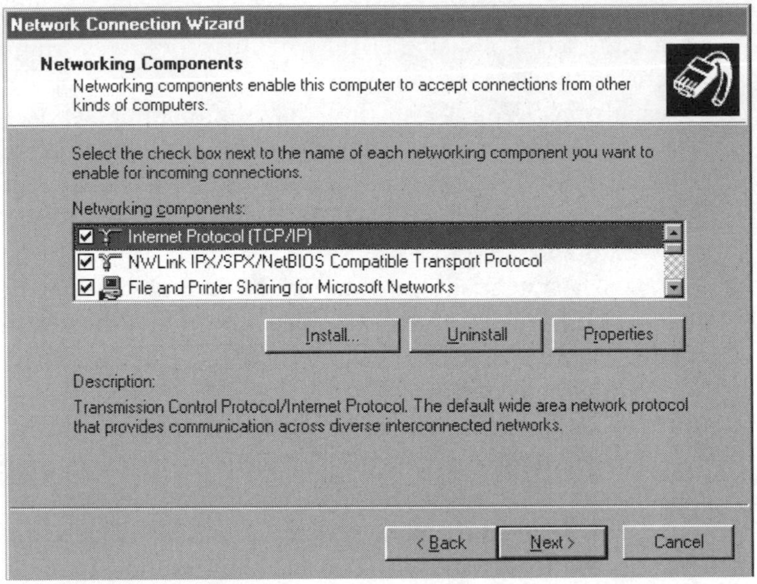

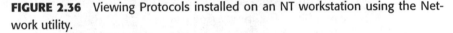

FIGURE 2.36   Viewing Protocols installed on an NT workstation using the Network utility.

the protocol assignments. Microsoft has been listening to system administrators in this area, and the evidence of that is in the structure of the Bindings page. Notice that there is now an option for Show Bindings For: entry. This is the first version of NT (Windows 2000) to include the ability to see bindings by adapters, clients, services, or protocols. This is particularly useful for system administrators in troubleshooting connections on a network-wide basis. If, for example, a system is not working correctly you'll be able to analyze the bindings associated with a given protocol and literally drill down to see what the connectivity issues are by device. Just like the options for configuring connectivity parameters in TCP/IP, the issue of configuring bindings can take hours to explore. We will go through how bindings work in detail throughout later chapters.

Power Management

One of the shortcomings of NT 4.0 was the problem of managing power on laptops. The letters to Microsoft must have been many and very specific because the end result-power management in Windows 2000 is comprehensive and easy to use. Figure 2.37 shows the contents of the Power Management Properties dialog box, showing the Power Schemes page.

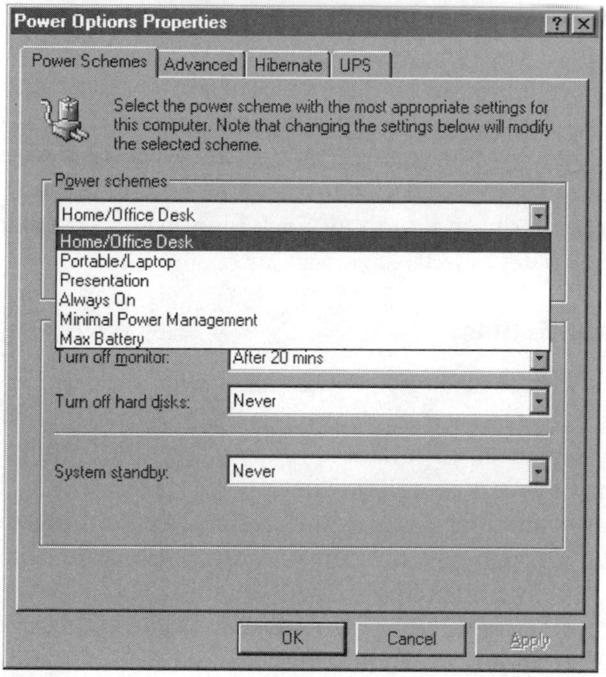

FIGURE 2.37 Extensive support for power management is available in Windows 2000.

From the Power Schemes page, there is a pull-down menu for configuring Home/Office Desk, Portable/Laptop, Always On, Presentation, Minimal Power Management, and Max Battery. In conjunction with these options there are options for configuring the turn-off times for monitors and hard disks.

The second page dedicated to Advanced functions includes a selection for having the power meter appear on the Taskbar. The third page, Hibernate, provides for your system to shut itself down and save its contents to your system's hard disk. This is very useful for laptops, and as a system administrator you can save the people you serve a lot of headaches by toggling on this option.

Printers

Even though it is located in the Control Panel, this is actually a jump point to the Printers folder that is located within My Computer. This jump point in Control Panel takes you to the contents of the Printers folder, and right after Windows 2000 is installed the Add Printer wizard is its only contents. Chapter 5, "Learning to Use Printers in Windows 2000 Professional," profiles in detail how to use these series of instructions. Installing support for a new printer using the Add Printer wizard is actually pretty easy to do, and as an administrator you'll be able to teach others how to get this accomplished quickly. Chapter 5 provides many useful tips and tricks on how to get printers up and running, including how to troubleshoot network-printing issues as well.

Regional Settings

If you work for a company that has a wide geographic coverage of customer accounts, and therefore regional offices around the globe, you'll get a chance to customize settings using the options in the Regional Settings Properties utility. The essence of this utility is to change the format of numbers, dates, time designations, and currency symbols to match the requirements of any supported international location. Included in the Regional Settings Properties utility are five pages. These include Regional Settings, Number, Currency, Time, Date, and Input Locales. What's different from this implementation in Windows NT 4.0? One major difference is the ability to read and write documents in multiple languages, configurable on the Regional Settings page. Figure 2.38 shows the Regional Settings page of the Regional Settings Properties utility.

Sounds

This applet defines sounds for specific events occurring on your system. There are five selections for Sound Schemes. These include Windows 2000 Default, Musica Sound Theme, No Sounds, Robotz Sound Scheme, and Utopia Sound Scheme. Any of these sound schemes can be selected from the Sounds page in the Sounds Properties utility.

System

This utility provides a comprehensive series of pages for defining the system parameters, options, and performance enhancements you can choose to include in a system's profile of characteristics. Windows 2000 Professional is the first iteration of the NT operating system to include direct access from the System applet to both the System and Device Manager wizards. This is useful from the standpoint of getting new and existing hardware up and running quickly. If you're a system administrator who

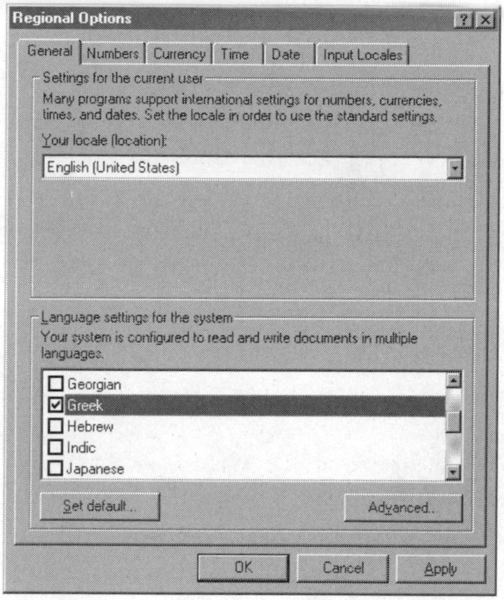

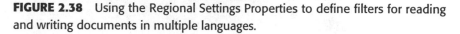

FIGURE 2.38   Using the Regional Settings Properties to define filters for reading and writing documents in multiple languages.

gets a lot of requests for troubleshooting why a system's performance is slow or seems to have incompatibilities with installed hardware, the System applet is where you should begin looking. Figure 2.39 shows one of the more advanced pages of the System applet, the Environment page. This is the page where you can define the attributes and environment variables that describe the set of parameters a given Windows 2000 Professional workstation uses.

Throughout this book we'll look at these variables on the Environment page in the context of the Registry, which is an application included with Windows 2000 that is used for defining system-level variables. We will cover the Registry in great detail later in this book. For purposes of our tour, click on each of the pages in the Systems applet. You can see that there are plenty of opportunities for getting higher performance from a Windows 2000 Professional-based workstation that is being installed or upgraded. Later we will cover the aspects of troubleshooting performance using the Performance Monitor and other tools just added to the Windows 2000 release.

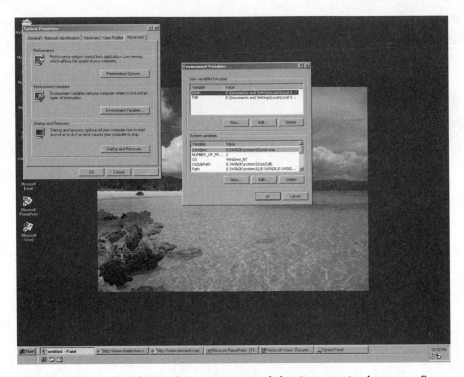

FIGURE 2.39 Using the Environment page of the Systems Applet to configure system parameters.

The next time you're in front of an NT system booting up, you'll see a complete list of all the operating systems loaded on that system. This is called **FlexBoot**. Originally developed on the UNIX platform, FlexBoot gives you, the user, an opportunity to choose the operating system you want to use for a given session. Notice how the FlexBoot is organized. How do you control the sequence of the operating systems listed? By using the Startup/Shutdown page of the System applet, you can easily customize the list of operating systems chosen by default under FlexBoot when a system first starts up. There are a multitude of reasons why this is a very useful tool; perhaps the best is that it gives you a chance to make the it easy for the the users you support to make choices. You can configure hundreds or even thousands of laptops, workstations, or even servers to revert to the NT operating system version you have labored over and installed. Figure 2.40 shows the System Startup option on the Startup/Shutdown page of the System applet.

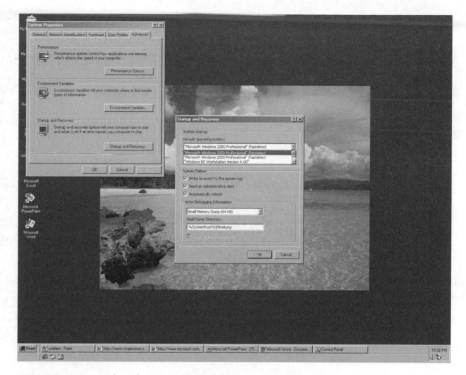

FIGURE 2.40　Using the Startup options to customize FlexBoot.

Telephony

This applet defines the parameters for getting a modem up and running. There are two pages included in the Telephony applet. Included here are options for configuring modem calls, using pulse or tone dialing, and the option of defining Calling Card numbers for charging telephone calls. You'll find the pages in the Telephony applet more focused on modem-like properties, with the second page of attributes, called Telephony Drivers, focused on a variety of provider connections for enabling an NT workstation to be used as a telephony server. Figure 2.41 shows the Telephony Drivers page of the Telephony applet.

UPS

This utility is used for configuring the options for the Universal Power Supply (UPS). A UPS is a device typically attached to the serial interface of

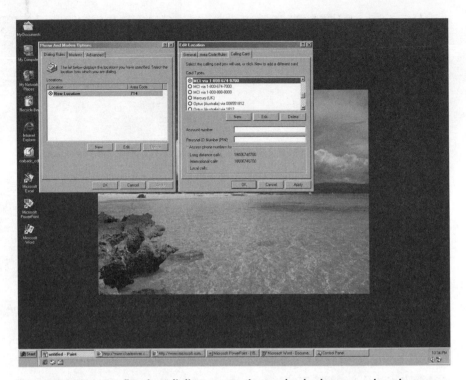

FIGURE 2.41 Configuring dialing properties and telephony services is accomplished using this applet.

a workstation. Its role is to configure the UPS attached to a workstation, ensuring that data is saved in the event power to a workstation is lost. There are options, for example, to have a power failure signal to be toggled on, to inform the user if a low battery is about to impact the UPS' effectiveness; this approach also gives the user the option to remotely configure a UPS shutdown. A UPS is particularly effective when using workstations in parts of the country where thunderstorms, electrical storms, and interruptions in system performance from forces of nature are common. A UPS is, of course, critical for use on servers where the data being stored and knowledge generated are critical to a company's business. Figure 2.42 shows an example of the UPS dialog box.

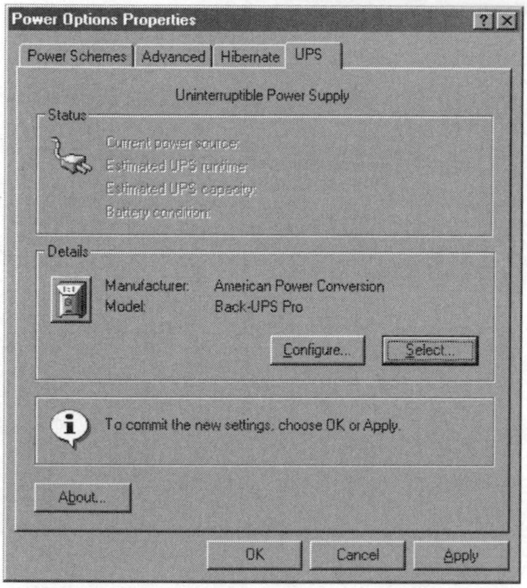

FIGURE 2.42   Universal power supplies are used for ensuring fault tolerance.

CHAPTER SUMMARY

What truly differentiates Windows 2000 Professional from previous versions of this operating system? The Active Desktop, the enhancements to the Taskbar, and the addition of customization capabilities in the Control Panel have not been possible before. The integration of Internet and intranet-based capabilities is stronger than ever—especially in the Internet applet that is found in the Control Panel. There are options within the Internet applet for configuring security aspects of Web surfing, toggling on the use of Java and scripting from the Active Desktop.

This chapter focuses also on the hands-on aspects of configuring a Windows 2000 Professional workstation desktop, and it includes a roadmap for getting around in the Control Panel. Many of the more detailed aspects of troubleshooting and working with Windows 2000 Professional are covered in later chapters of this book.

3 Customizing the Desktop and Exploring Commonly Used System Properties

There are a host of new changes within the graphically based tools of Windows 2000, many of which provide users with more flexibility than previously possible for managing networks and distributed environments. The Command Prompt, once a stand-alone emulation of an Intel 80486-based MS-DOS environment, now provides the functionality of an Intel Pentium-class system. This utility, the Command Prompt, has actually been included since Windows NT 3.5. In addition, many TCP/IP, network administration, and system maintenance commands can be run through the Command Prompt as well. Think of the Command Prompt as a command line utility for completing network administration tasks.

The intent of this chapter is to provide you with a thorough understanding of the role properties play in Windows 2000. If you are a systems administrator you will find this chapter a useful model for teaching others how to manage the features of Windows 2000 Professional.

INTRODUCING THE UPDATED COMMAND PROMPT

When you launch the Command Prompt window, it will remind you of the first MS-DOS-based computers you used in school or at your first job. It's an application based on the idea of emulating an MS-DOS environment within Windows 2000 Professional and Server. Why, you ask, would anyone want to revert to the file conventions and command lines that time forgot? Are the developers of Windows 2000 nostalgic? They are actually reacting to the needs that customers like you have reported as essential in an operating system.

Given a choice of whether to run MS-DOS applications on your own workstation, you would most likely opt for the higher-performance, multithreaded version of your favorite design and engineering applications instead of the MS-DOS version. Now put yourself in the position of your system administration colleagues at companies that used MS-DOS-based applications for communicating between various systems.

You will find that Windows 2000 supports MS-DOS applications either directly from the Command Prompt window or by clicking on the application from within Explorer. Interested in which provides the better performance on your MS-DOS applications? The Command Prompt window actually is a shade faster due to the lower overhead for the graphical interface of Windows NT.

Where do the applications actually run? Is Windows 2000 actually loaded on top of MS-DOS? Those users you support will have questions about where the application is actually running. The answer is in a protected memory subsystem, which is a key component of the architecture of the Windows 2000 operating system. Protected memory subsystems are specific to Windows NT, and they ensure that each type of application supported has its own memory address space to complete calculations. Each of the protected memory subsystems within Windows 2000 is built specifically to support compatibility with Win16, Win32, POSIX, and OS/2 applications. Supporting the protected memory subsystems that enable compatibility with these various applications is the Windows 2000 Executive, which is the centralized coordinator of all tasks within the operating system.

Using Properties to Customize the Command Prompt

You will find that many of the utilities included in Windows 2000 Professional have properties associated with them. The Command Prompt has a series of properties you can use to customize the font, colors, and layout, in addition to defining cursor size and command history. Using these properties you can make the Command Prompt easier to use.

These steps illustrate how to access the Command Prompt's properties:

1. Click once on the upper left corner of the Command Prompt window. Click once under the words Command Prompt, and a pull-down menu appears, as shown in Figure 3.1.

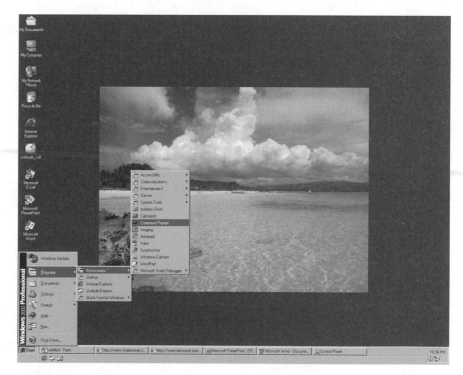

FIGURE 3.1   Accessing the Command Prompt's menu.

2. Moving down the list of commands on the pull-down menu, select Properties. The Command Prompt Properties dialog box appears and is shown in Figure 3.2.

3. Click once on the Layout tab across the top of the Command Properties dialog box. The Screen Buffer Size, Windows Size, and Window Position variables are all shown. Using these options you can also override the system's default locations for the Command Prompt window by deselecting the option, Let System Position Window.

4. Click once on Colors. This is the page of the Command Properties dialog box tailored to the color customization options for the Command Prompt window. Figure 3.3 shows an example of the Colors page of the Command Properties dialog box.

5. Click once on the Font tab. This is the page of the Command Properties dialog box that is used to define the structure of the characters used in the Command Prompt window. You can see

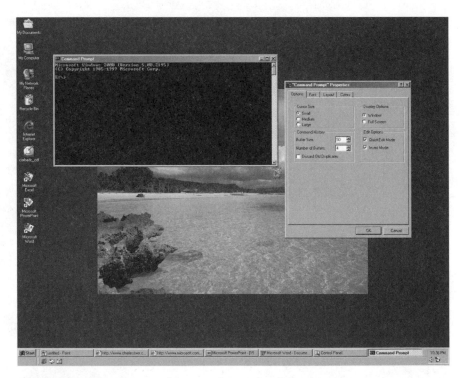

FIGURE 3.2   The Command Prompt Properties dialog box.

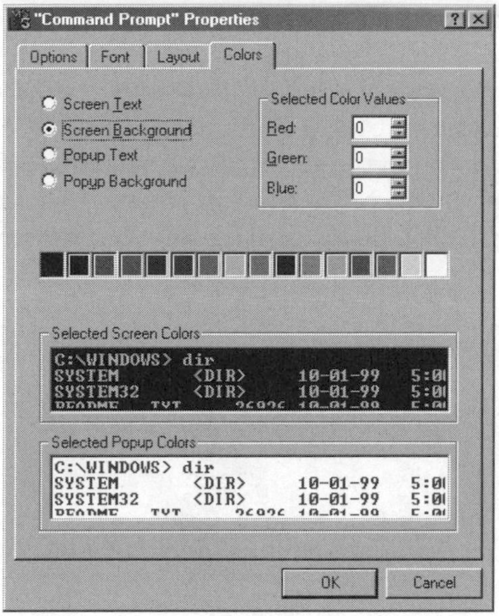

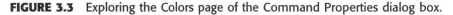

FIGURE 3.3 Exploring the Colors page of the Command Properties dialog box.

that a variety of options are available. Choose the one you find most readable.

6. Click once on OK. The Command Prompt properties dialog box closes, showing the Active Desktop.

What Can You Use the Command Prompt For?

If your organization has standardized on an MS-DOS application, you will find the Command Prompt a useful utility for getting these programs up and running on your workstation. More and more of the application providers who first began their companies with MS-DOS-based applications have ported them to the Windows 16-bit or Windows 32-bit programming architecture. Increasingly, the Command Prompt is being used as a mechanism for troubleshooting network connections, completing file transfers using the ftp (file transfer protocol) command, or using the Telnet command to log on to another workstation. True, some utilities have been developed for the Windows environment that make it possible to complete both ftp and Telnet commands through a graphically oriented interface, but many system administrators first learned how to work with networks using the command-line interface in both UNIX and mainframe systems.

The Command Prompt window also supports the CMD.EXE command that starts a new instance of the Windows 2000 command interpreter. A command interpreter is an application that displays the command prompt at which you enter in command syntax. You can also use the Command Prompt for creating logon scripts using shell script functions.

Networking Commands Compatible

As an administrator, you most likely spend a large percentage of your time working with the networks in your organization. Using the Command Prompt window gives you a level of communications flexibility that, quite frankly, is not possible using a series of graphical interfaces for completing the same tasks. The networking commands compatible with the Command prompt windows are as follows:

TCP/IP Commands
    finger
    ftp
hostname
    netstat
    ping
    rcp
TCP/IP Utilities
    lpq
    lpr
Networking Commands
    net help
    net computer
    net file
    net helpmsg
    net print
    net start
    net view

net config

net use

net time

These commands are described in detail in Appendix A, "Trouble-shooting TCP/IP Configurations," along with a comprehensive series of subsystem and native commands compatible with Windows 2000 Workstation and Server.

UPGRADED ACCESSORIES

Getting familiar with the WordPad and Paint features of Windows 2000 Professional can help you get everyday tasks done more efficiently than before. In the case of WordPad, you will find that many new features make this utility comparable to Microsoft Word.

Exploring the New WordPad Features

This is really a scaled-down word processor, and it is compatible (thankfully) with Microsoft's Word 6.0 and Office2000 compatible as well. You will find WordPad in the Accessories group, accessible from the Start menu. It's a very Word for Windows-like application that is useful for creating documents that you can save into Word documents' you can even import them into preexisting Word documents. WordPad bridges the gap between text editor and fully defined word processor successfully by integrating key elements of more advanced word processing applications.

Creating New Documents Using WordPad

Provided here are a series of steps for creating a document using WordPad. You'll find this a simple process, and one that is useful for creating note files for yourself, and then either mailing the file or saving it to disk and taking it with you. Here's how you create a document in WordPad.

1. From the Start menu, select Accessories, then WordPad. Figure 3.4 shows the path from the Start menu to the Accessories applet.
2. Launch WordPad. Unmistakable in its resemblance to Word, in WordPad notice that the toolbars are smaller with fewer commands and the menus shorter.

FIGURE 3.4   Finding Your Way to the WordPad applet.

3. You can enter text and emphasize it using Bold, Italic, and Under-line commands. You can also change the alignment of the text using the Left, Middle, and Right text alignment buttons. There is also an option for taking text and creating bullets from it. Figure 3.5 shows the Formatting Toolbar.
4. WordPad also has options for defining the finalized document that you'll be creating. Selecting Options from the View menu, the Options dialog box appears and is shown in Figure 3.6.
5. After creating your document, select Print. Just as with any of the other Windows-based applications you'll find that printing is iden-tically implemented in WordPad as it is in any other application

Importing and Exporting Documents from WordPad

One of the most valuable features of WordPad is its ability to import and export documents. This is very handy when you need to work with docu-ments on a new workstation that does not have Office97 or Windows 2000

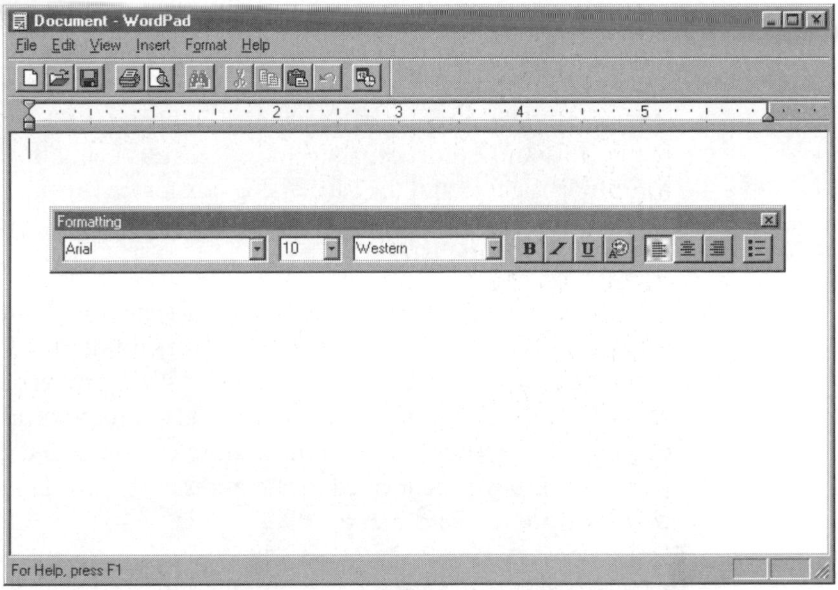

FIGURE 3.5 Exploring the Formatting Toolbar.

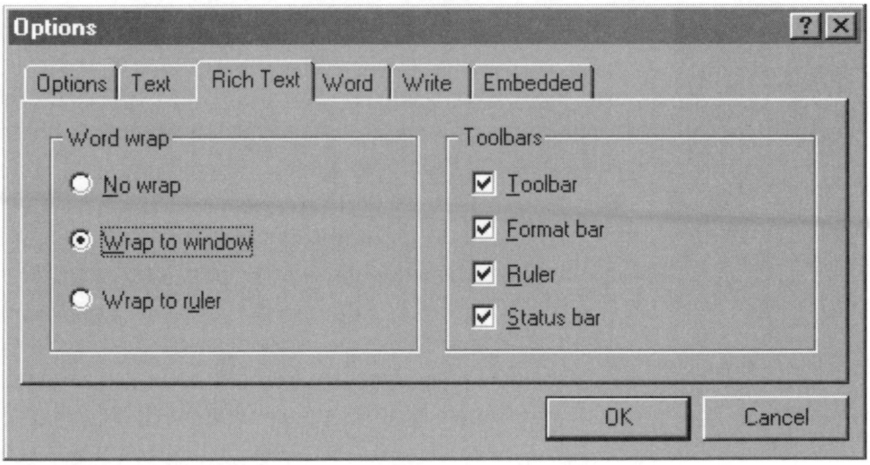

FIGURE 3.6 Using the Options dialog box for customizing destination files.

installed yet. If you need to import files, you will find that WordPad supports Microsoft Word, Windows Write, Rich Text Format (rtf), MS-DOS text, and Unicode. For exporting files, you will find support for Microsoft Word, rich text format (rtf), text, MS-DOS text, and Unicode. Accessing the import and export capabilities is as easy as using the Open dialog box for importing files and the Save dialog box for saving files in other formats.

What's New with Microsoft Paint

One of the more common uses for Paint is creating and editing bitmap, GIF, and JPEG files. You'll find that the version in Windows 2000 Professional has more flexibility in defining system properties, including the skewing and angling of images. There is also much improved support for color mixing and definition, plus a more customizable graphic interface. Paint continues to be located in the Accessories group, and it is accessible directly from the Start menu.

Creating Graphics in Paint

Meant mostly as a freehand drafting tool, Paint is easily learned and quick to use. You can use Paint to bring in a graphic that has already been created, a bitmap, GIF, or JPEG, for modification or color matching. Paint is also very useful for creating graphics. Paint, an applet (or small application), is intuitive and powerful enough for handling initial design efforts.

The process of creating a document in Paint is entirely up to the creative or graphic goals you have in mind. Let's take, for example, the task of creating a network topology diagram that defines the locations of systems on your LAN and their associated IP addresses.

1. Using the Start Button, navigate through the menus, first finding Accessories, then finding the Paint button.
2. Select Paint. The application launches. Figure 3.7 shows the Paint application after it has been opened.
3. Click once on the large A on the toolbar to the left. This is the tool for adding text to the Paint file you're creating. Clicking the A automatically shows the Fonts toolbar on-screen. By default, the Algerian font is shown, which is difficult to read at lower resolutions.
4. Using the pull-down Font selection bar, select Times New Roman (Western) instead. This font is easier to read at resolutions required on 14-inch monitors.
5. Click once in the workspace of the file so that the cursor is visible.

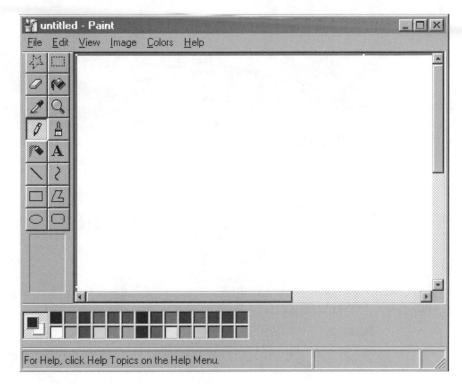

FIGURE 3.7 Introducing Microsoft Paint.

6. Type the title, **IP Address Definition**.
7. Next, click once on the rectangle tool and draw boxes, each representing a system on the network.
8. Click once on the text tool to select it.
9. Click one of the rectangles in the diagram, and stretch the text box until it covers an entire rectangular box.
10. Next, type in the IP addresses of each system. Figure 3.8 is an example of what this graphic would look like with two systems included in the diagram.

Importing and Saving Graphics into Paint

As with every Windows-based application, you can save your diagrams using the Save As... command from the File menu. You'll have the option of saving the file you have created in one of several formats, including bitmap, GIF, or JPEG. The majority of applications on the Windows platform have import filters that can accept each of these formats.

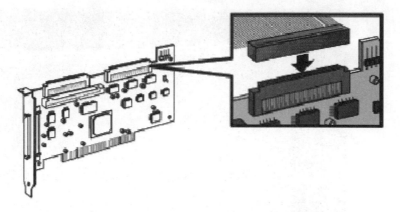

FIGURE 3.8   It's possible to create diagrams quickly using Paint.

The same file filters are provided for importing files into Paint for editing. You can import monochrome and color bitmaps, GIFs, and JPEG files.

SEARCH FOR FILES AND FOLDERS—WHAT'S IMPROVED?

Whereas in Windows95 there is simply the Find File dialog box, in Windows 2000 Professional there are more network-centric capabilities built into this command. For example, the Search command located directly on the START menu includes support for finding files or folders, finding a computer located on the network, finding a file located in a secondary directory, looking over the Internet for a file, and looking through personal phone books in your Outlook and e-mail folders. Figure 3.9 shows the location of the Find command.

How Find File Works

This application is unique within Windows 2000 in that is very command-like but also has a graphic interface associated with it. The FindFile command works just the same as the Find command did in File Manager from Windows 3.X and previous versions of Windows NT. This application gives you the opportunity to search for files by name, date created, or even application used to create the file. You can also search for files that contain a specific phrase or word, regardless of the file the target is located in. Figure 3.10 shows an example of the Find File dialog box.

FIGURE 3.9 Locating the Find command.

The majority of the time you're looking for a file, you know its contents but not its name. Say, for example, you have created a report that describes how "Hydra" works and its implications for your organization. You'd click once in the area titled Containing text: and enter the word "Hydra." Find File would then query every single file on your workstation, showing the search results. Figure 3.11 shows the text entered into the Find File dialog box to search for documents containing "Hydra."

Find File also includes options to locate files created during a specific time or date period, and it also gives you the option to search for all files that are a specific size. These options are configurable in the other two pages of the Find File command dialog box.

The Find File command is also integrated with the Active Directory, which means you can search for users, computer names, hardware and software resources.

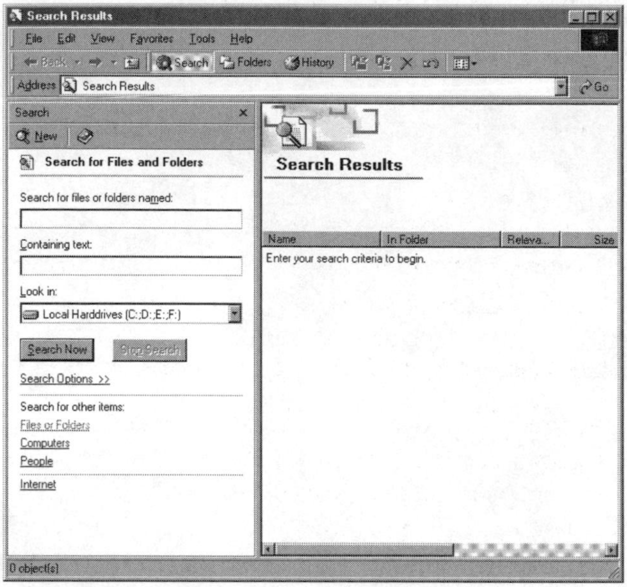

FIGURE 3.10 Introducing the Find File dialog box.

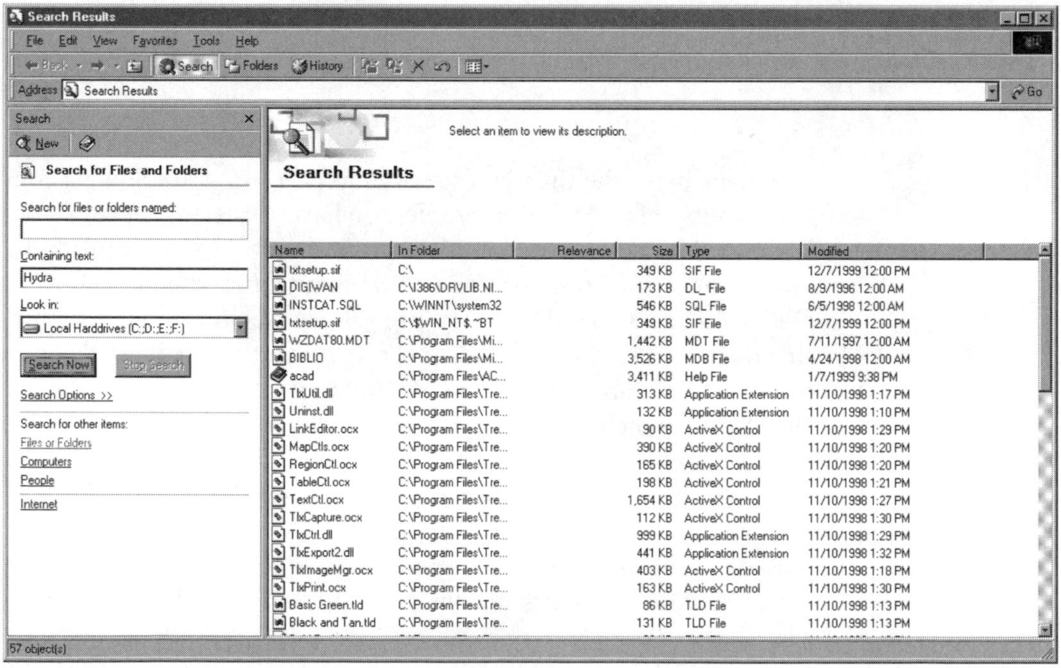

FIGURE 3.11 Using Find File to retrieve all files containing "Hydra."

EXPLORING SYSTEM PROPERTIES IN WINDOWS 2000 PROFESSIONAL

The emphasis in each subsequent version of Windows 2000 increasingly focuses on the properties of key resources within the operating system. This properties-centric approach gives you, the workstation user and administrator, more opportunity than ever before to customize the operating system with great flexibility. Throughout this section, we'll look at defining system properties in Windows 2000 Professional.

Desktop Properties and Their Options

Throughout Windows 2000 Professional you'll see significant effort has been made to make Internet access as seamless as possible. This is especially true on the Desktop, or as it is called in Windows 2000 Professional, the Active Desktop. It's called the Active Desktop because the Desktop itself can be configured to be the Web page of your choice. Chances are, if you are a member of a larger organization and have sufficient bandwidth to provide adequate refresh rates you will find this a useful tool. When Windows 2000 Professional is first installed, the Active Desktop is loaded by default. Many other customization options are available for the Desktop, and they are explained in this section.

Guided Tour of Setting Up Desktop Properties

First, let's start by seeing how easy it is to access the Desktop properties. You can either go through the Control Panel, which is located in the My Computer icon at the upper left corner of your screen, or access the Desktop properties by right-clicking on the Desktop. Many people prefer this latter approach as it saves time. Let's take a quick tour of the Desktop properties now:

1. Right-click on the Desktop. An abbreviated menu appears. Listed at the bottom is the word Properties. Scroll down the small menu, then select Properties. The Display Properties dialog box appears, as shown in Figure 3.12.
2. Notice that along the top of the Display Properties dialog box there are a series of page selections. Each of these selections represents another series of properties associated with the display. You can use the page titled Web, for example, to configure the Web site you want for your Desktop.

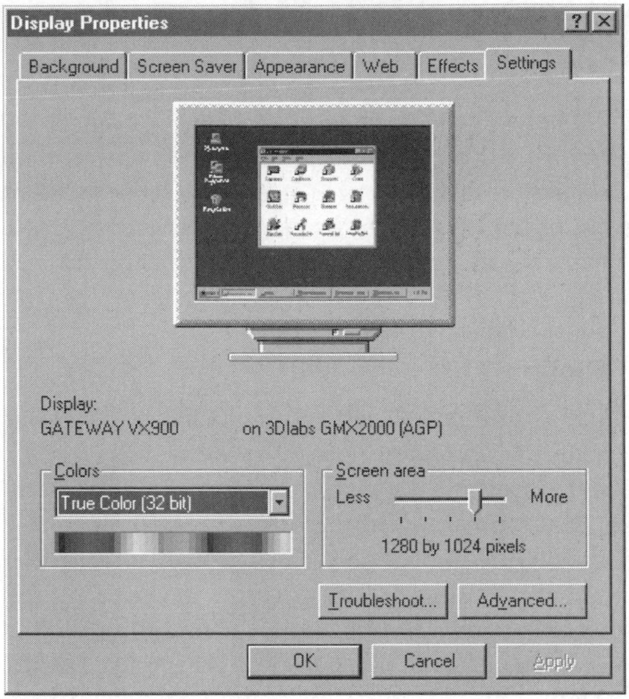

FIGURE 3.12   Use the Display Properties dialog box to customize the Desktop.

3. Click once on the Settings tab in the Display Properties dialog box. For many system administrators, this is the set of properties in this dialog box where they spend most of their time. Figure 3.13 shows the Settings page of the Display Properties dialog box.

4. Click once on the sliding bar in the Desktop Area section of the Settings page. The display monitor located in the top half of this page reflects dynamically the changes you make to pixel definition in the Desktop Area section of the dialog box.

5. Next, let's look at how you can change the Adapter device driver being used on your workstation. Let's say, for example that you're going to install a new device driver for your video adapter because you've just purchased a 21-inch monitor. You can first check to see which device driver is loaded, then change the device driver selection by using the options on the Settings page of the dialog box.

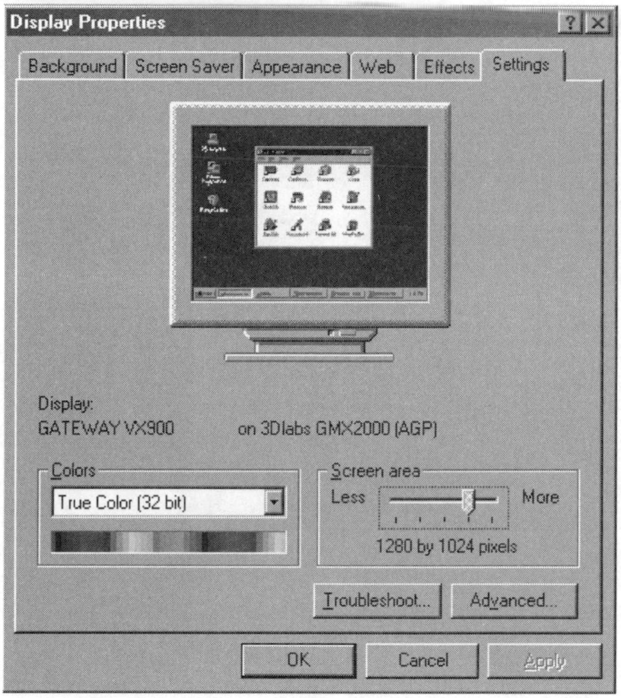

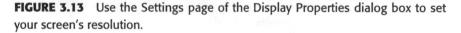

FIGURE 3.13 Use the Settings page of the Display Properties dialog box to set your screen's resolution.

6. Click the Display Type… button shown on the Settings page. Figure 3.14 shows the Display Type settings, options for changing adapter type, and an option for detecting the video adapter already installed. The Plug and Play functionality in Windows 2000 reliably senses how many and which graphics adapters are installed, and recommends a device driver for the specific graphics adapters being installed.

7. Click once on Change… in the Adapter Type dialog box. The Video Adapter wizard appears, which guides you through the selection of a video driver for your system. Figure 3.15 shows an example of the Video Adapter wizard.

8. Scroll through the list of drivers, then select the one that best fits your workstation.

9. Click once on Next within the Video Adapter wizard. The New Hardware wizard begins, looking for video adapter(s) installed in

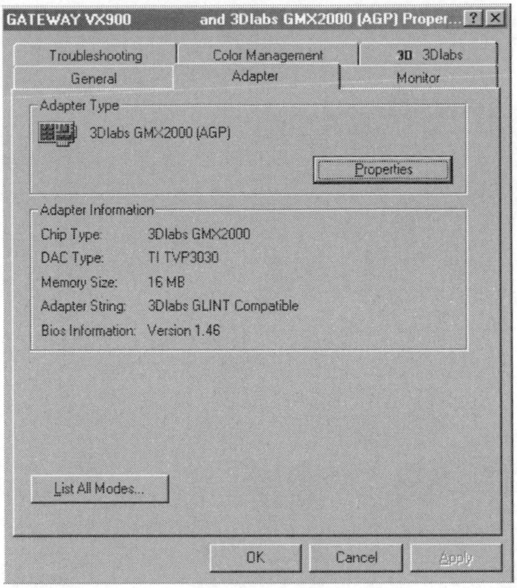

FIGURE 3.14 Changing video drivers in the Display Type dialog box.

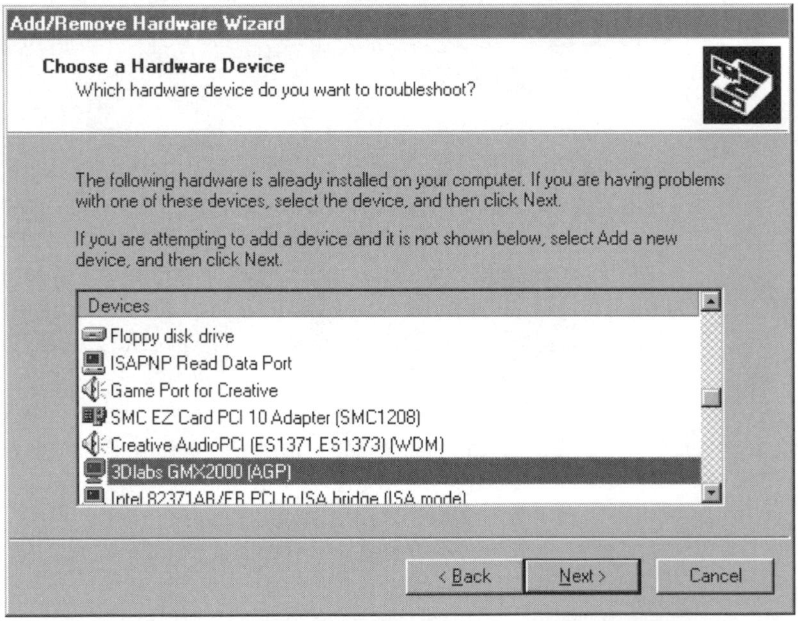

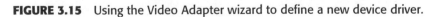

FIGURE 3.15 Using the Video Adapter wizard to define a new device driver.

your workstation. Figure 3.16 shows the Add New Hardware wizard that is launched from the Change… option selected in the Display Type dialog box.

10. Follow all steps of the Add Hardware wizard. You will end up back at the Display Type dialog box.

11. Click Cancel. The Display Type dialog box closes, showing the Display Properties dialog box.

FIGURE 3.16   Using the Add New Hardware Wizard for installing a new video adapter.

12. Click the Test button. A warning dialog box appears telling you that the screen will be temporarily re-displayed.

13. Click OK to have the dialog box disappear and the test completed. The test begins with the screen being filled with color boxes and patterns to test the resolutions supported by your video adapter and monitor. After five seconds, your screen returns to the Desktop, showing the Display Properties dialog box.

14. If the test completed successfully, click OK. The Display Properties dialog box closes, showing the Desktop in the resolution you have selected.

How Shortcuts are defined from the Desktop

Look at the Desktop of your workstation, or any system running Windows95 or Windows98. You will notice a small icon with an arrow coming up from the lower left corner. This is a *shortcut*, which is a direct link to an application or executable that is frequently used. Shortcuts are very useful for streamlining access to commonly used applications. Let's take a quick tour of how to create a shortcut on the Windows 2000 Professional Desktop:

1. Right-click on the Desktop. A small menu appears. Figure 3.17 shows how to find the Shortcut menu from the abbreviated Desktop menu.

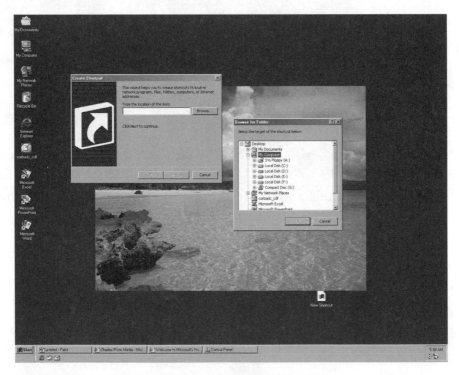

FIGURE 3.17 Creating a shortcut from the NT Desktop.

2. Selecting the Shortcut option on the abbreviated Desktop menu, the first screen of the Create Shortcut wizard appears. You can also create a Shortcut by dragging an icon directly onto the Desktop. You can also drag the icon with the right mouse button to create a shortcut.

3. Click once on Browse. The Browse dialog box is useful for finding the application or executable you want to run. Figure 3.18 shows what your screen looks like at this point.

4. Using the options in this dialog box, find the application to which you want to create a link. Figure 3.19 shows the Browse dialog box with Microsoft PowerPoint selected.

5. Click OK. The path you have defined through your selections in the Browse dialog box is now placed into the Create Shortcut wizard for definition with a shortcut. Figure 3.20 shows an example of what the wizard looks like.

6. Click once on Next. The final page of the Create Shortcut wizard is shown.

7. Click once on Finish. The Shortcut is then finished and ready for use directly from the Desktop.

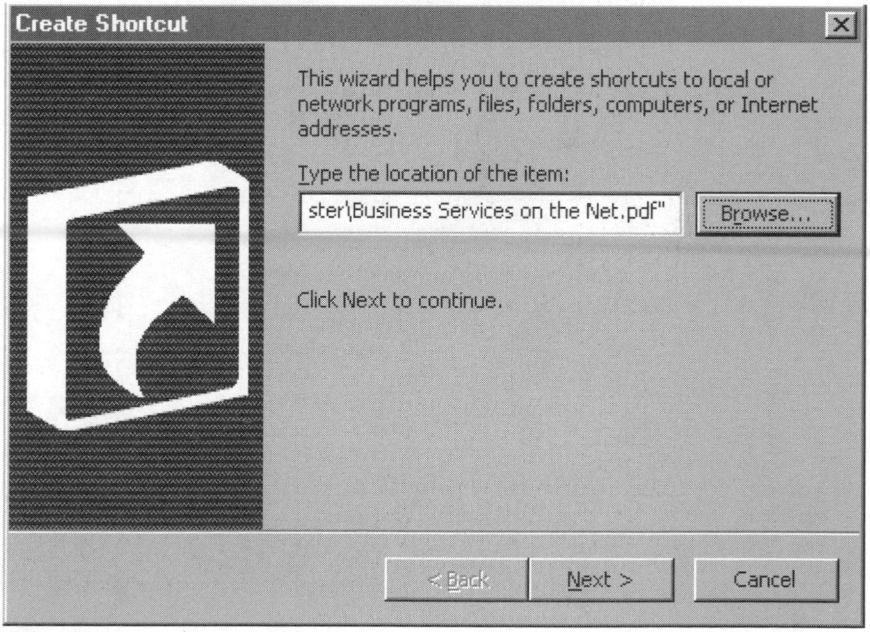

FIGURE 3.18   Using the Browse option in the Create Shortcut wizard menu.

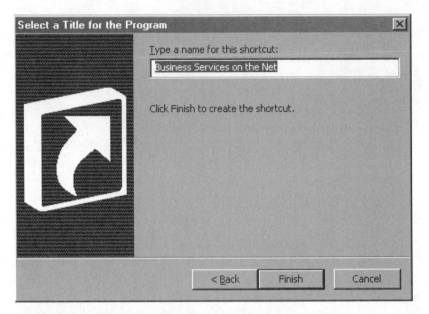

FIGURE 3.19 Finding the application to which you want to link using the Browse dialog box.

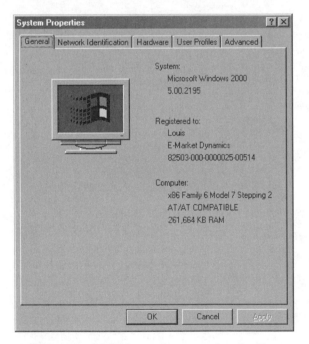

FIGURE 3.20 Finishing a shortcut using the Shortcut wizard.

Desktop Properties and Their Many Uses

In summary, the Desktop properties provide a wealth of customization options for you both as a user of Windows 2000 Professional and as an administrator. A book could easily be written covering the myriad option combinations and their implications for user productivity. Take some time to explore the Desktop Properties dialog box when you get a chance. It is a rich feature set that includes options for configuring Web site integration directly to your Desktop, which is a first for any network-based operating system.

Defining System Properties in the Systems Applet

You will find that many of the applications you work with in Windows 2000 Professional make their own entries in the file system. Many use the existing system profile you have created for defining parameters specific to an application's performance. This is especially true for many applications that use a temporary or TEMP subdirectory. In addition, NT looks at the options configured in Hardware and User Profiles for specific parameters controlling how the workstation needs to respond to specific requests. The System Properties application is integral to getting these profiles defined, in addition to setting performance parameters for how your Windows 2000 Professional will respond to each application's requests for resources.

Included in the System Properties application (or applet) are properties pages defining Hardware Devices, Hardware Profiles, User Profiles, General System Information, Performance options, Environment variables, and Start-up/Shutdown options. Throughout this section we will tour selected options on the pages of the System Properties dialog box. The Hardware Devices page includes two jump points, the first for starting the Hardware wizard and the second for starting the Device Manager. Figure 3.21 shows the System Properties dialog box.

Touring the Performance Aspects of System Properties

Whether you are a systems administrator or a power user of Windows NT, getting the most performance you possibly can from your workstation is undoubtedly a chief concern for you. This is especially true if you are working in engineering and other technically challenging professions where performance can make the difference between getting home before 6 p.m. or after 10 p.m. Let's take a quick tour of how Windows 2000 uses the options defined on the Performance page of the System Properties dialog box. Use these options to rev up your system's performance.

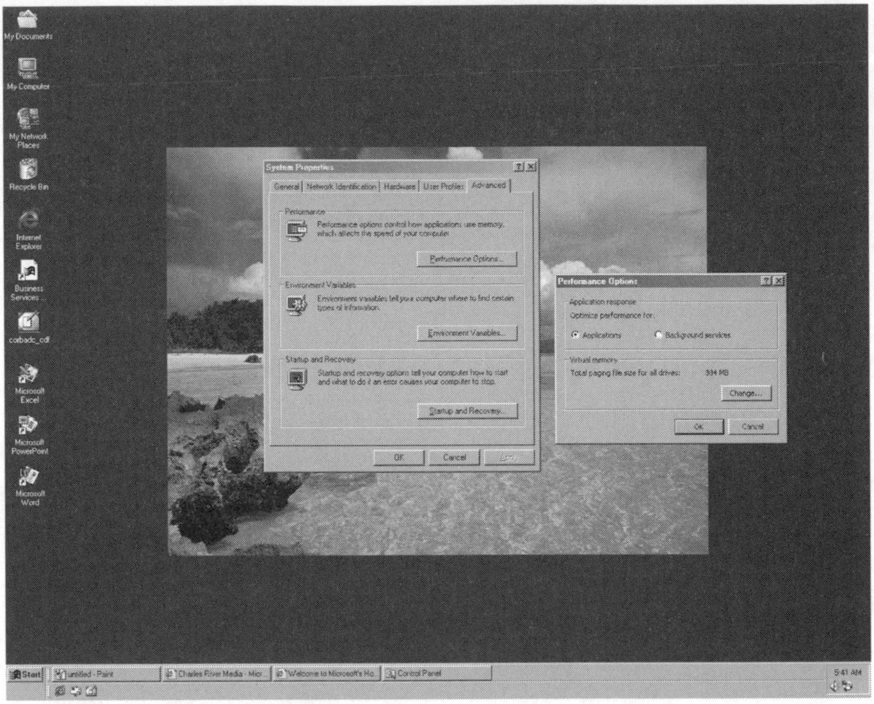

FIGURE 3.21   Touring the System Properties dialog box.

1. First, let's get to the System Properties dialog box. Double-click on My Computer in the upper left corner of the Desktop.
2. Double-click on the Control Panel icon. You will see many applets in this area, listed in alphabetical order. Look for the System applet, which is a blue icon in the shape of a desktop PC with a monitor on it.
3. Double-click on the System applet.
4. Click once on the Performance page of the System Properties dialog box. Figure 3.22 shows the Performance page of the System Properties dialog box.
5. Windows 2000 uses virtual memory: Tasks are swapped from RAM to disk and back, increasing the total memory space available to all tasks running. You can optimize this process as follows. Click once on Change... next to the Virtual Memory option. Figure 3.23 shows the Virtual Memory options page. This page is used to define the page file system for a selected drive. You can also use this

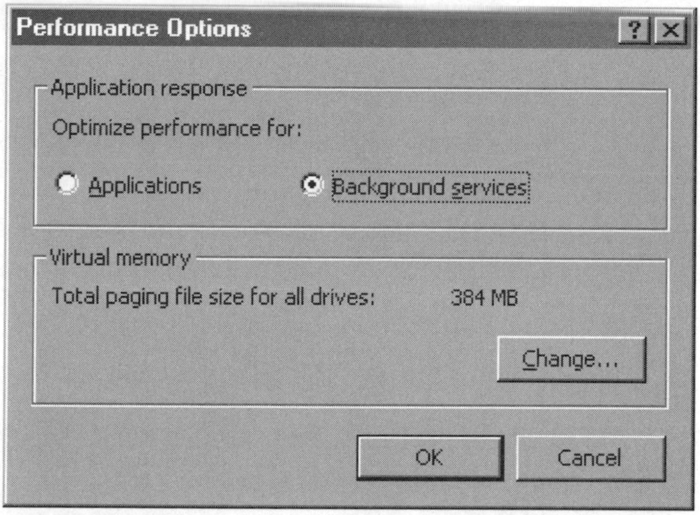

Performance Options

Application response

Optimize performance for:

○ Applications   ⦿ Background services

Virtual memory

Total paging file size for all drives: 384 MB

Change...

OK   Cancel

FIGURE 3.22   Using the Performance page in the System Properties dialog box.

option to define the total page file size and the size of the registry. You can use the options in this dialog box to define the size of the virtual memory in terms of disk space. While Windows 2000 calculates this automatically, you can increase it in the event you are using larger applications concurrently.

6. Click OK after setting the paging file size. The System Properties dialog box is again shown.

7. Notice the sliding bar in the first location of the Performance page. This is a relative measure of foreground application performance. Windows 2000 is a multitasking, preemptive operating system, so it can differentiate between foreground and background applications. Clicking once and dragging the indicator all the way to the right maximizes the application's performance, which is running in the foreground.

8. Click once on Apply. Your selections are now applied to the system.

You can also right-click on My Computer, then choose Properties from the Shortcut Menu to create a shortcut overall.

NOTE

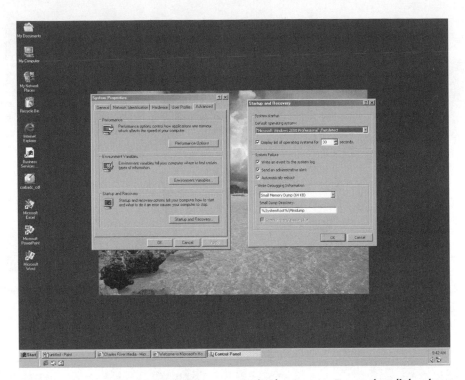

FIGURE 3.23   Using the Performance page in the System Properties dialog box.

CONFIGURING STARTUP AND SHUTDOWN PROPERTIES

Another area within the System Properties dialog box you'll find very useful is the Startup/Shutdown page. The purpose of this page is to define the System Startup operating system selection and to specify how long the FlexBoot option is shown on-screen. There are also a series of options you can select for enabling recovery of system errors. Figure 3.24 shows the Startup/Shutdown page of the System Properties dialog box.

FlexBoot is the listing of all operating systems found when you first power on a workstation. Windows 2000 shows all operating systems on the system or workstation, giving the user the option to select which one to use. The default operating system selected during FlexBoot is chosen from the Startup: option field, and the length of time the FlexBoot option is shown is by default 30 seconds, as shown in the Show List for: entry. You can change this to any value you choose.

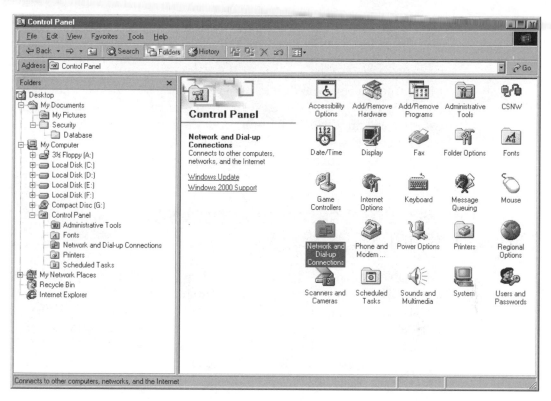

FIGURE 3.24   Configuring the System Startup and Shutdown characteristics.

Let's briefly tour each of the options in this dialog box:

Startup.   This is a pull-down menu that lists all operating systems installed on the computer you are using. Using this option, you can configure either Windows 2000 Professional or another operating system to be the first one selected during the start-up sequence.

Show list for.   This is the time period that FlexBoot will show the list of installed operating systems.

Recovery options.   In the event your workstation has either a soft or hard error, you can configure recovery services to write an event to the system log, send an administrative alert via e-mail (or to a pager if your system is configured with the proper options), write debugging information to a File, or automatically reboot. In many instances, system administrators who have many workstations to support choose to have the system automatically reboot itself. This makes it possible to clear any soft errors without having to run to the system itself.

After selecting the options in this dialog box, click once on Apply, which is the default selection in the active dialog box. Your selections are then recorded. When finished with the options in this dialog box, click OK. You will find that Applications Today is already interacting with System Properties without your intervention. Consider using the System Properties for diagnostic purposes most of the time.

USING THE NETWORK SETTINGS APPLET FOR DEFINING NETWORK PROPERTIES

Being able to share files, printers, and even processors through network connectivity is one of the most commonly cited reasons why organizations invest in networking operating systems like Windows NT. Networking is so integrated in Windows 2000 Professional that there are options found throughout this operating system for network configuration. You will find the Network applet in the Control Panel. Figure 3.25 shows the location of the Network applet in the Control Panel.

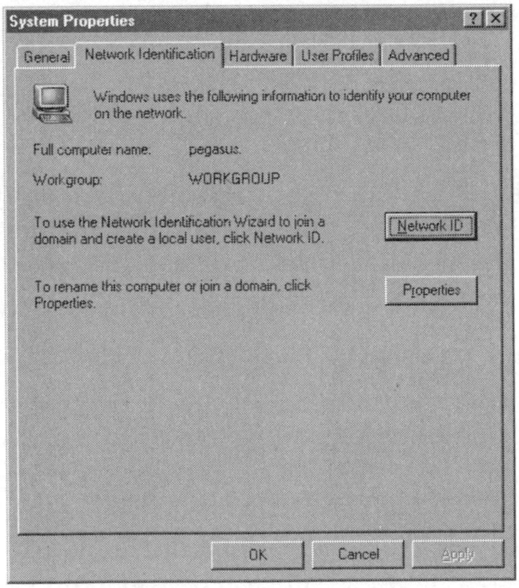

FIGURE 3.25   Finding the Network applet in the Control Panel.

The options within the Network applet include installing network services, defining network bindings for each of the network connections on your workstation, defining network protocols, and loading client services for your workstation. Let's take a tour of the Network Properties dialog box.

1. Double-click on the Network applet. The Network applet opens.
2. Click once on the Identification page of the Network Properties dialog box. Figure 3.26 shows the Identification page.
3. Both the Computer Name Workgroup names defined on this page are used by TCP/IP, Domain Naming Services, and even the Windows Internet Naming Services (WINS) for displaying this specific system's identity across the network. These names are defined during system installation, and they can be modified by clicking once on the Change... button.

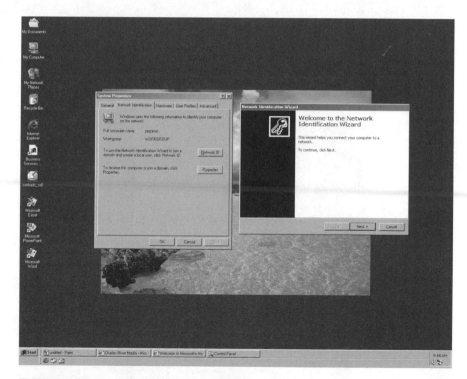

FIGURE 3.26 Introducing the Identification page of the Network Properties dialog box.

4. Click once on the Protocols page of the Network Properties dialog box. The purpose of this page is to define the protocol(s) installed on your system. Figure 3.27 shows the Protocols page.

5. It's also possible to add protocols to your workstation, making it compatible with future networks. Clicking once on the Add button displays a list of the protocols that are supported in the baseline configuration of Windows 2000 Professional. Figure 3.28 shows the Select Network Protocol dialog box with the protocols supported list.

6. Click once on the protocol you want to load. The selected protocol is then loaded. Once completed, the Protocols page then lists the protocol you selected for installation. Figure 3.29 shows the protocols installed.

7. Click OK. You will need to reboot your workstation to have the new protocols broadcast and/or sensed by the network (depending on the network protocol installed).

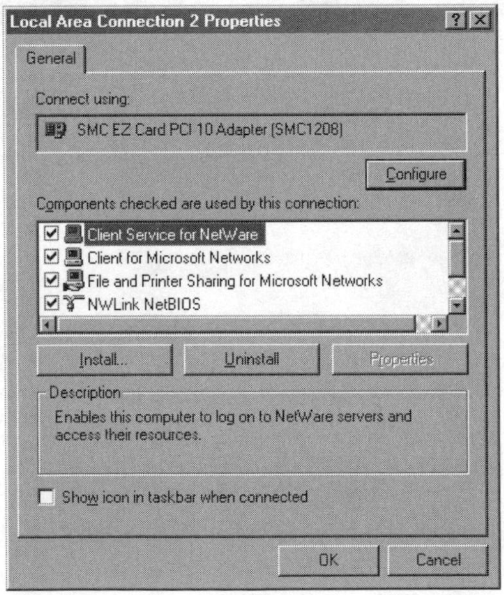

FIGURE 3.27 Determining which protocols are installed on your workstation.

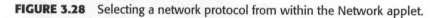

FIGURE 3.28   Selecting a network protocol from within the Network applet.

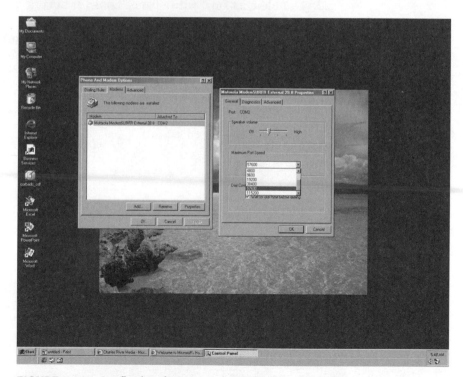

FIGURE 3.29   Confirming that network protocols are installed.

REMOTE CONNECTIONS AND THEIR PROPERTIES IN WINDOWS 2000 PROFESSIONAL

The Internet is making the entire issue of connectivity more urgent and necessary than ever before. NT's capabilities in this area are full of properties that can customize connections to Internet service providers, intranets, and Remote Access Servers (RAS) in an NT environment. The properties in Windows 2000 Professional serve to make Internet access more efficient and transparent.

Accessing and Changing Modem Properties

Touring modem properties begins in the Control Panel, where the Modem applet is, represented by a small yellow phone icon. Let's get started with a tour of how to change the basic modem properties here:

1. Double-click on My Computer.
2. Double-click on Control Panel. Look for the phone-shaped yellow icon.
3. Double-click on the Modems icon. Figure 3.30 shows the contents of the Modems Properties dialog box.

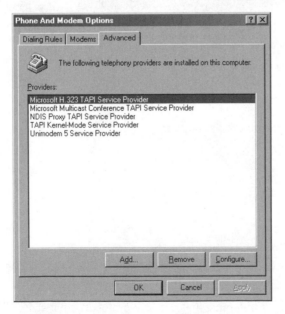

FIGURE 3.30 Introducing the Modems Properties dialog box.

4. Click once on Properties. For the specific modem selected in the initial Modems Properties dialog box, the Properties dialog box is shown. For purposes of this example, a Motorola ModemSURFR External Modem is listed. Figure 3.31 shows the properties associated with the modem selected.

5. Click once on the Connection page of this dialog box. This page shows the communication parameters associated with the given modem profiled. This page is specifically used for setting data bits, parity, and stop bits in addition to Call Preferences.

6. Click the Advanced... button on the Connection page. This presents the Advanced Connections dialog box, which is used for defining error and flow control, specifically the type of protocol used for handling connections between your workstation and the remote system.

7. Click OK. The Modem Properties dialog box is again shown.

Throughout the previous tour of the Modems Properties dialog box, you can see where you would enter the key parameters for handling communications connections with the host(s) of your choice. Many commu-

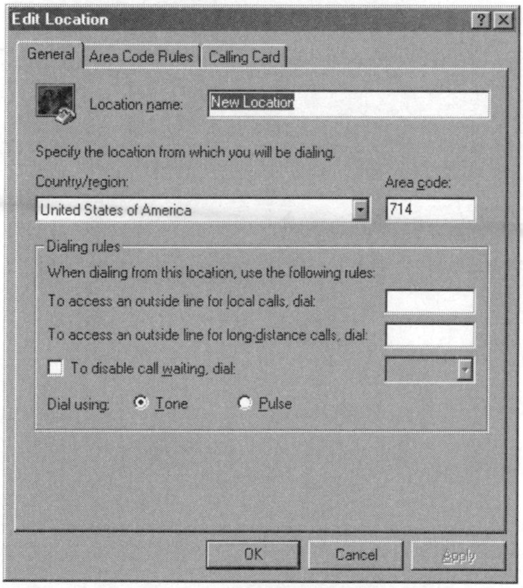

FIGURE 3.31 Every modem installed in NT has a set of properties associated with it.

nications programs today will reset these values during initialization, saving you from having to configure them manually. Nonetheless, it is a good idea to know how to change these properties if you need to.

Configuring Dialing Location Properties

Let's now look at how to configure the location you'll be dialing into. Because you are already in the Modems Properties dialog box, let's take a tour of how to get the Dialing Properties set up for the various locations you will dial out to on a regular basis.

1. Click once on the General page of the Modem Properties dialog box. Look at the lower portion of this dialog box for the Dialing Properties button.
2. The Edit Location dialog box appears, and it is shown in Figure 3.32. The intent of this dialog box is to define locations you are dialing from and the number sequence to access an outside line. You can also use options in this dialog box for configuring calling cards as well.

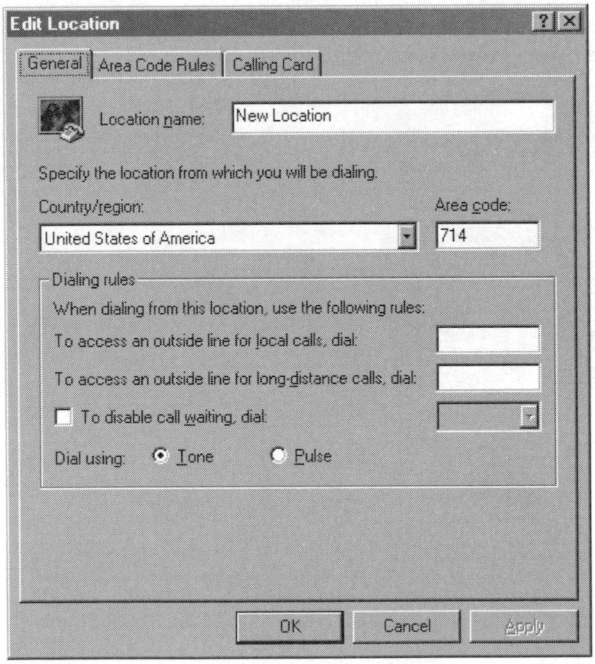

FIGURE 3.32   Configuring Dialing Properties.

3. Be sure to enter an 8 or 9 (depending on your phone system) for placing long distance calls.
4. Click either on the tone or pulse dial, given your preference
5. With all the options set in this dialog box, click OK. The Modem Properties dialog box again appears.
6. Click once on OK in the Modem Properties dialog box to close it. Your Desktop is again shown.

CHAPTER SUMMARY

The extensive use of properties for defining system events is more pervasive than ever in Windows 2000 Professional and Server. The use of properties is comprehensive, and it can have implications for the relative ease of use, performance, and communications capabilities. The intent of this chapter is to present a tour of the more common properties configured in Windows 2000. An entire book could easily be written on all the combinations of properties in Windows 2000 Professional and their implications for system performance and connectivity. As a system administrator or power user, you'll find that the properties in Windows 2000 give you flexibility to meet the requirements you have for analyzing system-wide performance. In the case of system administration, the wholesale configuration of systems is possible by first creating a customized profile using the customization and properties tools defined in this chapter.

4 | Installing Applications in Windows 2000 Professional

E ven before Windows 2000 was first introduced, development kits were available for creating 32-bit applications that would take advantage of the speed increases possible on the Windows 2000 platform. Microsoft's efforts to get software companies to port to Windows 2000 have been successful. A survey of software companies conducted by International Data Corporation showed that there is a 150% compound annual growth rate in the anticipated number of titles being ported to the Windows 2000 platform. With any luck, the applications you and your organization use are part of the en masse migration to Windows 2000 from UNIX, Windows, or DOS.

Inevitably, certain applications will not be ported to Windows 2000 due to a variety of factors. First, many of these applications were first developed on the DOS platform and have long since been out of a maintenance and upgrade cycle. Your organization may have a utility or calculation tool that was originally developed for DOS and no one ever had the time to port it to or rewrite it for the Windows or even Windows 2000 platform. Second, your organization may be dependent on a communications application, for example, that runs great on UNIX and has a Windows client version, and due to the strength of UNIX demand in the software vendor's customer base it doesn't feel the need to migrate to Windows 2000. Third, if your organization is part of a government agency and has made significant investments in UNIX and POSIX, any new operating system must support applications originally developed on these platforms.

Microsoft learned from talking with customers that many organizations are dependent on applications based on other platforms. No matter how sophisticated, intuitive, and responsive an operating system becomes, there is always the need for compatibility with the previous generation of applications that customers have been using. Operating systems serve as the foundation element of compatibility in a workstation.

In this chapter you'll learn how to continue using these applications on Windows 2000 Professional.

USING WINDOWS 2000 PROFESSIONAL WITH APPLICATIONS

Windows 2000 Professional is a 32-bit operating system that is distinguished from Windows95, Windows98, and Windows 3.X by its integration of multithreaded support of applications and its ability to preemptively multitask applications in and out of memory to ensure consistent performance. Another differentiating feature of Windows 2000 is its ability to run each application in a memory-protected subsystem to ensure reliability. These protected memory subsystems are assigned to a given application once they are launched. There are subsystems in Windows 2000's kernel to support the native 32-bit applications, sometimes called Win32-compatible applications, in addition to Win16, POSIX, and OS/2 subsystems. What about DOS? Windows 2000 uses an emulator that is accessible from the Command Prompt application to provide compatibility for DOS-based applications. Subsystems are included in the Windows 2000 operating system structure to ensure reliability with these various applications. The Windows 2000 subsystem overview is shown in Figure 4.1.

Windows 2000 Professional supports 32-bit applications in native mode, where the applications can make calls into the Windows 2000 Executive for resources. When a 16-bit application is launched, it is first loaded into memory, then assigned into the Win16 subsystem for ongoing processing. This subsystem approach to segmenting applications ensures a high level of reliability in Windows 2000. In the event that a 16-bit application stops functioning or freezes, the remaining applications on the workstation can continue running.

Using this subsystem approach also makes the request for hardware resources transparent to 16-bit applications. When a 16-bit application requests hardware resources they are taken by the Windows 2000 Executive and then queued up for a given hardware resource along with requests

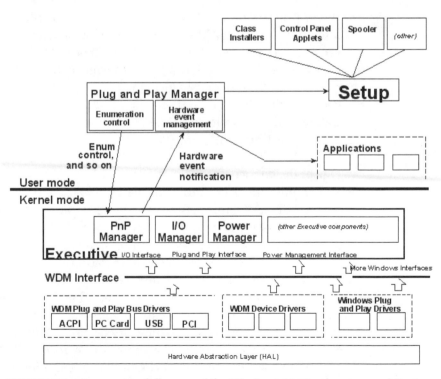

FIGURE 4.1 A conceptual diagram of the Windows 2000 subsystem architecture.

from other subsystems. Windows 2000's subsystem takes the requests for resources and translates them into 32-bit counterparts. Once the task is complete, a response is then reformatted into a 16-bit compatible response, which is in turn sent back to the application occupying the 16-bit subsystem. This entire process is just like going to another country and having an interpreter provide you with what a native is saying and vice versa. As you would suspect, 16-bit applications do pay a performance penalty in Windows 2000 environments due to the overhead of having to translate requests for resources back and forth. The term "VDM" in Figure 4.1 stands for Virtual DOS Machine, which is the term applied to the resource threads associated with the Win16 subsystem. An essential part of the Windows 2000 architecture is the WOWEXEC, or Windows on Windows. This area of Windows 2000 makes it possible to have Win16-compatible applications running on a Windows 2000 Professional-based workstation or server.

What Is WOWEXEC?

When using Windows95, I like to have several applications open at the same time—typically Microsoft Outlook, a variety of text and graphics programs, even a Web browser or two. The trouble begins when one application suddenly either runs out of memory or just stops working, period. The entire system then locks up, and often I lose portions of a file I've been working on. You no doubt have had the same experience. It's frustrating, and it can cause you to be cautious about using several applications at the same time in Windows95. Inevitably you end up rebooting the system to clear the memory.

One of the key benefits of the WOWEXEC application is the ability to run 32-bit applications in a protected memory space. This feature is one of the primary differentiators between Windows NT and Windows 2000 relative to Windows 95 and 98, as the protected memory functionality ensures the application doesn't affect the operating system functionality overall.

Windows 2000 Professional's integration of subsystems is aimed at solving this problem. A Win16-based application that suddenly quits working is recoverable in Windows 2000. Why? Because of WOWEXECS. These are the memory segments in the Win16 subsystem where each Win16-based application separates itself and its commands from other Win16-based applications. The payoff of WOWEXEC is a significantly higher level of reliability in Windows 2000 for running 16-bit-based applications. The downside is performance, as each application using

WOWEXEC must go through the Win16 subsystem for resources to complete tasks. Also, because the kernel for Windows 2000 relies on the Win16 application being launched from your workstation to detect the type of application being used, WOWEXEC does not apply to Win16-based applications launched over a network.

When you launch an application in Windows 2000 Professional and Server, the operating system checks to see if it is a 16- or 32-bit-based program. If it's Win16-based, then WOWEXEC is initiated. How can you be sure WOWEXEC gets initiated for an application? By choosing to launch the application from the Run dialog box, as this is the only dialog box in Windows 2000 Professional that gives the user the option of running an application in a separate memory space.

By default, when a Win16-based application is started it is assigned a separate memory space. Using the Run dialog box you can choose whether you want to have the application assigned a completely separate memory space. In general, it's a very good idea to have this feature turned on if you plan on launching applications from the Run dialog box. It's important to recognize that Windows 2000 looks at the application being profiled in the path definition to see if it is a 16-bit application first, then makes the option of selecting separate memory space available. Each subsequent Windows 16-bit application launched becomes part of the WOWEXEC process, thereby protecting the remaining applications running in other subsystems of the operating system.

Comparing Single versus Multithreaded Applications

With the increase in Win32-based applications there is increasing interest in the performance gains of single-threaded versus multithreaded applications. A single-threaded application is typically 16-bit, developed using the Win16 Application Programmer Interface (API) guidelines. For a software company to have a Windows-compatible application, it needed to write its application using the Win16 API. A key assumption behind writing applications for Windows 3.X environments was the availability of single threads or requests for processor and memory resources. Each application could generate a single thread or request for resources at a time. If a user running the 16-bit version of Word, for example, had several documents open at the same time, a single thread is created for the applications' resource needs. As a user opens more and more applications in Windows 3.X, the overall performance becomes very slow; the system would eventually stop running as there would be no more resources available.

Conversely, a Win32-based application is based on the Win32 APIs that have been so aggressively promoted by Microsoft as multithreaded. Due to the commands included in Win32, applications are capable of taking advantage of the multithreaded aspects of the Windows 2000 operating system. This API also provides support for the flat memory model of Windows 2000 and for protected memory mode functioning. A 32-bit application also has the ability in a Windows 2000 environment to make native calls for resources in the operating system, ensuring faster performance. You'll notice the different in performance between a Win16-based and Win32-based application.

What then are the key differences between Win16- and Win32-based applications?

Win16-based applications are native to the Windows 3.X environment. Due to the integration of the Win16 subsystem, VDMs, and WOWEXEC, they are also compatible and can be used with Windows 2000 5.0 Workstation. Their performance will be influenced by the need to have resource calls translated from Win16 to Win32 and back again.

Win32-based applications or 32-bit applications are native to Windows 2000 and are structured so as to have direct access to Windows 2000 resources, including protected memory subsystems, multitasking, and the multithreaded aspects of managing resources. All things being equal, it's better to choose a Win32-based application for your work and recreational needs as it is tailored specifically to the strengths of Windows 2000.

Software companies that have completed migrations of their Win16-based applications to Win32 have reported speed increases from 40 to 60 percent with their Win32-based programs. Just as car companies say "your mileage may vary," your system's performance may vary as well. There is a definite sped increase in multithreaded, Win32-based applications over their Win16-based counterparts, mainly due to the decrease in overhead.

From a practical standpoint, be sure to look for the Microsoft Compatibility Logo on the software packaging. The development of this program is in response to the widely varying results of companies providing applications for this platform.

INSTALLING WINDOWS 3.X APPLICATIONS IN WINDOWS 2000 PROFESSIONAL

Windows 3.X applications are the Win16-based applications we have been discussing, and they are installed in Windows 2000 Professional using one of the approaches outlined here. One the alternatives, using the Programs wizard, is new and appears in the Windows 2000 product family for the first time in Windows 2000 Professional. The other approaches—using the Run dialog box, launching Setup from Explorer, and using the Add/Remove Programs Properties tool—have been in previous versions of Windows NT.

Using the Run Dialog Box

Installing applications using this approach is very similar to using an MS-DOS prompt to launch an application via a path definition or using Windows 3.X dialog boxes (remember those?) for getting an application up and running. Let's take a quick tour of how you can install an application using the Run dialog box. Be sure to have the CD of the application you want to install handy. If your application also requires a diskette, be sure to have both inserted into their respective drives before starting this process.

1. Select Run from the Start menu. By default it is the first selection on the Start menu above the Log Off selection.
2. Click once on Browse to find the application you want to install. For purposes of this example Adobe Acrobat is being installed.
3. From within the Browse dialog box, click once on Open. The path for the application is then placed in the Run dialog box. Because this application is 16-bit based, the selection for running in a separate memory space was available as an option and is selected. Figure 4.2 shows the Run dialog box being used to install Acrobat.
4. Click OK. The executable highlighted in the Run dialog box is initiated and the application is installed.

Launching Setup from Explorer or File Manager

You can use either Explorer or the File Manager to install applications in Windows 2000 Professional. To install from Explorer, follow the steps provided here:

1. Right-click on My Computer, located in the upper left corner of the Desktop. A menu appears, showing a variety of options.
2. Select Explore… from the abbreviated menu. Explorer appears, as shown in Figure 4.3.
3. Using the options in Explorer, navigate to the install icon for your application.

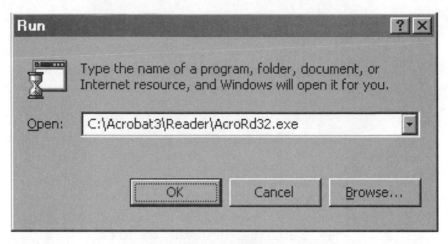

FIGURE 4.2 Using the Run dialog box to install Acrobat.

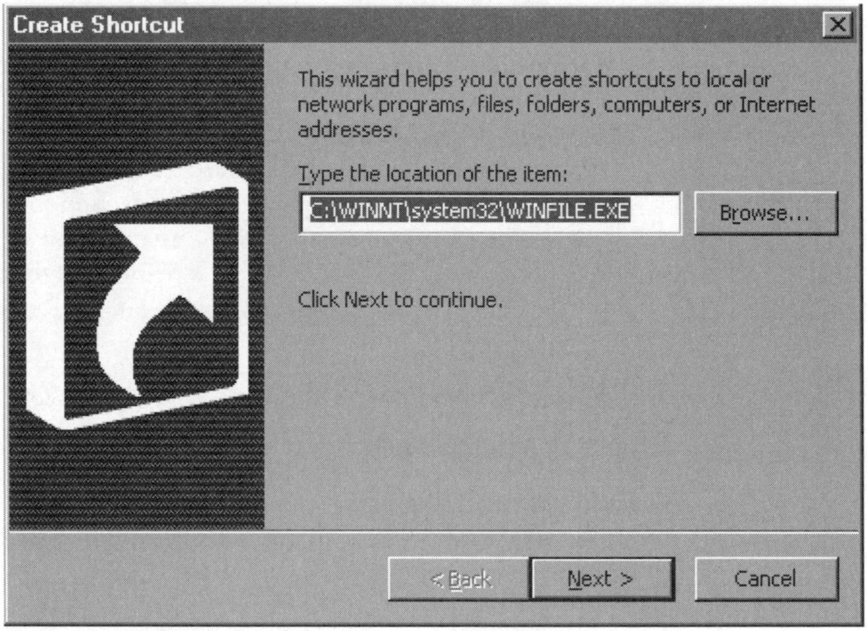

FIGURE 4.3 Using Explorer to install applications.

You can also use File Manager for installing applications. If a large number of users in your organization are using Windows 3.X still and you are planning to migrate them to Windows 2000, you can assist them in their transition by installing a shortcut to File Manager and then using its interface for installing applications. The principle of installing applications under Explorer is the same as using File Manager; it's just a different interface. The steps provided next show how to first create a shortcut to File Manager and then how to launch applications using its interface. If you've installed applications in Windows 3.X you'll find the File Manager intuitive.

1. Right-click on your Desktop. An abbreviated menu appears.
2. Select New, then Shortcut from the menu. The first page of the Create Shortcut wizard appears, as shown in Figure 4.4.
3. Click once on Browse. Select a file or application you use the majority of the time while using Windows 2000 Professional.
4. Click once on Next. This page of the Create Shortcut wizard prompts you for the name of the shortcut. Type the name of the application you are creating.

FIGURE 4.4 Creating a shortcut to NotePad.

5. Click once on Finish. The Shortcut is then created and appears directly on the Desktop.

Using the Add/Remove Programs Properties Tool

Another approach for installing Windows 2000 applications is to use the Add/Remove Program Properties Tool found in the Control Panel. As the name suggests, this approach makes use of a properties-oriented dialog box that leads into a wizard-like series of screens that guide you through installing software. First introduced in Windows NT 4.0 Workstation, the Add/Remove Programs Properties Tool is useful in that you can also use its features to define which Windows components you want to install or de-install.

The steps listed here show you how to install an application using the Add/Remove Programs Properties Tool.

1. Double-click on My Computer. The contents are shown, which include the Control Panel.
2. Click on the Control Panel. The contents display on-screen and are shown in Figure 4.5. The Add/Remove Programs Properties Tool is highlighted in the illustration.
3. Double-click on Add/Remove Programs Properties Tool. The Add/Remove Programs Properties dialog box appears, as shown in Figure 4.6.
4. Click once on Windows 2000 Setup. This shows the second page of this dialog box, which includes a list of the categories of utilities, tools, and applets included with Windows 2000 Professional. You can use the options on this page of the dialog box to either install or uninstall components delivered with Windows 2000.
5. Click once on the page titled Install/Uninstall again. Notice that along the bottom portion of the dialog box there is a list of the programs that can be automatically removed from Windows. These applications have uninstall scripts associated with them.
6. Click the Install… button, located in the upper portion of the Add/Remove Program Properties dialog box.
7. Click once on Next. The second page of the Install wizard is shown, which is used for browsing for the installation program or installation script for the application you want to install.
8. Click once on Browse. The Browse dialog box is then shown; it is useful for navigating to find the installation program you need to use.

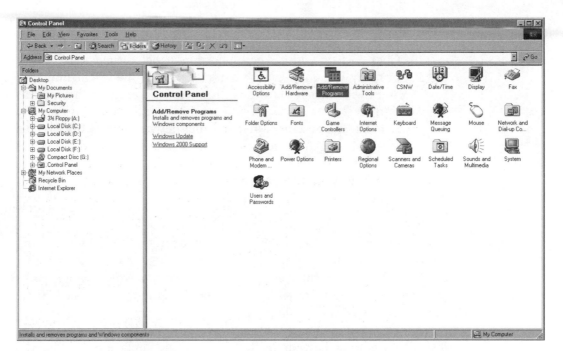

FIGURE 4.5   Finding the Add/Remove Programs Properties Tool in the Control Panel.

FIGURE 4.6   Using the Add/Remove Programs Properties dialog box to install applications.

9. Click once on Open. The selected program is then inserted into the Browse page of the Install Wizard dialog box.

10. Click once on Finish. The application is then installed. In the case of many Windows-based programs starting the installation program will actually start the introductory screens for the application, as is the case with Site Server. This intro screen is typically called a "splash screen" in that it splashes across the workstation's screen during the installation of the program.

Using the Add Program Wizard

This last approach to installing applications centers on a new series of tools available for the first time in Windows 2000. The Add/Remove Programs wizard is found in the Control Panel and has a series of selections and corresponding paths for adding, upgrading, repairing, or removing existing programs from your workstation. The approach of having multiple paths from within a wizard selectable also can be found in the revamped Hardware wizard in Windows 2000.

The steps outlined here show how to use the Add Program wizard for installing a new application. In this instance, the application will be Microsoft's Site Server.

1. Click once on the Add/Remove Programs icon in the Control Panel. The intro page of the Add/Remove Add Programs Wizard is shown. Click once on the Add New Programs selection on the left side of the Add/Remove Programs Wizard. Figure 4.7 shows the Add New Programs page of the Add/Remove Programs Wizard.

2. Click Add New Program, the first radio button on the intro screen of the Add Program wizard.

3. Click once on CD. The Add Program wizard prompts you for the location of the application you want to install. Figure 4.8 shows this second page of the Add Program wizard. For purposes of this example the application will be loaded from a compact disc (CD), so the first entry is selected.

4. Click once on Next. The third screen of the Add Programs wizard prompts you for the type of media being used.

5. Click once on CD. For purposes of this example assume you're loading software from a CD-based product.

6. Click once on Next. The Add Software wizard checks which application(s) are on the CD being used for installation.

FIGURE 4.7 Using the Add New Programs page of the Add/Remove Programs Wizard.

FIGURE 4.8 Defining the location of the applications being installed.

7. Click once on Finish. The application begins to install, showing the first of several splash screens for completing installation.

INSTALLING AND USING MS-DOS APPLICATIONS

You can run your MS-DOS applications in Windows 2000, and you can actually get more done than would have been possible using a dedicated PC with the fastest processor, highest amount of memory, and best disk subsystem. Why? Because the multitasking nature of Windows 2000 gives you the flexibility of starting and ending MS-DOS applications in parallel with both Win16- and Win32-based applications running concurrently.

How do you use MS-DOS applications in Windows 2000? Recall that to use MS-DOS applications in Windows 3.X you'd have to use the PIF Editor, or Program Information Editor. The PIF Editor was a tool used for defining environment variables and system requirements. The PIF Editor was somewhat cumbersome to fine-tune, and the property-centric approach to using resources in Windows 2000 is also reflected in the series of properties that are customizable for each MS-DOS application. You'll notice that the properties dialog box associated with an MS-DOS application has six pages, while the comparable properties dialog box for Windows-based applications has only two pages. The reason for this is that each of the MS-DOS applications, once defined to the Windows 2000 operating system, are considered to be dependent on the Command Prompt window for use.

The series of steps here profile how to define properties of for an MS-DOS application so that it can be used in Windows 2000 Professional. How do you find out if an application is MS-DOS based? By looking in Explorer and toggling on Details from the View menu of Explorer. The description of applications will show which are MS-DOS applications, which are specific data file types, and which are simply applications.

Here are the steps for getting to the properties of a given MS-DOS application:

1. Right-click on My Computer.
2. Select Explore... from the abbreviated menu. The Explorer launches.
3. Select Details from the View menu. The list of applications is shown in the Explorer's windows, with definitions detailing which are MS-DOS applications.

4. Right-click on the MS-DOS application you want to install. The properties dialog box for that application is shown. In this example, it's the PKUNZIP Properties dialog box. It would also be the name of any MS-DOS application being installed.

This page is useful in that it defines the file name and type, in which file folder the file being profiled is located, the size of the file being profiled, and whether the file is compressed. The date the file was created is shown, in addition to the last time modified, accessed, and written to. There is also a definition of the attributes associated with the file.

5. Click once on Program. The purpose of this dialog box is to confirm the paths to the file, the full name of the application, the working directory that the application uses for completing tasks, any shortcuts associated with the MS-DOS application, and shortcut keys associated with launching the application. There's also an entry for the size of the window to launch when the application is run.

6. Click once on Properties.... The intent of this dialog box is to show the Windows 2000 PIF File locations. These are the AUTOEXEC.NT and CONFIG.NT files that serve as the Program Information Files in Windows 2000 Workstation. These files are used for launching DOS applications just as AUTOEXEC.BAT and CONFIG.SYS were used for running applications. In the case of Windows 2000 launching MS-DOS applications, both the AUTOEXEC.Windows 2000 and CONFIG.Windows 2000 files are initiated and run with every request. You can change the options in these two files to more finely tune MS-DOS applications to run on Windows 2000.

7. Click OK.

8. Click once on the Change Icon. The Change Icon dialog box is next shown, which gives you the option of defining another type of icon to represent your application.

9. Click once on the icon you want to use, then click OK. Notice at the top of the Program page that the icon you just selected is shown.

10. Click once on the Font page.

11. From this dialog box you can define the type of font that will be used in the DOS Command Prompt window. You can choose between bitmapped or TrueType fonts for the DOS window. Many people prefer the TrueType font as it scales directly to the size of

the screen. You can decide which font type and matrix size shown is best for you by using the options in this dialog box.

12. Click once on Bitmap, then the TrueType font selection, then Bitmap again. Notice how the Window preview: segment of the dialog box changes in size to reflect the type of font technology chosen.

13. Click once on an entry in the Font size: entry. Notice how the size of the font and the size of the window in the Window preview: area change correspondingly.

14. After experimenting with these options, select the one that best fits your preferences and then click Apply. The selections made will then be applied directly to the MS-DOS application for which you are defining parameters.

CONFIGURING MEMORY OPTIONS TO SUPPORT DOS APPLICATIONS

Using the options in the PKUNZIP Properties dialog box defines the most important aspect of any DOS application's performance in Windows 2000, the use of memory. The Memory options page defines how conventional, expanded, and extended memory are used in conjunction with the DOS application being configured, which is in this case is PKUNZIP.

Because of the number of possible options available on this page, each major area is defined in detail here. As you read through these options keep in mind the DOS applications you'll want to run and decide how you'd like to configure system memory.

Exploring Conventional Memory

Included in this area of the dialog box is the option of defining parameters for both Total and Initial environment memory classifications. Let's briefly look at what each of these does.

> **Total.** This refers to the total number of kilobytes that PKUNZIP needs to work. By default, this option is selected as Auto. This will allocate just as much memory as possible to ensure that your MS-DOS application runs. If you have time to test each memory range, you can do so by using the variety of options available. In general, it's a good idea to let this value stay at Auto.

Initial environment. This defines the number of bytes (not kilobytes) reserved for the initialization files on your workstation. Files that are part of the initialization sequence include COMMAND.COM and AUTOEXEC.Windows 2000. Because this variable defines the initial amount of memory allocated to Windows 2000's start-up process it's a good idea to have this set to Auto all the time as well. Setting this too low will cause your workstation to crash.

Protected box. By default this is not selected, but it's a good idea to select this option with all DOS applications. Selecting this option creates a dedicated memory partition into which a DOS application is loaded and runs. Not selecting this option opens you up to the potential of having an errant DOS application crash your entire workstation.

Configuring Expanded (EMS) Memory

At present, many DOS applications are not using expanded memory. It's best to leave this option set to None. In the event that a customized application your organization uses has built-in support for expanded memory, use the pull-down dialog box to define the memory you'll need to run it.

Configuring Extended (XMS) Memory

This memory specification was created to provide DOS applications with memory above the 1024K DOS limitation. Many recent applications support extended memory, so be sure to check the DOS application's manual for compatibility with XMS. When using this option keep in mind that CONFIG.SYS is loaded high on Windows 2000 installation, so selecting Use HMA may be irrelevant to your application. Be sure when using this option to specifically select a memory amount that is within your workstation's physical limit becausea number exceeding available memory will cause the application not to load.

Configuring DPMI Memory

This is managed extended memory that you can allocate to an application. Unlike Extended Memory, this type of memory can be set to a value higher than what is physically installed in your workstation, and the application will still run. This type of memory is somewhat similar to using a protected memory structure, yet it stops short of providing the dedicated memory partitions that are essential for reliable application performance.

EXPLORING SCREEN OPTIONS

The purpose of this page is to define the settings used when the Command Prompt window opens and runs a DOS application. This option page gives you several options for configuring the Command Prompt window. Let's take a look at the options you can use for defining screen properties using this page of the PKUNZIP Properties dialog box.

Let's take a look at the options available in the Screen options.

Choosing between Full-Screen or Window

Keep in mind that many DOS applications were developed for CGA, EGA, or even VGA screen resolutions, and they will modify the size of the Command Prompt window to reflect the standard defined in the software. Don't worry if you intend to run the DOS application in a single window; you can easily resize the window once the application is up and running.

Selecting Window Appearance

By default, both Screen options are selected. It's a good idea to leave both selected, as the first option, Display toolbar, provides a Toolbar across the top of the Command Prompt window. The second option, Restore setting at start-up, saves the options you select during a window session and reinitializes the window to your latest set of preferences. It's a good idea to keep both of these selected.

Defining Performance Attributes

You can also use the Screen options page for increasing the performance of your applications. The options are Fast ROM emulation and Dynamic memory allocation. Here are brief summaries of each of these performance-enhancing options in the Screen options page.

> **Full ROM emulation.** One of the system management aspects that DOS became famous for was going directly to the BIOS for hardware definition and resources. If your organization has standardized on a DOS application that bypasses the operating system and goes straight to the BIOS to complete tasks, you'll find that DOS application performance is accelerated when this option is selected. Instead of going directly to the BIOS, this option provides your DOS application with Windows 2000-based resources and tools, the majority of which are

32-bit based and perform faster than a direct hardware call to the workstation's BIOS anyway. Even if your DOS applications make calls directly to the BIOS, be sure to select this option to increase performance.

Dynamic memory allocation. If your DOS applications display the majority of their functions in text mode with a few in graphics mode, you'll want to select this option. In fact, if any portions of your DOS applications use graphics displays, be sure to select this option. This option serves to inform Windows 2000 to use system memory for both text and graphics screens.

FINISHING UP ON PROPERTIES

There are other pages as well in the PKUNZIP Properties dialog box, and you can explore them to see how the options they contain can further customize the appearance of your DOS applications. You can, for example, look in the Miscellaneous options and explore the General options in greater detail.

Once you have selected all the options you need, click once on Apply, then click OK. The changes to the PKUNZIP Properties dialog box are then recorded and made ready for the next time you use the DOS application.

CHAPTER SUMMARY

One of the main design goals of Windows 2000 is the ability of previous-generation applications to run reliably. To this end, Microsoft has created a series of subsystems that use protected memory partitions to ensure reliability and consistent performance. This chapter explores the key concepts of how the Windows 2000 architecture has been created with the goal of compatibility in mind and how the subsystems support Windows 16-bit, 32-bit, POSIX, OS/2, and DOS-based applications. Beginning with an overview of the organization of the Windows 2000 kernel and progressing through the approaches used to ensure compatibility with the industry's previous-generation applications, this chapter provides a roadmap by which you can teach others and navigate through the intricacies of how multiple applications run on Windows 2000.

5 | Learning to Use Printers in Windows 2000 Professional

INTRODUCTION

After getting an entire network up and running, you realize that setting printers up presents a challenge. The marketing communications department needs to get the latest ads for your company printed from their workstations to the color printer down the hall. Finance wants to get their reports by 9 a.m. every Monday for the meeting with the general managers, and development wants to use the new high-speed laser printer. In short, almost every job in a networked environment, the deliverable for many working people is the paper their thoughts and efforts are captured on. Printing is a big deal in any business since it is the method by which many professionals have a chance to deliver their expertise literally in black and white. Printing is here to stay, and the challenge of making this a resource for those you support or for a network you are responsible is the topic of this chapter.

The world of printing has changed rapidly during the last five years, and with the advent of Windows-based printing architectures there has been an increasing emphasis on network printing as well. Sharing printers once meant taking a extra-long Centronics cable and passing it to your co-worker; today it means sharing the printer over, in the majority of cases, a TCP/IP-based network. The decentralized aspect of printing has continued to force application developers, including Microsoft, to modify their approaches to presenting printed pages from their applications. Today in Windows 2000 Professional, just like previous versions of Windows NT, each printer specifically has a queue of its own, complete with capabilities for managing print requests or jobs. Each printer in Windows 2000 can also have multiple logical names, which makes the process of creating separate printer identities on a network easily accomplished. Lastly, there is extensive support for internet-based peripherals. Novell pioneered the ability to print directly to a peripheral, and Microsoft's implementation in Windows 2000 is explained in detail in this chapter. This is a hands-on chapter that guides you through the intricacies of setting up, installing, managing and troubleshooting printing in Windows 2000 Professional.

PRINTING FROM WINDOWS 2000 PROFESSIONAL

Beginning with the first version of Windows, there has always been a Print Manager, which served as the centralized resource of all output devices, including printers. You'll find that instead of the Print Manager, there is now a Printers Folder, which contains both an Add Printer Wizard and icons representing the various printers you have created. The approach to printing in NT 4.0 and now in Windows 2000 Professional is to focus on the customization options inherent in each output device. Printing from applications is the same using Windows 2000 Professional as was the case with any previous version of the operating system, yet the approach to installing or creating printers and managing them is different. This chapter provides you with a thorough description of how these new features that pertain to printing work.

Printing from Windows NT involves several steps. First, the application that you request the print command from takes the document, graphic or spreadsheet and spools it to the intended printer. *Spooling* is the process of writing the contents of a document to a file on disk. This file is typically called a spool file. Operating systems spool files so that the data being printed can be sent to the printer at a printable speed. Many times you'll send multiple documents to the printer, and due to the imaging speed of the printer, you'll need to wait from several seconds to a few minutes for the documents to be produced. Documents waiting to be printed are actually stored in a *queue*. In developing a property-centric approach to printing in NT 4.0 and now with refinements in Windows 2000 Professional, there is the strong focus on managing queues of each printer. It's really the same series of steps that happened in the Print Manager that resided in Windows 3.0, 3.1, Windows NT 3.1 and 3.51. The only difference is now you have more options as a user to view and control the sequencing of print tasks or jobs in a given printer's queue. Figure 5.1 shows

FIGURE 5.1 Touring the Printing Features of Windows 2000 Professional Begins in the Printers Folder.

an example of the new graphical interface to two key printer utilities in Windows 2000 Professional: the Printers Folder with the Add Printer icon and an example of a printer queue.

Exploring the New Printing Features and Their Benefits

Printing has become more intuitive in Windows 2000 Professional Workstation. Here are the major enhancements or benefits to printing capabilities:

- Many of the printer drivers delivered by Microsoft have additional features included that make it possible to customize printing tasks more precisely than has been possible in the past. Support of additional features is more robust in the Windows 2000 Professional than before.
- Any printer connected to a network where NT is being used can have multiple logical names, meaning you can customize as many individual queues as you want for a printer, depending on the type of document you want to produce. You could for example have a HP Color LaserJet in your office where you create one queue specifically for transparencies, another for draft monochrome reports, and a third for high-resolution project management charts.
- Given the new graphical interface to how printer identities are stored, it's much easier to figure out which printer to use for a given task. Printers are stored as icons within the Printers Folder.
- One of the key benefits of print spooling, namely that the background printing gives control at the application back to the user faster than with a non-spooling print architecture.
- There are now more ways to get to the Printers Folder than before. You can for example get to the Printers Folder from the Control Panel, Explorer, Start Menu, or from within My Computer.
- Continuing from Windows NT 4.0, The Add Printers Wizard now includes options for configuring the printer's driver to take into account which type(s) of client workstations will be printing to it. For example, you can during the installation of a printer tell the device driver that other users on the network are running specific operating systems other than Windows 2000 Professional, and the printer driver compensates during printing for the variation in

formatting of data from those diverse client systems. This makes the issue of cross-platform connectivity transparent to both the users and the administrators of Windows 2000 Professional -based print networks.

- Using the options inherent with each printer, you can modify the print sequence of documents, even delete selected documents or purge the print requests of an entire printer.

- Printers are multi-threaded applications, which makes their performance significantly faster than their Windows 3.X counterparts. You'll find that the printing task, once requested, is spooled quickly into the printer queue, and the application is off-loaded from the task of using memory to support the print request.

- Sharing printers over a network is easier than before as all properties, including whether it is a local or network printer, are available directly from menus available on the queue window.

- It's possible to secure a printer and define user rights by class of user. You can vary how administrators, power users, or everyone (who is everyone else logged in) uses the printer. Using these options you can give administrators the rights to change printer properties, and give everyone else permission to print.

- If you have multiple documents going to several printers at once and want to see when they are completed, you can open up each printer's window and watch the progress of each print request.

- You can select the option in each printer to send you e-mail once printing has been completed, and also you can have cover sheets printed between jobs as well. This second point is very useful in a networked environment.

- You can also view printing documents in HTML format using Web View. This options requires that Microsoft Web Services also be installed and running on your Windows 2000 Professional.

- Windows 2000 Professional can function as a print server in small network environments where there are ten or few users. Many work groups and organizations use a Windows 2000 Professional as a small print server in addition to mainstream application. The Windows 2000 Professional print architecture can operate in either a client or server role. For over 10 users, Windows 2000 Server is recommended. Windows 2000 Professional supports up to 10 users per Microsoft's licensing guidelines.

Creating a New Printer

In Windows, many people called it installing a printer driver or getting a printer installed. Now in Windows 2000 Professional through the use of the Add Printer Wizard, it's called creating a printer. In getting a printer up and running in Windows 2000 Professional, it's a matter of following the steps of the Add Printer Wizard. At the end of this process an icon is going to appear in the Printers Folder that represents the printer queue. Let's start with a step-by-step guide on how to create a new printer in Windows 2000 Professional. Incidentally, these steps also apply to Windows 2000 Server as well.

Creating a New Local Printer

From the Desktop, double-click on My Computer. Figure 5.2 shows the contents of My Computer. Notice that now Printers have a folder specific to themselves, and is delineated with a printer icon on the top of it.

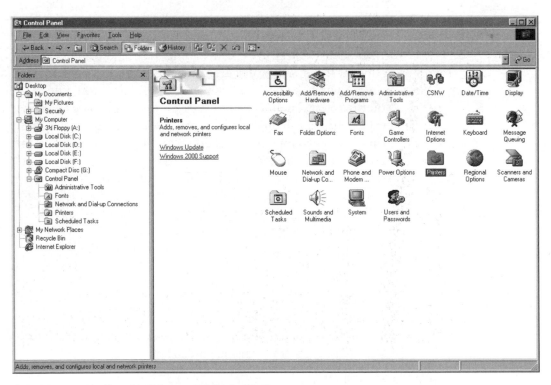

FIGURE 5.2   Finding the Printers Folder in My Computer.

Double-click on the Printers Folder. You'll find the Add Printers Wizard icon and any other printer icons representing those printers already installed.

Click once on the Add Printer icon. The first screen of the Add Printer Wizard is shown, and appears in Figure 5.3.

FIGURE 5.3 The Add Printer Wizard guides you through creating a printer.

The purpose of this first screen is to define if the printer being installed is going to be used for either local printer (My Computer) or over a network (Network Printer Server). By default, the option of My Computer is selected.

Click once on Next. The second page of the Add Printer Wizard is shown. This page defines which of the printer ports you'll be using for communicating between your workstation and the printer. The selection of LPT (the Centronics parallel interface) is already configured for use with a QMS 860 printer. You can see from Figure 5.4 that there are several other options, including Add Port....

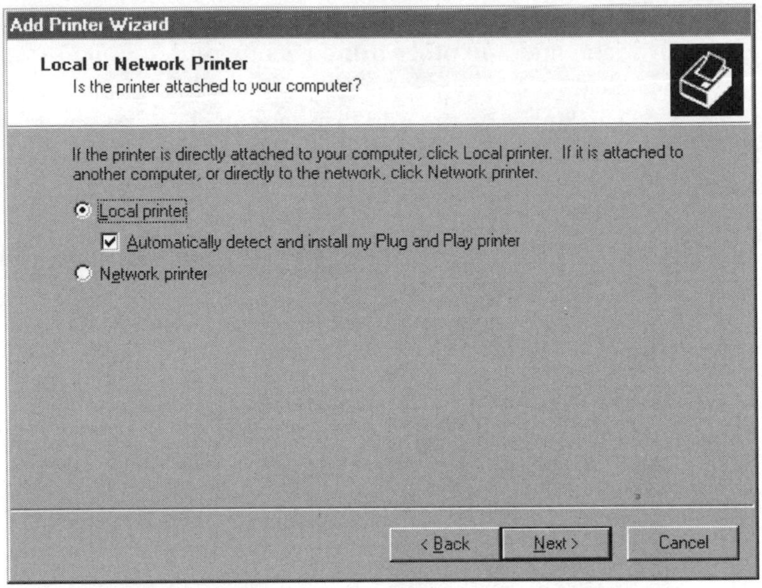

Add Printer Wizard

Local or Network Printer
Is the printer attached to your computer?

If the printer is directly attached to your computer, click Local printer. If it is attached to another computer, or directly to the network, click Network printer.

◉ Local printer
   ☑ Automatically detect and install my Plug and Play printer
○ Network printer

[< Back] [Next >] [Cancel]

FIGURE 5.4   Defining the port you'll use for printing.

TIP

Network Printing Tip
When installing a printer either on a network or on a print server, click once on Add Port... at this point and select the Print Monitor for use with the printer. A Print Monitor is the network-based component of a print driver, making it possible for other workstations on your network to see the printer after it is installed for shared use. The Add Port... dialog box is shown in Figure 5.5.

Click on the port you want to use to connect your printer to your workstation. Since there is already a printer on LPT1, the selection of LPT2 is made.

Click once on Next. The page of the Add Printer Wizard is shown that defines which printer driver you want installed to enable printing. Figure 5.6 shows this page of the Add Printer Wizard.

Scroll down the list of manufacturers to the left side first, then select the model of printer you are using. Figure 5.7 shows an example of selecting the HP 6P.

Device drivers are increasingly being delivered over Internet, and as a result the installation routines know where to look for them during the in-

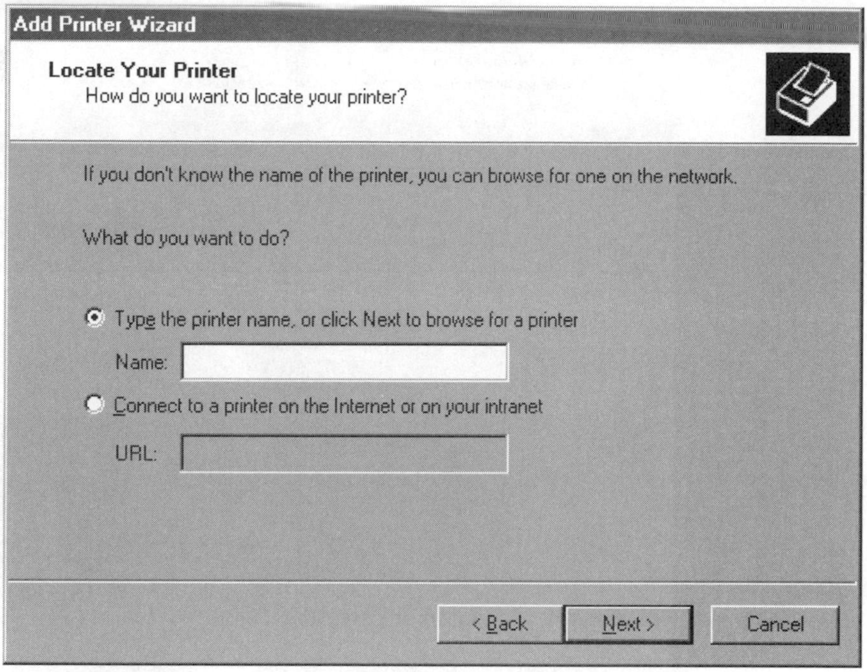

FIGURE 5.5 Configuring a Network Printer Port using Print Monitors.

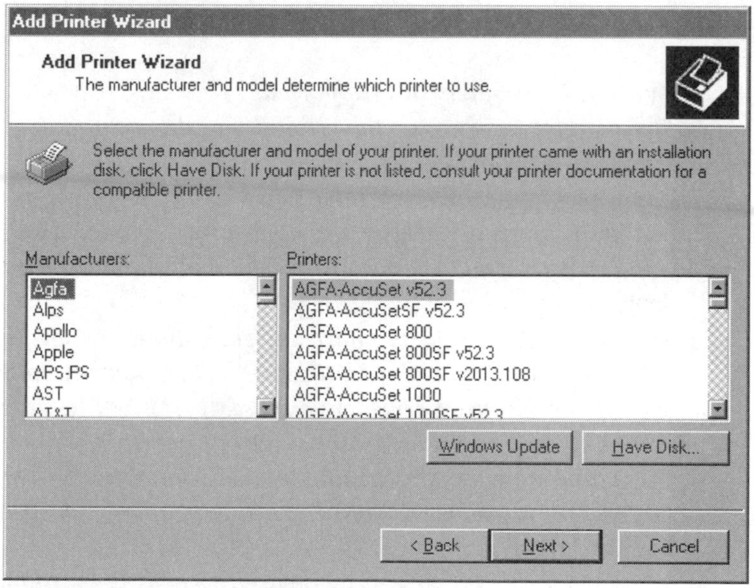

FIGURE 5.6 Selecting a printer driver.

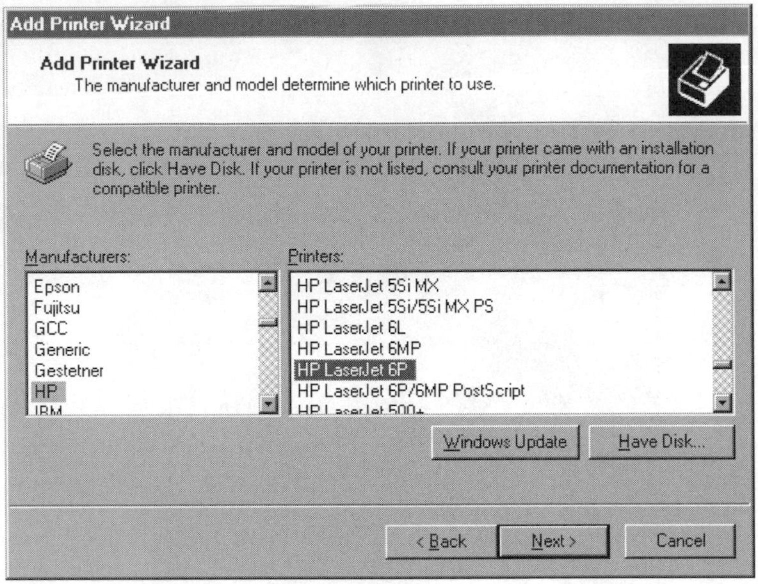

Add Printer Wizard

Add Printer Wizard
The manufacturer and model determine which printer to use.

Select the manufacturer and model of your printer. If your printer came with an installation disk, click Have Disk. If your printer is not listed, consult your printer documentation for a compatible printer.

Manufacturers:
- Epson
- Fujitsu
- GCC
- Generic
- Gestetner
- HP
- IBM

Printers:
- HP LaserJet 5Si MX
- HP LaserJet 5Si/5Si MX PS
- HP LaserJet 6L
- HP LaserJet 6MP
- HP LaserJet 6P
- HP LaserJet 6P/6MP PostScript
- HP LaserJet 500+

Windows Update   Have Disk...

< Back   Next >   Cancel

FIGURE 5.7   Setting up the HP 6P.

stallation process. In the case of device drivers delivered on diskette, you'll need to select the Have Disk option for the specific dialog box to initiate the driver installation.

TIP

What if my printer isn't supported?
If you've looked through the list of device drivers and don't see your printer listed, you can take these steps:

Check to see if the printer you're using has any emulations included within it. Many of the laser and inkjet printers today have either HP or IBM emulations included within them. Use the printer driver your printer emulates.

If your printer doesn't have any emulations included in it, then find the web site for your printer's manufacturer. There will hopefully be a listing of device drivers on the Internet for your use.

If the manufacturer's website doesn't have a device driver for Windows 2000 Professional that is specific to your printer, get the e-mail address of the product manager for your printer, and write to ask that they support it.

Click once on Next. The device driver is then applied to the printer being created in the Add Printer Wizard, and the next page appears. This page takes the name of the printer's device driver and applies it to the printer's queue. You can name a printer anything you wish, providing it fits within the 31-character field shown in this page of the Add Printer Wizard. Figure 5.8 shows the naming page for the printer.

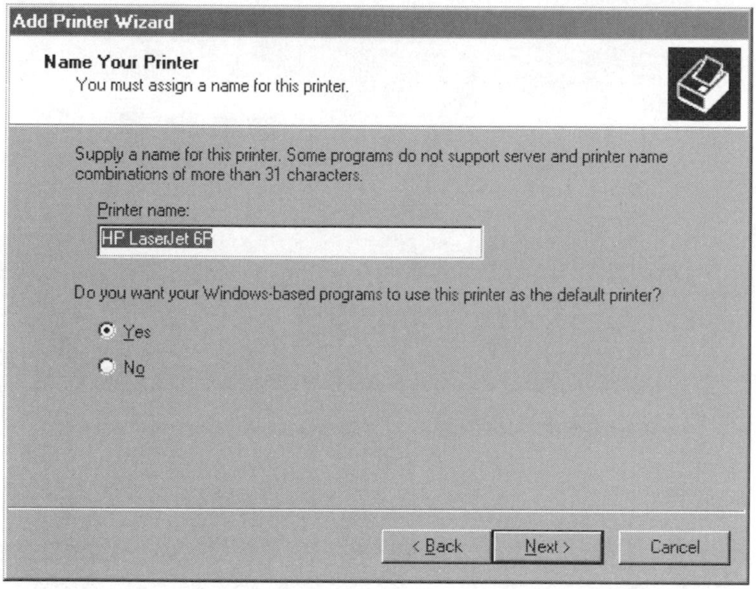

FIGURE 5.8   Defining the printer's name in the Add Printer Wizard.

Click once on Next. The last page of the Add Printer Wizard prompts you with a request to print a test page. The default setting is Yes.

Click once on Finish. The device driver is either loaded from CD, a network location, or from a hard disk location you have defined previously.

After the printer's device driver is loaded, then the test page is sent to the printer. Once the page has been sent, a dialog box is shown, prompting you if the test page was printed correctly. Figure 5.9 shows confirmation page.

If the page printed, then click once on Yes. If the page did not print, then the printing help file is shown with jump points that can be used for troubleshooting the problem.

With the HP 6P created, it now appears in the Printer Folder. Figure 5.10 shows the HP LaserJet 6P in the Printer Folder. The check mark next to the printer means it is the default printer for all printing tasks.

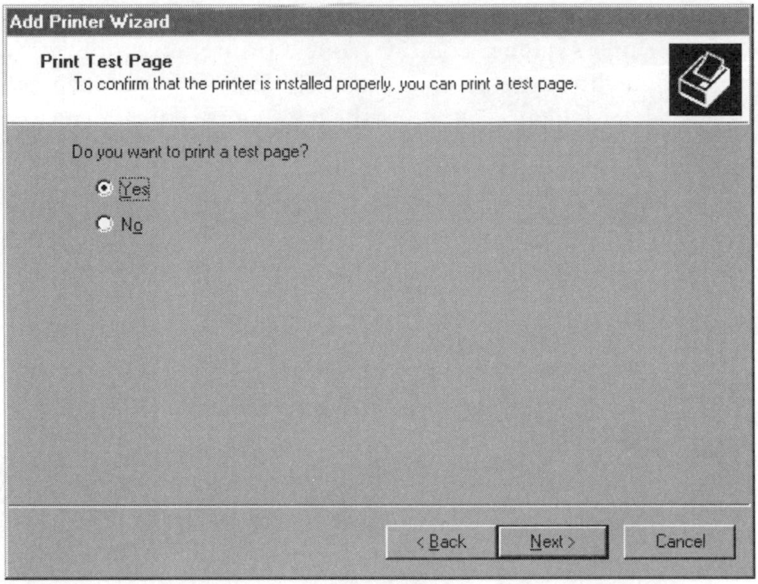

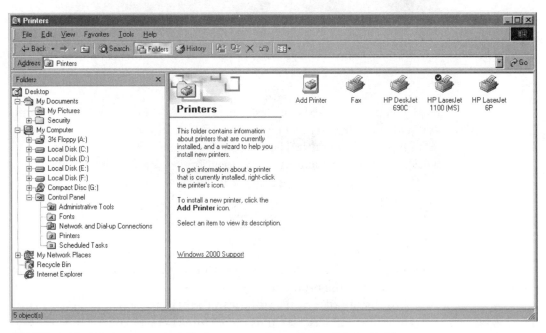

FIGURE 5.9   Confirming the test page printed successfully.

FIGURE 5.10   Printers once installed appear as icons in the Printer Folder.

Making the Printer Accessible over the Network

As the organization you're a member of continues to grow, there's more and more people needing to print. The LaserJet 4MPs are on order, and in the meantime your co-workers need to get their reports out. You decide that the best approach is to help them out by sharing your printer over the network. Here's the steps involved in taking a printer previously configured for local use and making it available to others who are members of your work group.

- Open up the Printer Folder.
- Double-click on the HP LaserJet 6P.
- From the Printer Menu, select Sharing... Figure 5.11 shows this step.

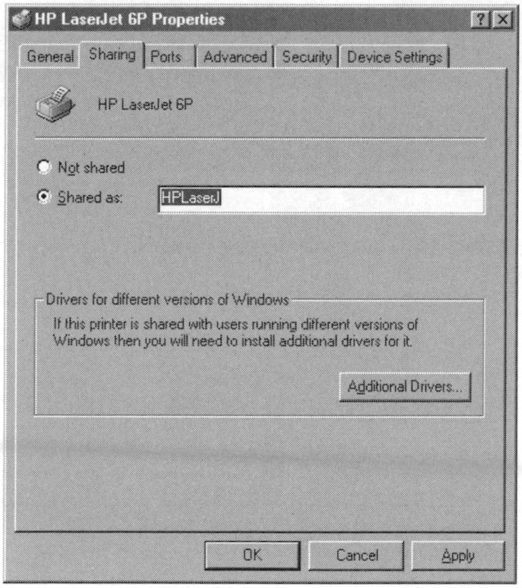

FIGURE 5.11   Sharing a printer created originally for local use.

The HP LaserJet 6P Properties dialog box is shown with the Sharing page selected. Using the options on this page you can toggle between Not Shared and Shared status for your printer, and assign a name for it as well. Figure 5.12 shows this dialog box with the printer name filled in.

Click once on OK. The printer is now shared for others in your work group to use. Figure 5.13 shows the results of sharing printers graphically.

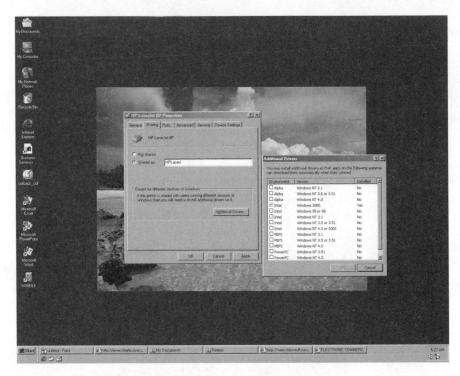

FIGURE 5.12 Using the Sharing Properties page.

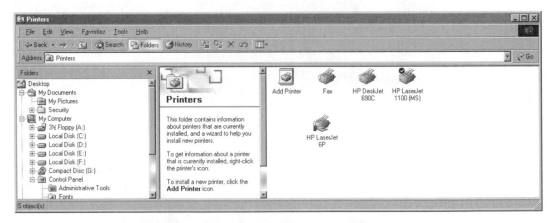

FIGURE 5.13 Confirming printers are shared in the Printer Folder.

Notice that under both the QMS and HP printers, there is a hand supporting them. This hand represents that the printer is shared.

TIP

Shortcut to Printer Sharing
You can also access printer sharing through the abbreviated command menu that appears when you right-click on top of the printer's icon you want to share. This abbreviated menu includes an option for jumping directly to the specific printer's properties menu, where you can toggled print sharing on or off.

Creating Multiple Printer Identities for the Same Printer

One of the strengths that Windows NT's printing architecture has is the ability to have multiple printer definitions or sets of properties associated with a single physical printer. This is very useful for being able to select different print queues for specialized tasks. To do this, follow the series of steps for creating a printer earlier in the chapter, being sure to apply a title to the printer that defines what the printer queue was specifically created for. You'll find this is useful for delineating between shared, networked printer queues and a localized queue that is going to the same printer.

Exploring Printer Properties

Once a printer is created, you may want to change the options associated with it as the needs of your organization change. Presented in this section is a definition of how you can go about changing the properties and options for your printer once created. Remember that you don't have to do this every time you print; rather, it would be a more efficient use of time to create separate printers, each reflecting the most commonly used properties and options you use. While the Add Printer Wizard made the process of creating a printer very easy, it didn't show you the full extent of the device properties available for, in the case of the example, an HP LaserJet 6P.

If not already in the Printer Folder, double-click on My Computer and then click once on the Printers Folder.

You can either click once on the HP LaserJet 6P queue and access Properties from the Printer Menu, or right-click on the HP LaserJet 6P icon and select Properties from the abbreviated menu. Figure 5.14 shows the latter approach.

The HP LaserJet 6P Properties dialog box appears, and is shown in Figure 5.15. Notice that this dialog box has several pages, all tabbed at the top.

FIGURE 5.14   Accessing a printer's properties through the shortcut menu.

This dialog box has been designed to provide accessibility to several of the features device drivers can control during printing.

Click once on the Device Settings Page. These are the characteristics or features of the printer that the device driver for this printer support, which makes it possible for you to customize them for your specific needs.

Clicking once on the Installable Options area of the Device Settings page, there are options for configuring this specific printer queue to take advantage of the memory installed in the printer. Figure 5.16 shows the Printer memory option being at 2MB, which is the factory default, and also shows the range of options selected for configuring once memory has been included in the printer.

Many printers also have a form-to-tray assignment. Click once on the first entry in this area on the Device Settings page. You'll see that there are a range of tray settings displayed on the lower portion of the page.

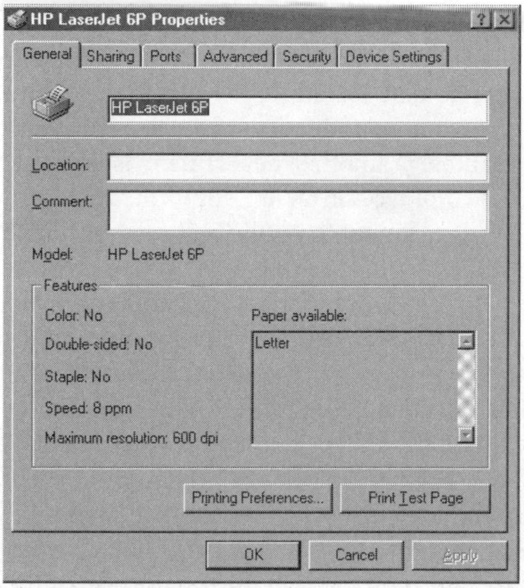

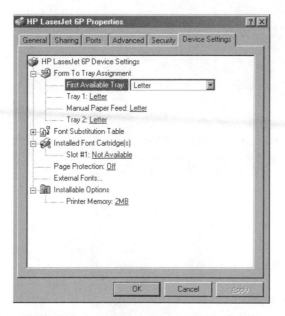

FIGURE 5.15   Introducing the Printer Properties dialog box.

FIGURE 5.16   Exploring the Installable Options on the Device Settings Page.

Setting Security Properties

Many printers are located in rooms where confidential accounting and finance information is kept, including payroll records. Being able to limit access to printing from confidential areas so that sensitive data is not inadvertently printed in another department is essential to preserve the confidentiality of employees in an organization. Many times the marketing communications departments have the best color and high-resolution laser printers, with prints costing up to $3 or $4 each to produce. Limiting access to these printers is a good idea so that for example an invitation to new club opening doesn't mistakenly end up in your company's catalog. Here are the steps for setting up security on your printers. Alternatively, if you shared a printer on your desk and find that now the entire company is using it, you can use these steps to only provide printing services for your work group.

Click once on the Security tab. The Security page is shown, and appears in Figure 5.17.

Click once on Administrator in the top portion of this dialog box. Administrators in general need to have the option of managing both the printer itself and documents in case there is a problem with printing.

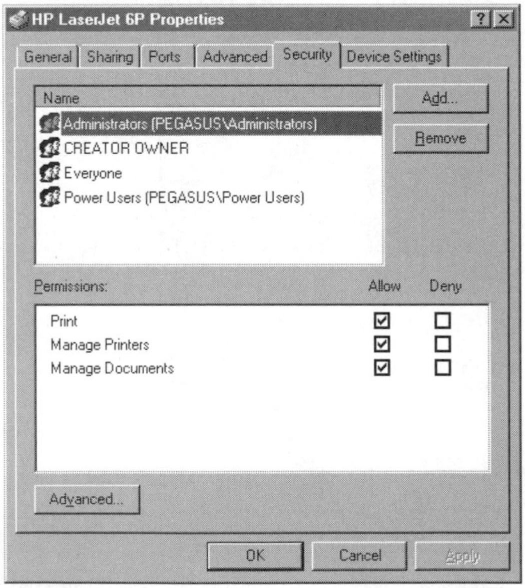

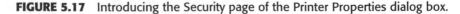

FIGURE 5.17 Introducing the Security page of the Printer Properties dialog box.

Click once on Power Users. These are members of the network that can share printers and directories. They are one level below administrators in the security levels of NT. Notice that by default this group has print capability only.

Click once on Everyone. As the title suggests, these are all the remaining members of the network you belong to. It's a good idea to select printing only for this group, unless the Everyone group is small and includes the members of your immediate work area.

Adding New Users & Permissions

As your organization grows you'll want to add new users to the security profile for the printer they are going to use. You don't have to do this for every single new user who joins your organization and users the network. These series of steps apply to an entirely new class of users, for example, guests who are logging onto your network and needing to print. Let's look at the series of steps for adding Guests to the Security profile for the HP LaserJet 6P we are using for this example.

With the Security page of the Printer Properties dialog box showing, click once on Add... button located along the top of the Security page. Figure 5.18 shows the Add Users and Groups dialog box which is used for defining which new groups will be added to the security profile for the printer.

Click once on the Guests entry in the Names: portion of the Add Users and Groups dialog box.

Click once on Add... The Guests entry is added to the profile, and appears in the Add Names: entry below. Figure 5.19 shows the results of adding Guests to the security profile for this specific printer.

Click once on OK. The Guests entry is now entered into the groups that can access this printer. Figure 5.20 shows the results of adding the Guests entry to the security profile.

By default, the Guest group has print privileges only. Clicking once on OK saves the security settings you have defined and closes the Properties dialog box.

Scheduling Printer Availability

At certain times of the day a very popular networked printer can have a queue that resembles a freeway at rush hour with job after job queued up waiting to get to its destination (the printer). As a system administrator you can manage when a printer or even a specific printer's queue is available to

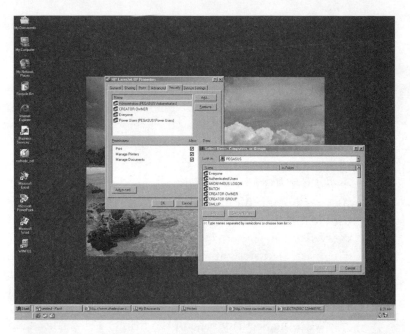

FIGURE 5.18   The Add Users and Groups dialog box.

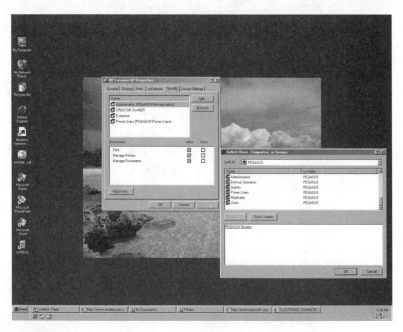

FIGURE 5.19   Adding Guests to a printer's security profile.

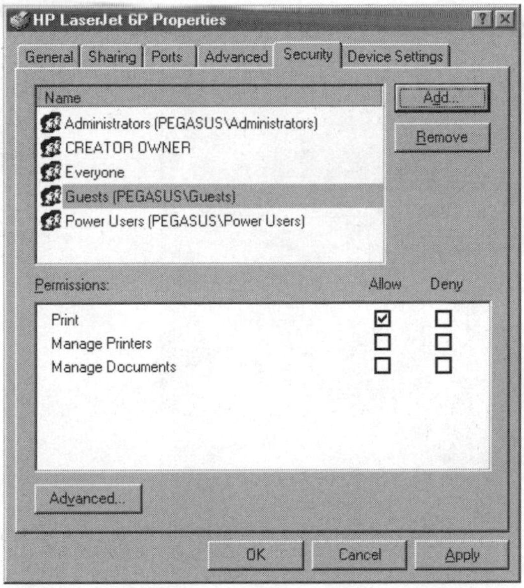

FIGURE 5.20   Guests are now added to the security profile for the printer.

others on a network. This is accomplished using the Scheduling page of the Printer Properties dialog box. Figure 5.21 shows the Scheduling page of the Printer Properties dialog box.

All printers when first created are available. You can also prioritize the printer you're using between 1 and 99. The higher the number, the higher level of precedence these specific printer queue takes over other printer definitions using the same physical printer. You can also use the option of spooling printer documents completely before being sent to the printer, or conversely send the print requests directly from your workstation to the printer.

What happens to a print request submitted when the printer is not available? The print request stays in the printer's queue until the system becomes available. When a printer's availability is on, all print requests are completed. The security aspects of printers in Windows 2000 does not differentiate between individual users. The security model of NT printing takes an entire group of users and restricts or grants access. If you want to make a printer available 100% of the time for administrators, you can create a second printer that share that specific printer's name with the administrators in your organization. While the properties associated with Windows 2000 printing do not provide for changing priorities on an individualized basis,

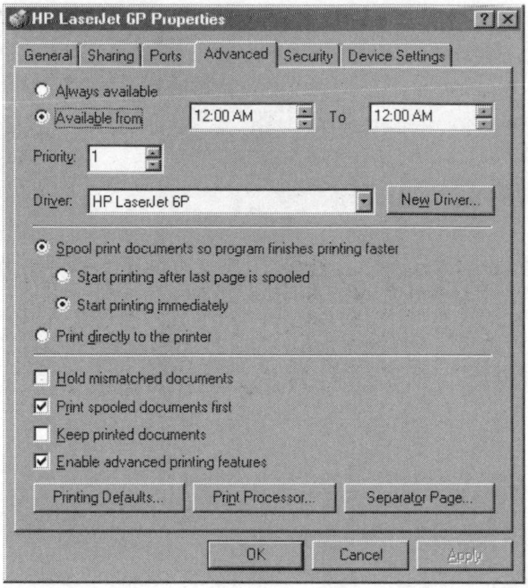

FIGURE 5.21 Introducing the Options for Scheduling Printer Availability.

you could feasibly create a separate series of queues, watch with times and properties that are specific for the needs of a given group.

Setting Print Job Priorities

What if your top-of-the-line Color LaserJet is suddenly getting more print requests that a wedding invitation printer gets in June? The Answer: create a shared printer queue for the high-use group with a priority of 99, and create a second queue for others that may not have as high a demand for color printing at 1. This serves the group that requires color printing well; they can get their work done faster with the higher priority queue.

Remember that when setting priorities the higher the number is the higher the priority. The default value for the sliding scale on the Scheduling page is 1, with 99 being the highest.

Configuring Separator Pages

When several people are sharing a common printer, figuring out which document is whose can be challenging, especially on busy days when each person has a project due. Separator pages have been developed specifically to solve this problem. They are used for delineating which print job is whose. Separator pages are also used for delineating between print jobs with different characteristics as well. You can create your own separator page using Notepad or use one of the several included in NT. To select one of the separator pages included in Windows 2000, go to the General Tab in the Properties dialog box. The Separator Page... button is located on the lower half of this properties page. Figure 5.22 shows the General page.

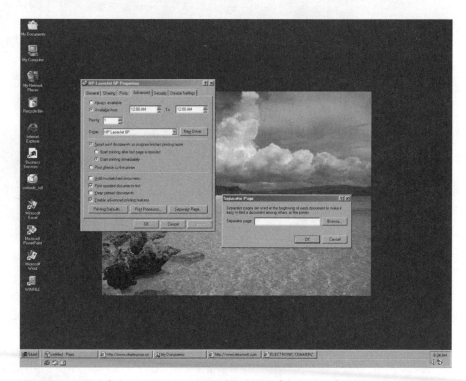

FIGURE 5.22   The Separator Page option is on the General Tab page.

With the General tab selected in the Printer Properties dialog box, use the following steps to have one of the four separator pages selected that are included in Windows 2000 Professional.

Click once on the Separator Page... option on the General Tab page. The Separator Page dialog box is shown, and appears in Figure 5.23.

Click once on Browse. By default this browse dialog box takes you to the winnt/system32 subdirectory where the four separator pages are located. These are pcl.sep, pscript.sep, sysprint.sep, and sysprtj.sep. The file

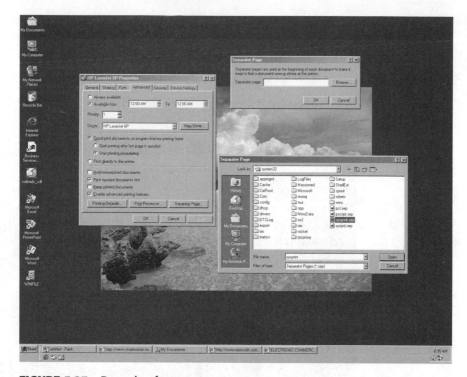

FIGURE 5.23   Browsing for a separator page.

extension.sep defines a separator file. Since the HP LaserJet 6P is running PCL, that page is selected.

Closing out the Printer Properties applies the separator page to the Registry of your workstation. If your workstation is not the print server, you'll need to complete these steps on the system that is the print server, because the Registry stores these values and applies them to every print request generated from this device driver.

TIP

Creating Your Own Separator Page
Using Notepad or any other text-based application capable of generating ASCII files you can create your own separator page. Here's how.

After opening up Notepad, enter a control character on the first line of the file. This character is referred to by the operating system as the control character, and informs NT this a file that will be used for functions, alleviating it from being read as a document. You can enter any non-alphabetic character. Consider using the # sign as the control character. This character goes at the top of *the file in the first position. You'll also see articles describing this process using the $ sign. Either the # or $ sign is fine to use as both work.*

Configuring a Network-Based Printer

With estimates of over 90% of the Windows 2000-based workstations being used in network environments from leading manufacturers including Dell and HP, it's understandable that many of the higher-end monochrome laser and color printers are primarily used via networks. In addition to the benefits of being able to amortize the cost of a higher-end printer over more users when it's networked, the actual performance of printers has continually increased in an effort to provide more responsive performance to users. Networked printers are now the standard for many organizations, and there is a definite trend towards including TCP/IP support in the form of device drivers for Windows NT, NetWare and in the case of Token Ring support, the IBM SNA architecture. Windows NT's integration of network printing support is comprehensive in Windows 2000 Professional. This section explores those features and how you can harness them in serving those in your organization.

What's New in Network Printing

In addition to the Active Directory and enhanced support for client printing (as will be shown in the hands-on tour of setting up a networked printer), Windows 2000 Professional includes these new features:

- Internet Printing. Enhancements made to the Windows NT printing architecture make printing from the Internet seamless. Client workstations can access shared printers either over a corporate Intranet or over the Internet. URLs can point directly to printers. Users can query as to the status of print tasks over the Internet as well, in addition to installing printers over Intranet and Internet connections. Drivers are also installable directly over the Internet as well.

- Simplified User Interface. A new web view of the Printers folder and print queues, with direct links to further information and technical support. There is also an enhanced Add Printer Wizard interface as has been demonstrated in this chapter. There are enhanced printer properties in each driver and an enhanced command set developers can call from in the development of their own device drivers. There's also a New Add Driver Wizard in the Server

Properties dialog box for installing additional printer drivers for down-level clients or non-Windows platforms. Driver properties are also available to administrators for managing printing tasks at a server level.

- Directory Services Support For Printing. The Active Directory treats every shared printer as an available resource in the Printers Folder. The Active Directory feature is best known for file sharing but also works with printing and imaging devices.

- Enhanced security and auditing features include the ability to add or delete classes of users to individualized queues.

Creating & Connecting to a Network Printer

Let's get started with the steps required for connecting to and creating a network printer. You'll find that these steps parallel in some respects the sequence of events in the Add Printer Wizard for creating a printer for localized or personal use. One of the major differences is in the definition of drivers for the client systems located on your network that have an operating system different than Windows 2000 Professional. Let's get started with the steps to creating a network printer.

Double-click on My Computer from the Windows 2000 Professional Desktop if it is not already open.

Click once on the Printers Folder, which opens it and shows it on-screen.

Click once on the Add Printer Wizard. The first page of this Wizard is shown. This is the page of the Add Printer Wizard where you decide if the printer you're creating or connecting with is managed by your computer or connected to a network printer server. Figure 5.24 shows the first screen of the Add Printer Wizard with the Network printer server option selected.

Click once on Next. The second page of the Add Printer Wizard is used for selecting the networked printer you want to connect to. Figure 5.25 shows this second page of the Wizard, and is designed to make it easy to navigate

Click once on the Microsoft Windows Network entry to view the shared printers on the network you are a member of. You can also see there is an entry for Windows 2000 Internet Printing. These are printers shared via Internet connections.

If you don't see the printer you want to connect to within the Microsoft Windows Network entries, type in the path to the printer in Printer: entry of this second screen of the Add Printer Wizard.

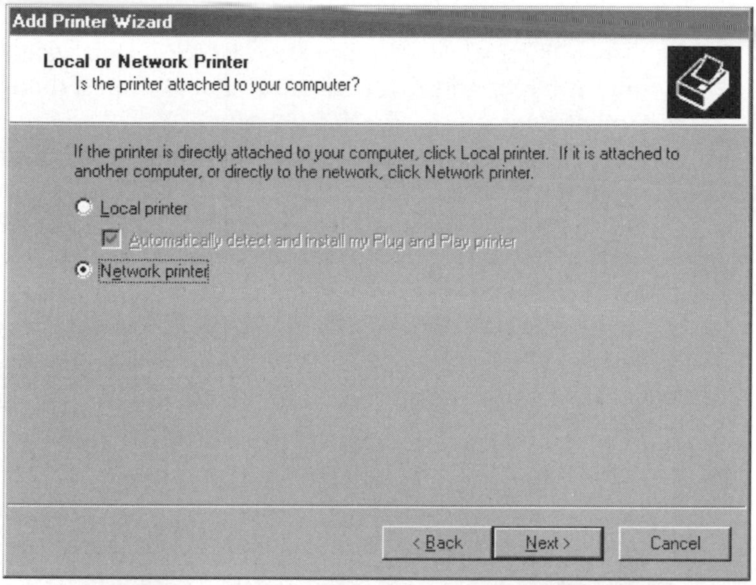

FIGURE 5.24   Using the Add Printer Wizard to connect with a networked printer.

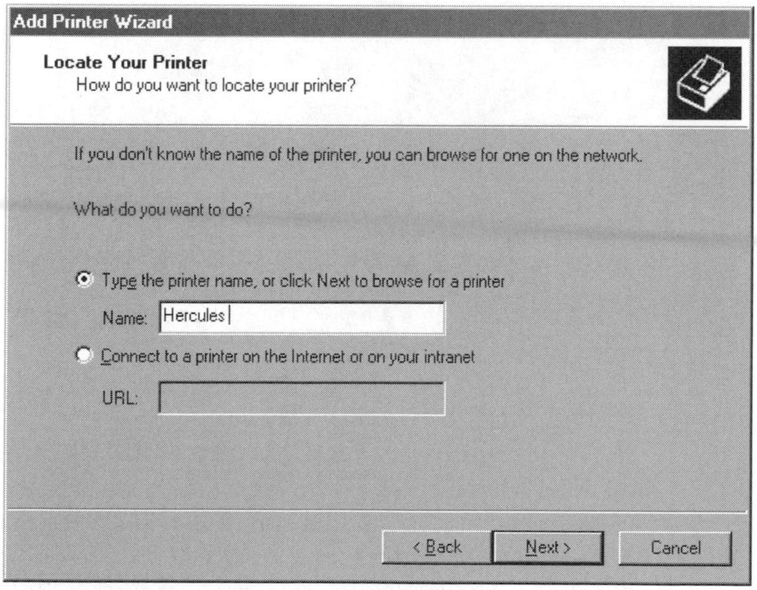

FIGURE 5.25   Using the Add Printer Wizard to connect with networked printers.

Once the printer is found, the third screen of the Add Printer Wizard is shown. This screen prompts you to see if you want to designate the printer you're connecting to as the default printer. If this is the primary department printer, leave the default value at Yes. If you're using these series of steps to connect with a printer which will be used only under special circumstances, then click once on No. Figure 5.26 shows this final page of the Add Printer Wizard when a networked printer is being created.

FIGURE 5.26 Defining if a networked printer will be used as the default printer selection for your applications.

Click once on Finish. The connection to the networked printer is now complete. It's represented as an icon in the Printers Folder. You can tell this is a networked printer with network cabling beneath the printer selected with the small hand underneath it. Whenever a resource has a hand underneath it that designates the resource as shared.

Testing the Connection

How can you be sure your workstation can now "see" and use the networked printer over the network? Simple. Open up any application, choose a sample document, and select Print... from the File Menu. Be sure to make sure the printer you want to test is defined in the Printers: entry of the printer dialog box. Send a document or drawing that is most representative of the work you'll be doing. If you happen to use AutoCAD or PhotoShop, send one of the larger files to make sure the memory configuration of the printer can also support the printing you typically do. It's a good idea to do this in off-hours to make sure others' documents don't wait behind you in the print queue.

Using Printers in a Networked Environment

Getting work done very often involves getting a hard copy of the work you have been completing. Increasingly, this means using a networked printer to get your work done. Using the hands-on activities in this section you'll be able to more effectively manage a networked printer and help others in getting their work done as well. If you're an administrator you'll find these steps a quick refresher course on concepts first introduced in Windows NT 4.0.

Using Universal Naming Conventions to Make the Connection

One of the key aspects of the printing architecture is the fact that printers are referred to using Universal Naming Convention (UNC) notation which is //servername/sharename.

For a printer named Sparta located on a server called Colussus, the UNC equivalent would be //PEGASUS/SCRIBE. You'd use this path for defining a shared printer in the Add Printer Wizard dialog box in the case of looking to connect with a shared network printer. Figure 5.27 shows the entry of a UNC-based definition of a shared printer in the Add Printer Wizard.

How Client/Server Printing Works in Windows 2000

One of the core strengths of Windows 2000 as an operating system is its reliance on a client/server approach to printing, in addition to file and application sharing. This approach to printing makes it possible for you to continue working on a document, even if the document is potentially hundreds of pages long. The benefit of client/server printing is that the

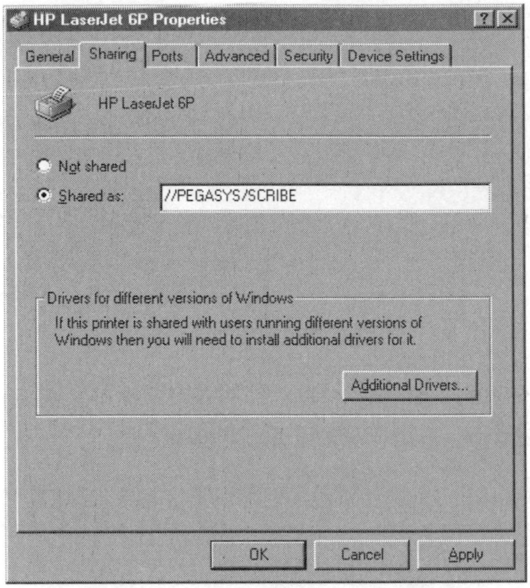

FIGURE 5.27   Using DNS To Identify A Shared Printer.

individualized client workstations offload printing tasks to a server for completion. Why not just have each individualized workstation process the print requests and then submit the print requests? Windows 2000 is a multi-tasking, multi-processing operating system, and should be able to handle printing as a background task. Why not just keep printing local?

When you are printing to a printing from your workstation to a local printer, that's exactly what is happening. If you happen to have a color printer that takes quite a bit of memory to print a document or graphic, your workstation's applications may have a slightly slower performance during print time. This is due to the fact that each thread for the applications you are running are competing for memory. A device driver handling an extensive color printing task will also generate a thread, in effect a request for memory and resources, just like every other application. Taking the print task and sending it to a server for processing frees up resources on your workstation so applications can complete their tasks. This is the essence of why client/server printing is so popular.

What are the other reasons why printers are shared? First, there's the cost issue. Being able amortize or share the costs of a printer across multiple users can make an HP LaserJet 4M affordable for smaller companies, as opposed to purchasing separate laser printer for everyone in the department. Second, the option of being able to enhance printing through the use of definition files and raster image processing utilities is also available when servers are used for printing. Third, queuing of larger documents from your co-workers and your workstations is possible in a networked environment. Fourth, there is a performance increase for each workstation off-loading prints requests to a printer server. Fifth, through the advances made in network operating systems, printer sharing and the integration of the benefits of client/server computing into printing is changing how printers of all types are made and sold.

Managing Printers in Windows 2000 Professional

Changing Document Printing Order

If you're the owner or administrator for a printer (more on these subjects later in this section) you can actively manage the documents in a printer queue using the commands located in the Document Menu. If your printer is attached directly to your workstation, you most likely have administrator privileges. Let's take a look at each of the entries in the Document Menu. This menu is delivered with every printer created in Windows 2000.

Figure 5.28 shows a HP printer queue with a document selected and the Document Menu extended. Each of these commands are briefly defined below:

Pause - Pauses the document from being printed. If you're in the middle of working on fixing a printer, you can use the Pause command in the Printer Menu to suspend printing to the printer.

Resume - Re-initiates the printing of the document.

Restart - Re-starts the printer queue after it has been paused.

Cancel - Stops the printing of the document.

Properties - Displays the properties for the selected document. Figure 5.29 shows the contents of the properties for the selected document, which happens to be from Microsoft Word. Notice that this properties dialog box has three tabs, with the Page Setup and Advanced being for the most part grayed as the selections for the documents' submission to the printer has already been made. The first page of the dialog box provides information on the priority of the document being printed, who the owner is, size of the document, number of pages, and the availability times of the printer.

FIGURE 5.28 Using the Printer Queue Options for viewing print jobs.

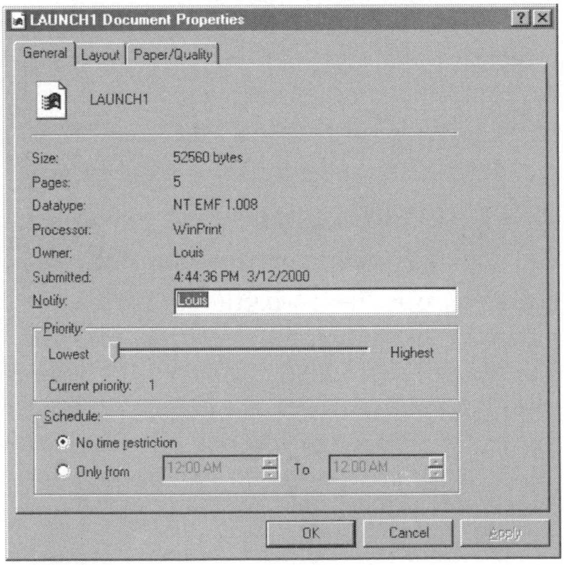

FIGURE 5.29 Viewing Document Properties from within a printer queue.

Updating Printer Queue Status

The nature of network printing is based on delivering print requests from a client workstation to a server. Many applications involved with printing assume that the communication will be one-way, with the printer queue showing the status of print requests for networked and shared printers. The printer queues in Windows 2000 are updated via network connections periodically, and depending on the network traffic, can be up to a minute. If you want to get the immediate status of a printer queue, press F5 with the printer queue selected. This immediately queries the status of the queue and displays the entire list of jobs.

Adding Owners to Printers

One of the truly differentiating aspects of Windows NT is its use of properties. In expanding the functionality of printing, Microsoft has integrated ownership options into printers from the first version of NT.

Ownership of a printer entitles you to changing its properties, defining security levels, modifying the printer order of documents, and changing the times the printer will be available in a shared mode. Printers created from within an Administrator account automatically take that account as the owner. You can add owners to a printer using the steps provided here:

From within the Printers Folder, right click on the Printer you want to add owners to.

Click once on the Security tab of the Printer Properties dialog box. Each of the groups who have access to the printer are shown in Name: area of this specific page. Figure 5.30 shows the Security page of the Printer Properties dialog box.

Click once on Add.... The Add Users and Groups dialog box is shown, and appears in Figure 5.31. The purpose of this dialog box is to add groups and individual users to the owners of the printer.

Click once on Replicators. This is a group that is responsible for system maintenance, including replicating print and application files in networked environment. At times during the support of a networked printer, the replication group may be the only staff onsite when a printing error occurs.

Click once on OK. The DNS entry for the Replicator group is now added to the Ownership list for the printer being configured.

Click once on Close. The properties dialog box is then closed.

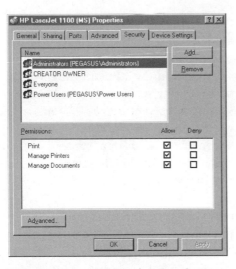

FIGURE 5.30   Viewing the Security Page of the Printer Properties dialog box.

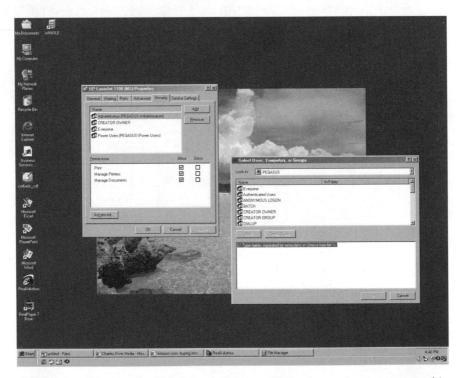

FIGURE 5.31   Using the Add Users and Groups dialog box to increase ownership of printers.

Re-positioning Print Jobs in Queues

If you're part of a group who owns printers or are the owner of a printer (in the context of Windows 2000 permissions) yourself, you can click and drag printer requests to any location in a queue you choose to. If you've ever used PowerPoint or any other graphically based application then you'll find the process of re-positioning print jobs in a queue something you can do quickly. Here are the steps for changing the order of print jobs in a queue.

- Open the printer's queue you want to change the order of print jobs in.
- Click once on the print request you want to move.
- While holding the mouse button down, drag the outline of the print request's name to the location in the print queue where you want it.
- Release the mouse button. The print request is then placed at that location.

Deleting Printers

Many shared printers have a specific purpose, and once you've connected to and used them to complete a project, you may want to delete a reference to the printer. You may also find that you have several profiles for the same printer, and want to trim down the number of printers you have defined. Here's how you can delete a printer from the Printers Folder. Once a printer has been deleted the device driver and font files associated with it stay on your workstation, in the event you want to re-create it. Follow the series of steps here for deleting a printer from the Printers Folder.

Right click on the icon of the printer you want to delete.

From the abbreviated menu of commands, selected Delete.

A confirmation dialog box is next shown. Click once on OK to confirm that you want to delete the printer. The printer is then deleted.

Troubleshooting Printing Problems

Supporting Printers in Windows NT–Getting a Device Driver

Going through the Add Printer wizard you suddenly find that there isn't a printer driver for your printer. You can solve this problem using the steps outlined here:

Using the manufacturer's web site search capabilities, see if they have a Windows 2000 device driver for your printer. Be careful to make sure the device driver you download is specifically for Windows 2000, as Windows95 and Windows98 drivers use different system-level resources than their Windows NT and Windows 2000 counterparts. Note that in many cases the device drivers for Windows NT specific tasks are incompatible with their Windows 2000 counterparts.

Check your user's manuals, or if you don't have them, check the printer manufacturer's web site for the emulation that is included in your printer. Select the printer driver for the emulation supported in your printer. Some of the advanced features of your printer may not work, as the commands to complete these tasks may be specific to your printer's command set.

If your printer doesn't have a device driver or emulation that is supported in Windows 2000, call the printer manufacturer and ask when a device driver for Windows 2000 will be available. Calling Tech Support may also yield an NT work-around that the company may have found to work with Win2K. Also, many times manufacturers will develop interest lists of customers who need device drivers. Once a device driver is complete, manufacturers will mail the device drivers to the people requesting them.

How to Test Your Printer with Windows 2000

The best approach to testing a printer in Windows 2000 is to try and print a test page from the Printer Properties dialog box. When a printer is first created in Windows 2000 a printer test page is submitted at the end of the wizard to check for communication. If you are just creating a printer and not getting any printing, try going back and re-installing the device driver. If your printer has already been installed and the printer just quits working, you can submit a print test page from the General page of the specific printer's properties dialog box. In the lower right corner of the dialog box there is a Print Test Page button that sends test data to the printer.

Solving Network Printing Problems

What about submitting a print request over a network, having the print request show up in the print queue, then suddenly drop out and disappear? This could be caused by a few factors, described here:

Disk space on workstation. Check to make sure there is enough disk space on your system to print the document. Documents take up to twice their size in disk space. If you have just a little free space on your hard disk you may find it difficult to print.

Disk space on server. Is the print server routing print requests up and running? Is there enough disk and system memory on the server? Is the disk running? Are there many other queues running on the server as well?

Memory. If your workstation is running low on memory either due to many applications being open or having a minimal amount of memory to run Windows 2000, you'll want to free up memory by either closing applications or getting more memory.

Connectivity. Is your workstation talking with the network? Check another network-based resource to see if you can get around on the network.

Printer. Is the printer powered on running? You might want to go over and re-cycle the power on the printer to make sure it is running OK and connected to the network.

Driver Options. Check the printer driver options using the printer properties dialog box to make sure the options selected match the physical attributes of the printer.

CHAPTER SUMMARY

Printing from Windows 2000 has progressed from Centronics parallel or serial connections to being able to print directly to network-based peripherals. The printing architecture for Windows 2000 now includes printer-specific properties that give you the option of tailoring printer queues for individualized tasks. Keeping with the option of having multiple logical queues being able to address a single physical printer, the addition of enhanced printer properties makes it possible to create individualized printer queues for specific tasks. Security features in Windows 2000 Professional also make it possible to ensure users who need a given printer's resources always have them available. In inclusion of web-based printing and configuration of printers via the web add value to the Windows 2000 Professional print architecture, making it capable of being used in larger, more diverse organizations.

6 Understanding TCP/IP in Windows 2000 Professional

Incoming TCP/IP Properties

Network access
- ☑ Allow callers to access my local area network

TCP/IP address assignment
- ⦿ Assign TCP/IP addresses automatically using DHCP
- ○ Specify TCP/IP addresses
  - From:
  - To:
  - Total:
- ☐ Allow calling computer to specify its own IP address

OK   Cancel

Microsoft, in attempting to gain a stronger presence in the enterprise operating system marketplace, has continually focused on how to effectively compete with UNIX and other already proven operating systems. The fact that Microsoft emanated from MS-DOS to a Windows-based GUI and finally to the robustness of Windows 2000 shows a progression to more fully recognize the needs of TCP/IP-based organizations. Microsoft's trails and tribulations at getting their network connectivity to the level of performance needed in organizations is best exemplified by their chosen implementation of TCP/IP in Windows 2000, which is the subject of this chapter.

In getting this large TCP/IP project moving, Microsoft learned about the needs of TCP/IP users in its own company by actively beta testing Windows 2000 Professional and all levels of Windows 2000 Server—including the connectivity and TCP/IP commands. Based on this extensive testing and integration effort, Microsoft added features to the NT TCP/IP protocol suite including Dynamic Host Configuration Protocol (DHCP) and the Windows Internet Name Service (WINS).

Now TCP/IP is considered the great equalizer, making it possible for network-based devices of all types to share all types of data, regardless of the operating system of the specific node. TCP/IP works in conjunction with the Java Virtual machine (JVM) software included in many network browsers, yielding a new ubiquity to information sharing not seen in the previous generation of operating-system-bound devices. Let's get started in this chapter with an overview of the TCP/IP protocol and move through the DHCP and DNS areas of NT's implementation.

A BRIEF HISTORY OF TCP/IP

Originally developed to meet the needs of universities, researchers, and even defense contractors to share data across broad geographic distances quickly, TCP/IP has progressed into an ever-increasing range of protocols. Defining how systems interact with each other through a common set of commands, protocols are the methods by which computers communicate with each other over a network.

The precursor to the Internet, ARPANET (Advanced Research Projects Agency Network) was originally created by the Department of Defense Advanced Research Project Agency. Established in 1969, ARPANET was primarily an experiment in large-scale packet switching. The foundation of what is today the Internet was created to accommodate the goals of the ARPANET project. First created as a research endeavor, the TCP/IP protocol is not slanted to a given vendor or manufacturer. This actually made ARPANET more ubiquitous and more nimble at responding to the needs of its users because the growth of the network could draw on the resources of multiple contributing companies and universities, and it did not have to be constrained by the technology of a single provider. ARPANET and now the Internet are successful because TCP/IP takes the best of multiple sources of technology, rather that just a single source of innovation.

How TCP/IP Fits into Microsoft's Networking Strategy

The role of TCP/IP in the product strategy of Windows 2000 is to create a foundation within each of the operating systems released from Windows NT 3.51 forward. This translates into several product generations today that have been used to translate users' needs into a TCP/IP protocol suite robust enough for the needs of corporate users to ensure connectivity between wide numbers of users.

Microsoft continues to use added functionality and differentiation within the TCP/IP command set to continually add value to the Windows NT product family. Sooner or later, you will encounter the need to configure TCP/IP in Windows NT, and the steps in this chapter will give you the tools you'll need. The role of TCP/IP in the Microsoft product strategy is to accomplish a variety of tasks. First, TCP/IP provides a solid foundation on which to build a connectivity strategy and drive the development of mixed or heterogeneous networks that include systems of many different operating systems yet share the commonality of TCP/IP as their method for communicating. Second, TCP/IP streamlines the integration of Windows 2000 Professional workstations and servers with the Internet. Many of you as system administrators will face the task of ensuring connectivity between servers and the Internet via routers and T1 lines from your own organization or via Internet Service Provider (ISP) connections.

TCP/IP BASIC CONCEPTS

It was with foresight that the first companies began building TCP/IP into their organization. UNIX provides TCP/IP as a standard feature, working to ensure that file transfers and connectivity are strengthened. UNIX was the first major operating system to include TCP/IP, and because of its ability to handle multiple sessions, and its appeal to the technical applications areas, UNIX and TCP/IP quickly became market standards. Why does TCP/IP continue to gain in acceptance? Because the founding concepts continue to meet needs and add value; the ongoing need for connectivity makes the protocol the first priority and the operating system secondary. Here are the reasons for TCP/IP's continuing gain in popularity within the NT user community:

Hardware independence. TCP/IP is the great equalizer when it comes to processors and network connectivity hardware. Instead of being limited to only a certain microprocessor, TCP/IP is truly hardware independent. The growth of network computers, or NCs, relative to lower-priced traditional PCs, points to the fact that the microprocessor, in specialized network applications, becomes secondary. The network protocols will in many cases allow these specialized devices to play the role of the operating systems. Nowhere will this become more prevalent than on the Internet. This hardware independence is best illustrated by growth of the Internet, where the HTTP protocol is independent of hardware platforms through the extensive development of browsers which enable Internet connections across a diverse series of hardware platforms.

Standardized addressing. Because any host computer on any TCP/IP network can be individually and specifically addressed, applications and programs using a specific IP address will be able to find the targeted system and know without a doubt that the data sent is intended for that specific system. This standardized addressing technique in TCP/IP also ensures that the assignment of IP addresses, either through static means with each individual system getting a discrete, separate identity or entire "pools" of PCs being able to draw from a range of addresses, will be consistent in meaning and structure. Later in this chapter you'll learn about Dynamic Host Configuration Protocol (DHCP). The essence of this protocol is the creation of an entire pool of IP addresses that are checked in and out, much the same way books and tapes are checked out of a library. In effect, a reservation can be made to get an IP address, and within the reservation, the time the IP address will be needed will also be communicated. Many of the Internet Service Providers that make their revenues based on offering access to the Internet use DHCP for assigning IP addresses to customers who want access to the Internet. The standardized approach to structuring IP addresses allows for the flexibility of creating protocols such as DHCP that make it possible to dynamically assign addresses from a range or pool, instead of having to construct addresses on a one-by-one basis.

Open standards. TCP/IP continues to gain enhancements because it is based on open standards that are available to anyone. The standards organization that manages the TCP/IP protocol is much more approachable, and it actually publishes proposed additions to the standard in the form of Requests for Comments, or RFCs. These RFCs are actually descriptions of planned enhancements to the standard, and anyone is welcome to comment on the content of these documents. Some are quite technical, but many are understandable for someone with a solid understanding of the OSI model, the essentials of TCP/IP, and the patience to read through them. TCP/IP's open standards are truly that; unlike other highly technical standards you don't need to be invited to comment. You can find these Requests for Comment at the following anonymous ftp sites:

ds.internic.net

nis.nsf.net

nisc.jvnc.net

ftp.isi.edu

wuarchive.wustl.edu

src.doc.ic.ac.uk

ftp.ncren.net

ftp.sesqui.net

nis.garr.it

http://www.cis.ohio-state.edu/htbin/rfc/rfc1213.html

http://www.ccs.neu.edu/home/modiano/TCP-IP.html

http://oac3.hsc.uth.tmc.edu/staff/snewton/tcp-tutorial/

http://www.weyrich.com/bookreviews/communications std 3.html

http://www.sis.pitt.edu/~mbsclass/standards/viar/TCP-IP.html

Application protocols. Another force driving the increased adoption of TCP/IP is the inclusion of applications based on this protocol that make sharing, transferring, and managing files on dissimilar systems possible. For example, you could use the ftp command to transfer files from a mainframe running TCP/IP to a Windows 2000 Professional workstation, where a spreadsheet would complete the analysis. After finishing the spreadsheet, you could ftp the file to a UNIX system, where a time series could be completed using the telnet command. All this could be accomplished from the Command Prompt window of the Windows 2000 Professional workstation you're using. You can easily trade files with mainframes, Macintoshes, and UNIX workstations of any operating system type—just as long as the TCP/IP application suite is present, the type of operating system is immaterial. TCP/IP, as you can see, levels the playing field between operating systems as well.

INTRODUCING THE OSI MODEL

In learning about network protocols and how they compare, it's useful to have a frame of reference or a structure to apply to the terms and concepts. The Open Systems Interconnect (OSI) model provides this framework for illustrating the components of the TCP/IP networking protocol. While the OSI Model is used pervasively for showing the differences between network commands and protocols, it's important to realize there are just as many variations in how protocols are structured as there are protocols—so the OSI Model is a good foundation on which to learn. Just as with any foundation of a house, the elevation can be quite different for each one. It's the same with network protocols—all are based on the foundation of the OSI Model, yet their elevations or appearance look different based on the needs of the given set of customers. In this section you'll first get an overview of the OSI Model, followed by an explanation of how the TCP/IP protocol fits into this structure. You'll also see how the OSI Model is organized to provide for data packets or datagrams to traverse the levels of the model to ensure communication and data reliability.

The Open Systems Interconnection (OSI) model was originally developed to meet the most basic yet essential elements of computer networking challenges—making a common basis for making diverse systems work together. In 1984, the OSI model was developed to serve as a common basis of reference for handling the communications tasks between diverse systems.

The OSI Model is organized into a hierarchy where data travels from the lowest point to the top, with each level either stripping away data elements needed for handling the data transaction or in the case of an outgoing message, adding data to ensure the targeted system receives the complete message. Figure 6.1 shows an example of the OSI Model.

The first two layers, physical and link, define the way data is physically transferred on the network. The network, transport, and session layers define the way data is processed before being passed to the operating system (or processed before being passed to the lower layers of the OSI Model). The presentation layer represents the operating system, and the application layer represents the workstation's network software. Above the application layer are the user interface and general applications that use network resources.

By breaking the network model into different layers, it's also much easier to see how independent yet linked physical attributes of network connectivity are relative to their logical counterparts. For example, the network protocols (TCP/IP, NetBEUI, and so on) are not dependent on the network interface card (NIC) or cabling, but instead work on any hardware

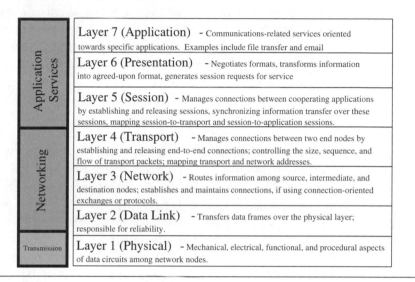

Introducing the OSI Model

Overview of OSI Model

Application Services	**Layer 7 (Application)** – Communications-related services oriented towards specific applications. Examples include file transfer and email
	Layer 6 (Presentation) – Negotiates formats, transforms information into agreed-upon format, generates session requests for service
	Layer 5 (Session) – Manages connections between cooperating applications by establishing and releasing sessions, synchronizing information transfer over these sessions, mapping session-to-transport and session-to-application sessions.
Networking	**Layer 4 (Transport)** – Manages connections between two end nodes by establishing and releasing end-to-end connections; controlling the size, sequence, and flow of transport packets; mapping transport and network addresses.
	Layer 3 (Network) – Routes information among source, intermediate, and destination nodes; establishes and maintains connections, if using connection-oriented exchanges or protocols.
	Layer 2 (Data Link) – Transfers data frames over the physical layer; responsible for reliability.
Transmission	**Layer 1 (Physical)** – Mechanical, electrical, functional, and procedural aspects of data circuits among network nodes.

FIGURE 6.1   The OSI Model provides a useful framework for comparing network protocols.

because the intervening layers translate and process the network traffic into the format these protocols can understand. Layering components of the OSI Model also helps make the network functionality transparent to the user interface and applications. Because the layers hide the protocols and physical hardware from the interface and applications, you can change protocols and hardware without making any changes to the interface or your programs to enable them to access network resources.

Let's look in more detail at each layer of the OSI Model. These are descriptions of each layer, beginning with the bottom-most layer and moving upward.

Physical layer. The physical layer defines the connection of one system to another via the network. At this layer, the network transmission consists of individual bits, with upper layers of the network processing and packaging bits into packet form. The physical layer does not define the media used, but it does dictate the methods by which data are transmitted over the network.

Data link layer. The primary purpose of this layer is to process *frames*, which are units of data by which the layer communicates upward in the OSI Model. A frame comprises a string of bits grouped into various fields. In addition to the actual data bits, the frame includes fields to identify the source and destination of the frame and an error control field that enables the receiving node to determine if a transmission mode has occurred. The technique used for checking if the frame has successfully been received relies on cyclic redundancy checking (CRC), where the data bits are compared using a parity-checking algorithm. This technique of ensuring data integrity was originally developed for hard disk drive controllers.

Other functions of the data link layer is to convert outgoing frames into individual bits for transmission and to reassemble incoming bits back into frames. In addition, this layer establishes node-to-node connections and manages the transmission of data between nodes, as well as checking for transmission errors.

This is also the first layer in the OSI Model where the specifics of protocols are defined. The specific data packets for Token Ring versus TCP/IP are defined for example. From this point forward in the OSI Model, differences in protocols are reflected in the packet definitions.

Network layer. The network layer is primarily responsible for routing data packets efficiently between one system and another on the network. The two network protocols included with Windows NT most commonly used in networks where routing is required are TCP/IP and NWLink, Microsoft's implementation of the IPX/SPX protocol. The IP protocol, which is part of the TCP/IP suite, is responsible for routing at the network layer. IPX, which is typically used in Novell NetWare environments, serves the same purpose. Microsoft and other vendors involved in network operating systems see routable protocols as a given that their implementations of TCP/IP must provide.

In addition to TCP/IP and IPX/SPX, Windows NT also includes the NetBEUI protocol. In the first versions of NT (3.1 and 3.51), NetBEUI was the default network protocol. Starting in NT 4.0, TCP/IP is the default protocol. The reason for this change is that NetBEUI was originally designed to provide efficient transport in small networks. NetBEUI can't be routed, however, which eliminates its use in large, complex networks where routing is often necessary. If you have a small, self-contained network, NetBEUI should be your protocol of choice—or at least one to consider in a peer-to-peer network. Otherwise, if you are going to route data from one location to another, you'll find TCP/IP the best choice for linking dissimilar systems or IPX/SPX if your entire network or the majority of it is Novell NetWare-based.

Transport layer. Although the network layer is responsible for routing, it doesn't perform error checking: It simply passes the packets back and forth. There's no guarantee that the packet will arrive with all its components, or if there are contiguous packets, that they will all arrive in the proper order. Some level of error checking is required, and the transport layer provides this reliability. The transport layer checks for errors and causes the packets to be retransmitted when an error in sending or receiving occurs. In essence, the transport layer acts as the first point of quality assurance in the definition of a transaction over the network.

Windows NT's two protocols that are pervasively used due to their routing characteristics are TCP/IP and IPX/SPX. These two protocols also provide reliable transmission due to their inclusion of additional features at the transport layer of the OSI Model. The TCP portion of the TCP/IP protocol is the component that specifically is responsible for error checking and placing packets in the correct order. SPX performs the comparable function in the IPX/SPX protocol.

Session layer. This layer also is responsible for adding reliable and secure communication between nodes of the network. When two network nodes or systems communicate, it's called a session. The session layer is responsible for establishing the communication link and the rules by which the session will be accomplished, including the transport protocol to be used. This layer also involves security authentication—to server nodes authenticating the client node's request through the security subsystem, validating what the client can and can't access.

Presentation layer. The function of this layer is to be an intermediary between the session layer and the application layer. The presentation layer translates the data from one format to another if required for the session and application layers to communicate. The presentation layer can also perform such tasks as data compression and encryption.

Application layer. This layer provides the interface between the network and user interface and applications. In general, the application layer provides a layer of abstraction between the user and the network, making the network essentially transparent. Because of the application layer, a program you're using to open a file can do so as easily across the network as it does locally.

UNDERSTANDING HOW TCP/IP ENABLES COMMUNICATION

The OSI Model provides a useful reference point for comparing the components of the TCP/IP command set. You're asking yourself, "How will this assist me in my efforts to serve my clients regarding TCP/IP?" The answer is that by having these foundation elements you'll be able to give them assistance and guidance for their most challenging aspects—those of getting systems to work in conjunction with each other.

Three processes are critical for having datagrams routed from the source computer to the destination host via the intended or destination network. These three processes are as follows:

Addressing. With an IP address, every network and every workstation have a unique identity, and they can be specifically communicated using their specific address.

Routing. This is the process by which messages pass through gateways to reach the destination workstation or network.

Multiplexing. This process entails combining transport protocol traffic into a single IP data stream.

IP Addressing

This attribute gives a network, workstation, or even an intranet its own identity within a large extranet or even from the Internet. It's been called the biggest single differentiator in the entire TCP/IP protocol suite. Many organizations prefer to have each workstation always have its own IP address. This is typically called "static IP addressing," and it makes a lot of sense, especially when the workstations will be staying in their locations. The DHCP protocol, in contrast, is a network protocol specifically developed for defining IP addresses just as a library dispenses books—on demand. Both the static IP addressing approach and DHCP use the same values for the IP address; these are just different ways of assigning them to workstations.

If you are running TCP/IP on a network that is not linked to the Internet, you can assign any valid IP address values that you want to your networks and hosts. Any computers that you will be connecting to the Internet must be on a network with an address that has been registered with InterNIC, the Internet Network Information Center. InterNIC is a clearinghouse for all Internet IP addresses, ensuring that host addresses are all unique.

InterNIC, however, does not register individual workstations, only networks. As a system administrator, you will use the series of IP addresses specific to your organization for defining a unique one for each workstation. The network address is part of the 32-bit IP address. The address class and the subnet mask determine which part of the address represents the network and which part represents the host.

Be sure to check out Network Solutions and their website for an example of how easy it is to reserve a URL of your own. Your URL is registered with InterNIC through the Network Solutions licensing relationship now in place. You can find Network Solutions at www.networksolutions.com and it is a good idea to get familiar with their services if you haven't seen them already.

IP Address Classes

InterNIC uses three classes of IP addresses, designating the specific class by the first three bits of the address. These classes vary depending on needs of the organizations receiving them. Here are the classes of IP addresses:

Class A. The first bit of a class A address is always 0, meaning that the first octet of the address can have a value between 1 and 126. Only the first octet is used to represent the network, leaving the final three octets to identify 16,777,214 possible hosts.

Class B. The first two bits of a class B network are always 1 and 0, meaning that the first octet can have a value between 128 and 191. The first two octets are used to identify 16,384 possible networks, leaving the final two octets to identify 65,534 possible hosts.

Class C. The first three bits of a class C network are always 1, 1, and 0, meaning that the first octet can have a value between 192 and 223. The first three octets are used to identify 2,097,151 possible networks, leaving the final octet to identify 254 possible hosts.

What's Subnet Masking?

If you have a network address registered with InterNIC, then the host to which you assigned that address can be used on the Internet. What if you want to set up numerous internal networks? Do they all need to be on the same InterNIC address? he answer is no; you can use subnet masks to differentiate other networks from the one visible on the Internet. Subnet masks are used for dividing your assigned network into subnetworks and assigning IP addresses based on the physical layout of your own internal network.

A *subnet mask* is a filter designed to register a value in the TCP/IP stack that tells the system which bits of the IP address represent the network and which represents the host. The subnet mask value is expressed in the form of a normal decimal IP address, but it is easier to think of it in its binary equivalent. A subnet mask is 32 bits in which 1s indicate the part of the address representing the network and 0s represent the host.

For example, if you have a small one-segment network and you obtain a class C address, then you can use it without modification. You would enter a subnet mask of 255.255.255.0 in all of your workstations and assign individual host values using the values 1 to 254 in the last octet of the IP address. The 255s represent binary octet values of 11111111, meaning that all of the first three octets are used to identify the network.

A more complicated example occurs when you are assigned a class B address and you have an Internetwork that is composed of many individual segments. It would be impractical to use the class B address as it is (with a subset of 255.255.0.0) and the number of hosts in a single sequence from 1 to 65,534, using the last two octets. Instead you would establish subnets that present the physical segment of your network. You would do this by applying the subnet mask of 255.255.255.0 on all of your hosts. The first two octets of their IP addresses would be values assigned by InterNIC. The third octet, however, would now represent part of the network address instead of the host. You can assign values to this third octet representing each of the segments on your network, and then you can use only the fourth octet to address as many as 254 hosts on each subnet. Subnet masking can get even more complicated than this, for example, when a mask is applied to only certain bits of an octet. Careful conversion between binary and decimal values is then necessary to ensure accurate configurations.

Subnet masking is something that must be organized on an Internetwork-wide basis. Be sure to check with your system administrator on the values of the subnet masks you should include.

IP Routing

The other side of getting messages delivered over a network is having the IP routing in place to ensure datagrams are delivered to the correct workstation or network. When the source and destination workstations are not on the same network, then one or more gateways are used to route packets to the correct network. A gateway is actually a hardware router and multi-homed Windows NT or Windows 2000 server in this instance.

Every TCP/IP system maintains an internal routing table that helps it make routing decisions. For the average host computer, located on a network segment with only a single gateway, the routing decisions are simple; either a packet is sent to a destination host on the same network, or it is sent to the gateway. As stated earlier, IP is aware of only the computers on the networks to which it is directly attached. It is up to each individual gateway to route packets on their next leg of the journey to the destination.

It may be the case that a host computer is located on a network segment with more than one gateway. Initially, packets addressed to other networks are all sent to the default gateway that is set as part of the host configuration. However, the default gateway may be party to routing information that is unavailable to the host, it may send an ICMP Redirect packet instructing the host to use another gateway when sending to a particular IP address. This information is stored in the host's routing table. Whenever the host attempts to send packets to that same IP address again, it consults its routing table first and sends the packets to the alternate gateway.

How can you be sure the routing table is correct and has the values entered into it that you need to support your Internetwork? Using the command line NETSTAT -NR or the ROUTE PRINT command, you can view the table. This command lists the IP addresses found in the table, the gateway that should be used when sending to each, and other information defining the nature and source of the routing information.

Depending on the destination host (which can be a network or an individual workstation or server) the routing implications can become elaborate. There are several TCP/IP protocols designed for use only by gateways to exchange routing information that allows them to ensure the most accurate and efficient routing possible, such as the Gateway to Gateway Protocol (GGP) and the Exterior Gateway Protocol (EGP). This presents the traffic for each individual host from flooding the entire Internet in search of a single network. As you can see from this example, the majority of TCP/IP routing is based on tables. Hosts and gateways each maintain their own routing tables and perform lookups of the destination address found in incoming packets. While the Internet at one time relied on a collection of central core gateways as the ultimate source of routing information for the entire network, this become impractical due to the tremendous growth of the Internet in recent years. Routing is now based on collections of autonomous systems called routing domains that share information with each other using the Border Gateway Protocol (BGP). Much of this is hidden from users, yet there are Web sites for showing the number of connections datagrams make as they traverse from the targeted host to your workstations and back again.

Getting to Know IP Multiplexing

After the host computer has received IP datagrams, they must be delivered to the transport layer protocol for interpretation and use by the targeted application service. On the sending workstation, the process of combining the requests made by several different applications into traffic for a few Transport protocols, and then combining the transport protocol traffic into a single IP data stream, is called multiplexing. Multiplexing is really the process of taking several messages and sending them using a single signal. At the receiving computer the process is reversed, in effect creating a de-multiplexing routine of steps for in-bound messages.

The numeric values assigned to specific protocols and application services are defined on the host computer in text files named Protocol and Services, respectively. These are located in the Winnt/System32/drivers/etc folder on a Windows 2000 Professional system. Many of the values assigned to particular services are standardized numbers found in the Assigned Services RFC. FTP, for example, traditionally uses a port number of 20 on all types of host systems. Port numbers are individually defined for each transport protocol. In other words, TCP and UDP both have different assignments for the same port number. The combination of an IP address and a port number is known as a socket. Be sure to keep the TCP/IP term socket with the Windows Sockets application development standard created by Microsoft, even though the Windows NT TCP/IP utilities were developed using this standard.

Integrating Microsoft TCP/IP into a Network

Two market forces that has led to the rapid adoption of TCP/IP in organizational networks are the growth of the Internet and the need for integrating Windows NT and now Windows 2000 into heterogeneous networks. TCP/IP is now very much the foundation of connectivity of the Internet, and it is effectively leveling the playing field of operating system competition.

Many organizations today have multiple network operating systems running at the same time. There are a multitude of scenarios that can lead to this condition, with the most common being an integration of network architecture and protocols that are mainframe-based, with both Novell

NetWare and Windows 2000 joined onto the total network as the number of users has grown and their corresponding needs have caused the first generation of network architecture to grow to meet evolving requirements. Another factor that drives organizations to have multiple network operating systems is users' need for specific client/server applications that are supported on Windows 2000.

Organizations today increasingly find that Novell NetWare has provided excellent file and print services, while Windows 2000, due to the pervasive support for Win32 APIs in the application development community, has the majority of 32-bit applications. Consequently, organizations have typically integrated Windows NT as the application server. What's different about Windows 2000 Professional is that file and print services have significantly been improved, giving a Windows 2000-based dual-processor workstation higher performance than its Novell-based counterpart.

Taken together, the needs of users for running client/server applications that use data from UNIX servers and print services from NetWare create a challenge for system administrators in troubleshooting problems. These separate protocols all cause redundant traffic and make it difficult to troubleshoot transmission problems over a network. Many administrators use TCP/IP as the unifying protocol in these kinds of scenarios, as TCP/IP is supported on all three platforms.

LEARNING TCP/IP CONNECTIVITY UTILITIES

In using TCP/IP as the unifying protocol in a network of heterogeneous operating systems, a series of connectivity and diagnostic utilities is always useful for checking connections and ensuring reliability of the network. Windows NT, in fact, contains its own versions of the standard TCP/IP utilities found on other platforms supporting this protocol. In global terms, TCP/IP utilities are separated into diagnostic utilities used for monitoring a network and its clients and connectivity-oriented utilities. Diagnostic utilities are described in the next chapter, "Configuring TCP/IP in Windows 2000 Professional." The connectivity-oriented utilities are described here.

Finger

While not commonly used between organizations today, many companies continue to have internal applications that still use this command. It's not typically found outside companies, but if you do work in an organization, which has this as a standard, be sure to get familiar with the syntax of this command. Finger is a command-line utility that displays information about user(s) logged onto a remote system. The remote system must be running the Finger service for this command to function, and the output of this command varies depending on the remote system being addressed. Table 6.1 lists Finger's parameters.

Syntax: finger [-l] [username] @hostname

TABLE 6.1 Parameters of Finger

Option	Description
-l	Displays information in long list (verbose) format.
username	When this isn't defined, all users on the remote system are listed. Use this option to check the status of an individual user.
@hostname	Use this option for defining the IP address or host name of the system you're querying.

FTP

Many times FTP is seen as a command, yet it is, in fact, a protocol that uses the TCP/IP connection to transfer files to and from remote computers regardless of the file systems being used at either end. FTP stands for File Transfer Protocol (FTP). FTP is the most pervasively used protocol within the TCP/IP suite as it is used for bridging diverse hardware platforms. This command was originated on the UNIX platform, and it is used pervasively in shell scripts for enabling communication between workstations and servers. Predominantly used for moving larger files between locations quickly and mostly transparently to users, FTP is the "glue" that holds together client/server applications that use UNIX as a server component with Windows NT as clients, and vice versa. FTP really embodies the goal of hardware interoperability between platforms.

You can use FTP both within shell scripts and interactively for moving files around a network, or even the Internet. You can, in fact, use a Web browser (Microsoft's Internet Explorer, for instance) to access FTP sites anywhere in the world, providing they are public-domain or you have the username and password to log in. Keep in mind that usernames and passwords are transmitted in clear text format, so anonymous FTP access is the best bet if you want to distribute information inside your organization using this command.

To connect from a workstation to another system via FTP, the destination system needs to be running an FTP server component. Microsoft delivers FTP as standard within Windows 2000 Professional. An FTP server for Windows 2000 Professional is delivered in the Microsoft Peer Web Services package, which began shipping in Windows NT 4.0 Workstation. You can configure a Windows 2000 Professional-based workstation as an FTP server using Microsoft Peer Web Services, making it possible to quickly post files and images for others to gather and use via the FTP command. In general, FTP is used from clients to larger file servers, where disk space greater than the capacities of workstations is available.

FTP is also delivered as part of Microsoft's Internet Information Server, and it ships with the Windows 2000 Server operating system. In the instance of configuring IIS for FTP use, FTP actually runs as a network service, allowing multiple users to connect with and use the server simultaneously. Nearly all UNIX operating systems run an FTP server daemon by default; FTP is often used for file transfers between UNIX and Windows NT systems. A daemon is the UNIX equivalent of a service under Windows NT. It is a program or utility than runs all of the time, and it provides resources that are available to any process that requests them.

Using the FTP command to connect with another system, you'll use the syntax provided in this command description. When logging onto an FTP site, you'll be prompted to enter your username and password. In the case of using anonymous FTP sites, your username needs to be anonymous, and the password you enter needs to be your e-mail address. Many of the anonymous FTP sites do verification checking of e-mail addresses, while a few do not.

After logging onto the remote system you can traverse the directories that have been made available through the definition of system security properties by the system administrator. If you're a system administrator, you'll find that Windows 2000's Peer Web Services can quickly configure an anonymous FTP site using the options in that service. Once logged onto a system using FTP, whether the host system is Intel or RISC-based is transparent. The only delineator of which hardware platform you are logged onto is the subdirectory structure. FTP is indeed transparent to the hardware platforms supporting it.

When you want to download files from an FTP server, you use the command GET. Conversely, loading files onto a server requires the PUT command. Some servers have restrictions on where files can be loaded, so be sure to check with the permissions set on the login and password you have before trying to load files to an FTP server. Keep in mind that the file names and commands themselves are case-sensitive, which makes sense given the fact that the commands included in this protocol bridge Windows NT (which is not case-sensitive in its file structure) to UNIX (which is case-sensitive in its definition of files).

Compared to the performance from a 14.4K or 28.8K modem line, the FTP command is by far the most efficient method for downloading files from the Internet. Providing the organization you want files from has an FTP site, you can get 2MB or greater files in a matter of seconds. Microsoft maintains a comprehensive FTP site at ftp.microsoft.com, which can be accessed through any Internet browser. Simply type in the FTP location in the place of a Web address, and the FTP site will be displayed. While the standard interface for the FTP command line is quickly becoming outdated due to the pervasiveness of using browsers for traversing FTP sites, utilities, both public-domain and offered for sale, are available that streamline the FTP process. Once FTP has been installed in Windows 2000 Professional on your system you can use the command line that follows for accessing and downloading or updating files to selected FTP sites. If you're using a public domain or purchased program for FTP functions, these commands are the basis for activities in those programs. Table 6.2 lists FTP's parameters.

Syntax: ftp [-v] [-n] [-i] [-d] *[hostname]* *[-s:filename]*

What about using the commands in the FTP protocol? Let's take a quick tour of the most commonly used commands during FTP sessions (see Table 6.3).

TABLE 6.2   Parameters of FTP

Option	*Description*
-v	Prevents the display of remote server responses to client commands.
-n	Prevents autologin upon connection.
-I	Prevents individual file verifications during mass file transfers.

TABLE 6.2 (continued)

Option	Description
-d	Displays debugging messages.
-g	Allows wildcard characters to be used in file and directory names.
Hostname	Defines the host name or IP address of the remote system to be accessed.
-tcp port #	Describes the port number being used for handling the ftp transmission.
-username	Changes the username that is being used for logging into the remote host.
-password	Changes the password that is being used for logging into the remote host.
-s:filename	Allows you to specify a test file containing a series of FTP commands to be executed in sequence. This parameter in effect launches a shell script of FTP commands to be executed on the remote system.

TABLE 6.3 Profile of FTP Commands

Command	Description
open hostname	Initiates a command session with a remote FTP host.
close	Terminates the current session (without closing the FTP command window or program).
exit	Closes the FTP program, returning you to the command prompt of your system.
ls	Lists all the files in the current directory.
ls -l	Lists full information for the files in the current directory.
cd /dirname	Changes to the different directory.
cd ..	Moves up one level in the directory tree.

(continues)

TABLE 6.3   (continued)

Command	Description
pwd	Displays the current directory name.
binary	Sets the file transfer mode on FTP to binary or bit-by-bit mode. It's a good idea to use this command by default, especially on image-based files.
ascii	Specifies that the file to be transferred is an ASCII file.
get *filename*	Transfers the specified file to the local system.
recv *filename*	Functions the same as get; transfers the specified file to the local system.
mget filename	Used in conjunction with wildcards in the command statement, this command transfers multiple files to the local system.
put filename	Transfers the specified file to the remote system.
send filename	Functions the same as put; transfers specified files to the remote system.
mput filename	Transfers multiple files to the remote systems. This command is typically used in conjunction with wildcards.
hash	Displays status of the current operation as files are transferred.
prompt	Toggles the use of prompts for each individual file during multiple transfers.
help	Provides the FTP command summary.

TFTP

Keep in mind that there is no user authentication services in TFTP. Further, TFTP does not provide for browsing of directories, as the UDP protocol is not connection-oriented. If you plan to use this command you'll need to know the exact file name and location of what you need to retrieve. Table 6.4 lists TFTP's parameters.

Syntax: tftp [- i] *host* [get] [put] source *{destination}*

TABLE 6.4   Parameters of TFTP

Option	Description
- i	Defines the file to be transferred in binary mode.
host	Replaces the host name or IP address of the remote system.
get	Transfers a file or files from the remote system to the local system.
put	Sends a file from the local system to the remote system.
source	Defines the name of the file to be transferred.
destination	Defines the location where the file is to be transferred.

Telnet

Telnet is a terminal emulation program, which makes it possible to log onto and use a remote workstation or server running telnet server services. Windows NT does not currently include a telnet server service. Many UNIX servers do have telnet server capabilities, and this command can be used from the Command Prompt window of a Windows 2000 Professional to log onto and use a UNIX-based server or any other server running a telnet service. The telnet implementation in Microsoft's TCP/IP supports DEC VT100, DEC VT52, or TTY terminals via emulation.

Using the options associated with the Command Prompt window, you can define the terminal and display preferences for the telnet sessions you plan to initiate from a Windows 2000 Professional. Terminal emulation using telnet occurs through the Command Prompt window. Once connected with the remote system, the entire session can be completed from within the Command Prompt window. Telnet then works in conjunction with Windows 2000 using the emulations available via the Command Prompt for communicating via the TCP protocol to other compatible systems. Table 6.5 lists telnet's parameters Syntax: telnet [*host*] [*port*].

RCP

The purpose of the RCP utility is to provide a command for copying files between a local system and a remote system that is running a remote shell server, rshd. Alternatively, you can direct two remote systems running rshd to exchange files between themselves using this protocol. Keep in mind this

TABLE 6.5   Parameters of Telnet

Option	Description
Host	Defines the host name or IP address of the remote system you want to log onto.
Port	Defines the port number in the remote system to which you will connect. When omitted, the value specified in the remote system's services file is used. If no value is defined in services, then port 23 is automatically used.

command is typically used for enabling Windows to UNIX communications and vice versa.

A text file on the remote system, called. rhosts, contains both the IP address and the corresponding user's name of the local system.

Syntax: rcp [-a] [-b] [-h] [-r] source1 source2 ... sourceN destination

TABLE 6.6   Parameters of RCP

Option	Description
-a	Specifies the file(s) to be transferred in ASCII mode.
-b	Specifies the file(s) to be transferred in binary mode.
-h	Allows files on a Windows NT system with file attributes set as hidden to be transferred.
-r	Copies the contents of all of the source's subdirectories to the destination (when both source and destination are directories).
Source	Specifies the name of the file to be transferred (and optionally its host and user names, in the format host.user:filename).
Destination	Specifies the name of the file to be created at the destination (and optionally its host and user names, in the format host.user:filename).

REXEC

The REXEC utility provides for batched, or noninteractive, commands that are executed on a remote system. This command needs to have rexec service running on the remote system to be available to Windows NT clients. Redirection symbols can be used to refer output to files on the local system (using normal redirection syntax) or on the remote system (by enclosing the redirection symbols in quotation marks, for example, ">>"). Table 6.7 lists REXEC's parameters. Like RSH, this command is typically used for Windows to UNIX connectivity.

Syntax: rexec hostname [-l *user*] [-n] *command*

TABLE 6.7 Parameters of REXEC

Option	Description
hostname	Specifies the host name of the system on which the command is to be run.
-l user	Specifies a username under whose account the command is to be executed on the remote system (a prompt for a user password will be generated at the local machine).
-n	Redirects rexec input to NUL.
command	Specifies the command to be executed at the remote system.

RSH

The RSH command is used for executing commands and options on remote systems. This command requires the remote system to have the rsh service running on the remote system. It has the same redirection capabilities as REXEC and the same.rhosts requirement as RCP. Table 6.8 lists the parameters of RSH.

Syntax: rsh *hostname* [-l *user*] [-n] *command*

TABLE 6.8   Parameters of RSH

Options	Description
hostname	Specifies the host name of the system on which the command is to be run.
-l user	Specifies a user name under whose account the command is to be executed on the remote system (a prompt for a user password will be generated at the local machine).
-n	Redirects rsh input to NUL.
command	Specifies the command to be executed at the remote system.

LPR

The LPR utility provides the flexibility of printing a file to a printer connected to a remote BSD-type printer subsystem that is running an LPD server. Table 6.9 lists LPR's parameters.

Syntax: lpr -S*server* -P*printer* [-J*jobname*] [-ol] *filename*

TABLE 6.9   Parameters of LPR

Options	Description
-Sserver	Specifies the host name of the system to which the printer is connected.
-Pprinter	Specifies the name of the printer to be used.
-Jjobname	Specifies the name of the print job.
-ol	Used when printing a nontext (PostScript) file from a Windows NT system to a UNIX printer.
-l	Used when printing a nontext (PostScript) file from a UNIX system to a Windows NT printer.
filename	Specifies the name of the file to be printed.

WHAT IS DHCP?

Let's say you are responsible for supporting 50 to 100 users of laptops who travel extensively and then come back into the office to check e-mail and download files for use at customer sites. How can you create a network that makes it possible for members of this group to get full TCP/IP access while in the office or even using Remote Access Services (RAS) from regional offices equipped with fractional T1 lines? Instead of choosing NetBEUI and settling for its limitations from both a performance and security perspective, how can you enable this group with TCP/IP access? Through the use of the Dynamic Host Configuration Protocol (DHCP) originally developed by Microsoft. This protocol truly functions just as a library checks books in and out—only this protocol checks in and out IP addresses just as a library checks out books. Best of all, this IP address mechanism is for the most part administered at the server level, alleviating ongoing configuration at the client level of TCP/IP configuration options. This means that once a laptop is configured to accept an IP address, the user at the remote location doesn't need to configure any other parameters.

The Dynamic Host Configuration Protocol, or DHCP, is an element of the TCP/IP protocol suite that enables you as a system administrator to automatically configure parameters such as IP addresses, subnet masks, and default gateways for remote systems that have sporadic access to a TCP/IP-based network. A DHCP Server, running Windows 2000 Server, manages these attributes responsible for handling a TCP/IP connection. Microsoft, working in conjunction with other companies, created this protocol to solve the dilemma of how to more easily solve the inherent difficulties associated with the assignment of IP addresses to workstations, large networks, and mobile users who need IP addresses when they dial into a network.

DHCP Server is included in the baseline of Window NT Server 4.0 and Windows 2000 Server. The DHCP Server consists of two components: a mechanism for tracking and allocating TCP/IP configuration parameters and a protocol that can distinguish when an IP address has been delivered to and acknowledged by the client. The DHCP standard, published in the Internet Engineering Task Force's RFC 1541, has been strongly influenced by the BOOTP protocol defined in RFC 951. BOOTP is a segment of the TCP/IP command set that was originally developed for use with diskless terminals needing TCP/IP compatibility.

BOOTP was first developed for these diskless workstations from the server-side to ensure TCP/IP connections could be made and the TCP/IP configuration settings delivered. While BOOTP is capable of assigning the same IP address to a diskless workstation or terminal, DHCP can dynamically assign and renew workstation configurations from a pool of IP addressees, solving many of the problems that have hampered the ongoing acceptance of the BOOTP protocol. Windows NT Server ships with integrated DHCP server modules that are compatible with the TCP/IP command set, regardless of the operating system of the originating system.

Why DHCP Was Created

With the widespread growth of large-scale networks and the integration of multiple needs within large organizations, and the need for uniquely identifying systems on a large, packet-switched network like the Internet, a dominant need to selectively identify each client quickly became a need. The approach to identifying client systems needed to be transparent to the hardware; it had to be an approach based solely on stable protocols that could traverse the hardware differences between systems.

Different types of networks have varying approaches to identifying and communicating with systems, which are part of their topology. An Ethernet or Token Ring LAN uses a unique MAC (Media Access Control) address hard-coded into each network interface card by the NIC's manufacturers. Because each manufacturer numbers its card sequentially, and because part of that MAC address consists of a code identifying the manufacturer itself, the device's address is unique not only on the local network where it is used, but on all networks everywhere. No other adapter uses that exact same address.

This hard-coded approach would be a pervasive solution if all users of the Internet had a compatible system that could communicate these protocols. For many networks, system administrators actually override the hardware address with their specific IP addressing approach. For the majority of Internet users today, there are simply no NIC cards installed in their systems, as they are using modems to dial up to ISPs and gain access to the Internet. This approach to connectivity to the outside world necessitated a network protocol thatcould effectively assign IP addresses dynamically from a "pool," or inventory of addresses, ensuring that the dial-up system has a unique identity.

Understanding How DHCP Works

When a client is configured to use the DHCP protocol, it acquires its IP configuration settings; it will attempt to communicate with the DHCP servers available on the local network each time it is rebooted or when the TCP/IP stack is reinitialized. This process relies heavily on the BOOTP protocol to initialize the entire process of finding the DHCP server on a network. DHCP runs on top of the User Datagram Protocol (UDP) for its communications, with clients transmitting to the DHCP Server and services using the RFC assigned numbers. Windows 2000's DHCP communications are defined through the use of the Remote Procedure Calls (RPC) application programming interface. All DHCP communications use the same packet format.

Let's walk through the major steps of how DHCP works. First, a DHCP client broadcasts to all DHCP servers on the network, using the BOOTP protocol to transmit its request. It is the responsibility of the routers and computers functioning as DHCP relay agents on the local network to propagate or forward the traffic generated by a client to other networks where additional DHCP Servers may be located. If the workstation looking for an IP address using the DHCP protocol does not find one, it will revert to the settings for the previous IP address that had been assigned in a previous session. When a client system does get an IP address assigned to it, it is called an IP lease, or the act of leasing an address.

If another client is actually using the options defined in the previous settings, an error message will be sent back to the client. The Windows 2000 Professional will in this instance repeat its attempt to contact a DHCP server until an appropriate response is received, generating an error each time the predetermined time-out values are reached.

A workstation configured as a DHCP client that has no previous configuration settings assigned to it by a server must start the client configuration sequence from the beginning. As an administrator it's important that you realize the circumstances that can leave a DHCP client in an unconfigured state. Here are the specific instances of Windows 2000 Professional configured as a DHCP client, which can be defined in an unconfigured state:

- The workstation has configuration options as defined from a DHCP client session whose lease has expired.
- The workstation bas just been configured and has yet to receive a DHCP client setting parameters options.
- The workstation has moved from its existing subnet to a new subnet that is unrecognized by the DHCP server that originally defined the IP lease.

■ The workstation may have explicitly release its hold on previous IP addressing options.

Installing DHCP Client Services

By default, when Windows 2000 Professional is installed, the DHCP protocol is also installed and made active. This protocol is designed to ease the complexities of configuring TCP/IP addresses and give client workstations the flexibility of having addresses assigned to them dynamically. This is particularly valuable when a laptop is running Windows NT Workstation and needs access to either an Intranet within a company or to the Internet via an ISP. After installing TCP/IP, and with it support for DHCP, every time Windows 2000 Professional is rebooted the TCP/IP protocol stack is reinitialized and the DHCP parameters are again communicated to the network.

USING DNS IN WINDOWS 2000

As with many of the advances made in the TCP/IP protocol, the Domain Name Services (DNS) was created during the 1980s to accommodate the growth of the Internet and the need for a reliable naming service. Just as the DHCP protocol was specifically developed to dynamically assign IP addresses through a leasing approach, DNS is a distributed name service and database system that spreads the load of system identification around the network by dividing it into domains. Domain names, rather than individual systems or workstations, are then are registered throughout the network. The unique name of the user and the domain it is a member of are the basis of the DNS databases that are distributed through networks based on this approach.

How DNS Naming Resolution Happens

DNS servers actually are centrally located hosts, where the names of individual workstations are compared and then cross-referenced for the originating system—making it possible to communicate with a targeted server. Requests for the identity of a workstation or node are sent to a DNS server, which then looks into the DNS database and returns the IP address corresponding to the targeted workstation. Its underlying strength is evident only when users try to resolve the names of computers located at remote Internet sites.

In the distributed architecture of DNS, there is no single listing of database that contains all the computers using the Internet. There is instead a collection of computers known as root servers that contain a complete listing of the domain names registered to individual networks. The entire DNS architecture is heavily based on a tree-like structure, emanating out of the root servers. The listings of root servers for each domain include the IP addresses of the DNS machines that have been designed as the defining servers for DNS-based traffic.

Despite its limitations the DNS architecture makes it possible for administrators to easily manage the network traffic specific to and relying on the DNS servers in Windows NT-based domains. Every DNS server uses its host tables as pointers for messages being sent from initiating the client workstation to the destination. By beginning at the root of the Internet tree, any host on any computer can be located by parsing the database from its root domain down to its host name and consulting the databases for each level.

The Role of the HOSTS File

In the context of DNS, the HOSTS file is the "deliverable" or the item where the host or workstation name and IP address are recorded. Since the inception of the TCP/IP protocol, the HOSTS file has actually been the clearinghouse or common place where name and IP address resolution are completed. The HOSTS file was challenged in its performance by the rapid adoption of TCP/IP networking and with it, the triple-digit compound annual growth rates of systems of all types needing an IP address identity. The HOSTS file, in smaller configurations, however, does provide an efficient name resolution approach because it is consulted before communications begins.

CHAPTER SUMMARY

Considering the growth of the Internet, and with it the explosive growth in the number of workstations of all types using the Internet, there is a continual need to evaluate and expand the capabilities of network protocols to assign names and IP addresses to these systems. The intent of this chapter has been to give you an overview of the single biggest connectivity force in the world of operating systems and connectivity today: the TCP/IP networking protocol. TCP/IP has leveled the playing field of operating systems and workstations of all types by making connectivity a given. No longer are workstations connected via proprietary operating systems, as Windows NT's tight integration of TCP/IP has made Windows 2000 Professional-based workstations capable of communicating with virtually any other system that is compatible with the TCP/IP protocol. The integration of naming conventions into the TCP/IP standard continues to improve the performance of these networks.

7 Configuring TCP/IP in Windows 2000 Professional

Incoming TCP/IP Properties

Network access

☑ Allow callers to access my local area network

TCP/IP address assignment

◉ Assign TCP/IP addresses automatically using DHCP

○ Specify TCP/IP addresses

From:

To:

Total:

☐ Allow calling computer to specify its own IP address

OK   Cancel

A n essential part of the Microsoft culture is to thoroughly deploy its own technology internally. That is the case with the Microsoft implementation of TCP/IP. This approach to product design makes the development teams accountable to their fellow employees for the performance of applications, and the development team's reputation is typically determined by the level of bug-free performance the team's applications provide. Talk about pressure! You have to face the people using your products the very next day after they are released. This, at Microsoft, is called "eating your own dog food" and is an essential part of the company's culture.

The integration of TCP/IP into Microsoft involved the migration of 35,000 network nodes globally. Through these experiences Microsoft realized that although the first implementations of TCP/IP were robust for smaller workgroups, they definitely needed work to scale accurately for the thousands of users within Microsoft's headquarters and regional offices throughout the world. As a result of their thorough internal testing, both the DHCP and WINS protocols were integrated into Windows NT 4.0 and now Windows 2000. Microsoft is being very careful to not market a proprietary solution, so the RFCs for the DHCP and WINS implementations have been presented to the Internet Engineering Task Force.

With the latest release of TCP/IP in Windows 2000 Professional and Server, Microsoft set out to make its interpretation standards-compliant and portable across platforms. Specifically, the design objectives as defined by Microsoft's design teams included making TCP/IP the following:

- Standards-compliant
- Interoperable
- Portable
- Scalable
- High performance
- Versatile
- Self-tuning
- Easy to administer
- Adaptable

The continual development of Microsoft's networking expertise has yielded further differentiation between its operating systems to specific command support within each operating system. Because networking is so pervasive, networking must be included in each level of operating system (Windows95/98, Windows 2000 Professional, and Windows 2000 Server) at different levels of functionality. Appendix A, "Comparing TCP/IP Implementations," defines in detail the differences between each of the operating system versions.

GETTING TCP/IP UP AND RUNNING

Due to the extensive graphical interface of Windows 2000 Professional, installing TCP/IP support is relatively easy. You'll find greater control over the configuration of options due to the extensive use of tabs within the

Network dialog box. These tabs give you an easier approach to configuring network options.

As in Windows95 and 98, the networking components of Windows 2000 are broken into three types: adapters, protocols, and services. These types form a stack corresponding to the overall network architecture; the services run on top, passing application data through the protocols, which code it for transmission via adapter.

At a minimum, you'll need to have network adapters and the corresponding adapter drivers installed in each Windows NT workstation. In the case of integrating Novell NetWare into your networks, you'll also want to have NWLink on each workstation. On the workstation you'll also want to install the NetBIOS Interface and the RPC Name Service Provider services so that each workstation will be able to see other network-based systems. Many of the network services are optional and must be explicitly installed from the Control Panel. Some Windows NT services can be run with different protocols; others are designed for use with a specific one.

HANDS-ON TUTORIAL: GETTING TCP/IP WORKING IN WINDOWS 2000 PROFESSIONAL

The real value-add of having a network operating system is the ability to integrate workstations of various operating systems into a single, cohesive network. TCP/IP is the network protocol of choice for ensuring the communication of workstation and servers—any computer that needs to reside on a network and share resources. Throughout this section, we'll explore how to configure Windows 2000 Professional for use in a TCP/IP-based environment, including the fundamentals behind each of the functions included in each of the protocols defined.

Installing TCP/IP

Because Windows 2000 Professional uses a browser-like shell for many of the configurable properties and attributes, the entire process of installing and customizing TCP/IP is relatively simple. In Windows NT 4.0 the Network dialog box was found in the Control Panel; Windows 2000 Professional's networking is found in the Add/Remove Programs applet, also within the Control Panel. Once you have selected the Add Network Software entry in the Add/Remove Programs application, you can configure a workstation for TCP/IP or any of the other networking protocols

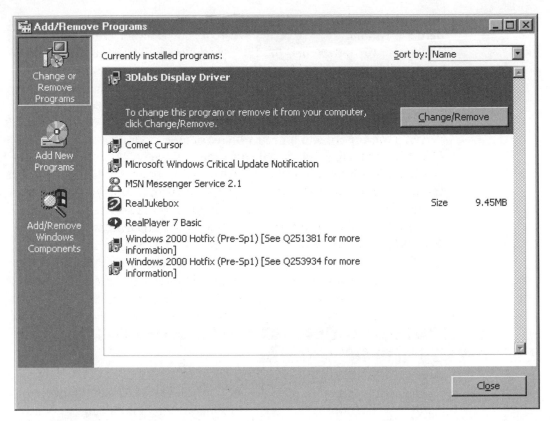

FIGURE 7.1 Windows 2000 Professional now includes support for loading software directly from the Add/Remove Programs dialog box.

supported in Windows NT. Figure 7.1 shows the Add/Remove Programs applet opened with the Add Network Software entry selected.

As in Windows95 and Windows NT 4.0, the networking components included in Windows 2000 Professional are broken down into three types: adapters, protocols, and services. These types form a stack corresponding to the overall networking architecture itself. The services run on top, passing application data through the protocols, which code it for transmission via the adapter.

Configuring TCP/IP for DHCP Support

You'll find that Windows 2000 Professional has significantly different navigational options for getting to the DHCP options available within the context of a TCP/IP network session. DHCP stands for Dynamic Host Configuration Protocol, and it is used for allocating Internet addresses dynamically as a system user dials in to use the network. DHCP is considered to be one of the strongest differentiators the Windows NT operating system has, as Microsoft's implementation is now three product generations old. DHCP relies on client/server architecture, as do other components of the TCP/IP command set in Windows 2000 Professional andWindows 2000 Server. DHCP is considered to be one of the command options of the TCP/IP command set. It is not a separate client, service, or protocol within Windows 2000 Professional; it is integral to the implementation of TCP/IP and is configured using the tools available for customizing the TCP/IP protocol definition.

The following steps profile how to get DHCP configured within Windows NT Workstation 5.0:

1. Access the Network Connections window. In Windows 2000 's the new graphical interface, the dialog boxes you are accustomed to using are one level lower in the overall navigational structure of this operating system. You can access the Network Connections window either from within the Control Panel or by right-clicking on My Network Places from the Desktop and selecting Properties. Either approach puts you in the Network Components window, as shown in Figure 7.2.
2. Right-click on the icon titled Local Area Connection, and select Properties from the abbreviated menu. Properties, the last selection on the menu, opens the Local Area Connection Properties dialog box, which displays the network components associated with the adapter listed in the Connect Using: section of this dialog box. Figure 7.3 shows the Incoming TCP/IP Properties dialog box.
3. Click once on the Internet Protocol (TCP/IP) entry of the dialog box.
4. Click once on the Properties… button to the right of the screen.

You can also control DHCP services from the command line, using the following NET commands:

NET START DHCPSERVER
NET STOP DHCPSERVER
NET PAUSE DHCPSERVER
NET CONTINUE DHCPSERVER

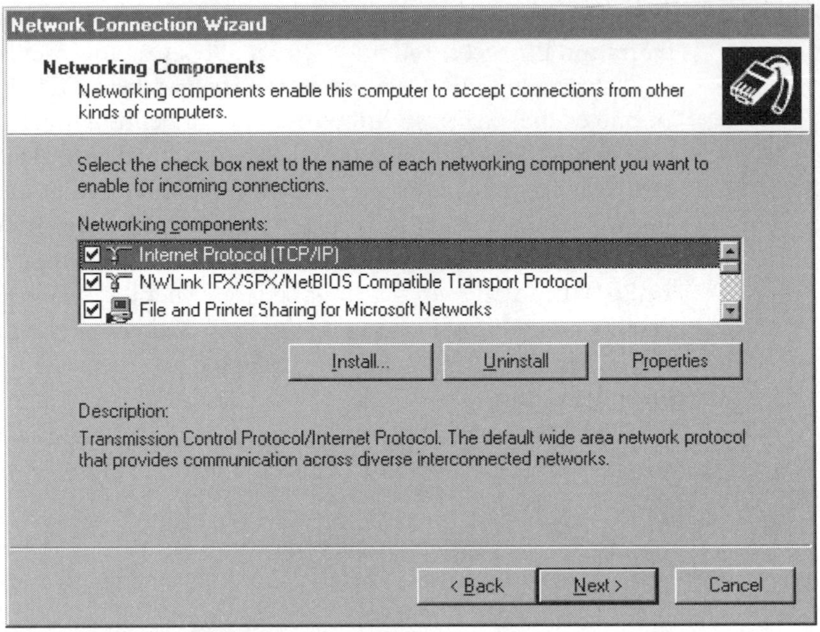

FIGURE 7.2 Using the Network Components window.

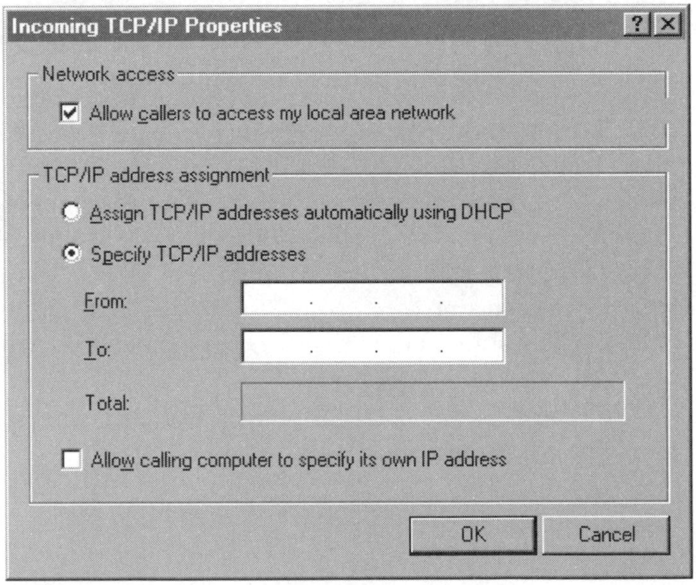

FIGURE 7.3 The Local Area Connection Properties dialog box provides access to network connections.

UNDERSTANDING DOMAIN NAME SYSTEM SERVICES

Created during the 1980s to handle the increasing set of name-resolution needs on the Internet, Domain Name Services (DNS) was originally run on only a small set of computers. The network on which DNS was first tested was the now-famous ARPANET, a network of systems that linked many of the nation's campuses together. The earliest versions of DNS relied on a host table to resolve computer names into IP addresses. At the time, it was possible for every connected computer to have a complete listing of all computers on the internetwork. Changes were e-mailed to the Network Information Center at the Stanford Research Institute, which maintained the list and made it available to all users. Each user of the APRANET system had only to download the latest version every few weeks to stay current with all the names and associated IP addresses on the entire network.

As more and more computers were included on the ARPANET network of systems, the entries to DNS tables defining each member at his or her associated addresses grew quickly. As the beginnings of what would eventually turn into the Internet, with millions of users across the world, ARPANET found that the volume of entries would quickly create a file that would take too much time to parse for address definition. Domain Naming System (DNS) is based on the logic of the first computers having an exhaustive list of every other system on the network. DNS has grown from the simple files used for arbitrating name resolution in ARPANET to a distributed name service and database system designed to spread the administration tasks around the network, dividing it into domains. Now, only domains are registered within the central repositories of the Internet. InterNIC is the corporation that in effect rents unique domain addresses throughout the world. Unlike the early days, as DNS was being created, when each individual system was defined within a DNS file, today only domain names are tracked centrally. The network administrator assigns a host name to each workstation. As long as the administrator avoids creating duplicate names within the domain, the combination of the host name (such as columlou) and the domain name (such as gateway.com) forms a fully qualified domain name or FQDN (Fully Qualified Domain Name), which in this example is columlou@gateway.com.

Quick Tip: It's easy to get confused by "domain" when used in conjunction with the Internet and "domain" as applied to a Windows NT domain. Although both terms represent groups of computers, an Internet domain is registered with InterNIC to identify computers belonging to a particular company or organization. A Windows NT domain is a collection of computers located on a single internetwork that have been grouped for administrative efficiency of management.

Understanding to the Differences between DHCP and DNS

DHCP and DNS, two approaches to managing network IP addresses, both retain in their respective databases the unique identities of the systems to which they are connecting. There are several differences between these protocols, which are profiled here:

Static vs. dynamic IP addressing. The first and most significant difference between DHCP and DNS is the method by which addresses are assigned. In networks configured with DHCP, the server assigns IP addresses by checking them out as a library checks out books to members. In the same way, a DHCP-based network uses the assignment technique to ensure that each system gets a unique IP address. This approach to assigning IP addresses is called dynamic IP addressing; as addresses are provided, each system on the network requires one. DNS-based networks are largely composed of systems that already have their IP addresses defined. The DNS entries within a DNS Server reflect the systems on the network, as a phone book reflects the people living in a city. DNS is then a network arbitration approach relying on IP addresses being already in place.

System identification during network operation. Within a DHCP-based network, each of the systems on that network uses an initialization sequence that alerts the DHCP Server that it is ready to be assigned an IP address. The initialization sequence on the client systems of the network uses a BOOTP protocol sequence to alert the DHCP Server that an IP address needs to be defined. This BOOTP protocol was originally developed in the era of diskless workstations, and it continues today with TCP/IP-based network computers. DNS uses an IP address that has been previously defined during a system's installation.

Content of databases. DHCP and DNS differ significantly in what is included in each's databases. In DHCP, the database acts as a central repository of IP addresses, which are then checked out on an as-needed basis. In the DHCP protocol you can also create client reservation-specific IP address, giving continuity to a specific client systems' connectivity definition with outside resources, for example. In a DNS database, the names and addresses of each system are recorded as they have already been installed within the client system.

INTRODUCING SNMP

If you're the system administrator for several dozen or even several hundred systems, you'll find that the Simple Network Management Protocol (SNMP) is a useful tool for managing the systems for which you're responsible. The SNMP protocol provides basic administrative information about devices attached to a network. SNMP is typically used by SNMP management software, which in turn takes advantage of the many desktop management initiatives (DMI) underway between Microsoft and Intel today. Specific to the needs of system administrators integrating Windows NT into their environments, SNMP is compatible with legacy DMI-based software tools. These include HP's OpenView, Netview, SNMPc, and several others.

What does SNMP do? Its primary function is to enable communications between systems on a network and a host system, most often a server, that has administrative tools installed to make it possible to monitor the overall health of the entire network by checking the performance and health of the client systems that make up the network. SNMP is pervasively used and included in many peripherals and components of all types including power conditioners, multiprocessor UNIX systems, and even motherboards themselves. One of the major developments in the SNMP arena is the integration of DMI-compliant firmware calls into the latest generation of dual-slot I and dual-slot II motherboards that are Pentium II, Pentium II Xeon and Pentium III compatible. Instead of relying on just the components to give you feedback on the overall performance of the system, you can get a baseline measure of system health from many of the motherboards available today in conjunction with the functionality of Windows 2000. In turn, as system administrator, you can manage the preventive maintenance on systems throughout your network. The integration of SNMP in Windows 2000 and the DMI-compliant features in many motherboards today give you a foundation for creating a comprehensive desktop management system. Software components can be loaded on top of SMNP to add definition and clarification to the systems being monitored from both the systems' components and the motherboard.

TIP

Quick Tip: It's a good idea to get SNMP Services running all the Windows NT-based systems you have, as this service specifically provides object/counter relationships within the Performance Monitor, which makes troubleshooting system performance that much easier. Why? Because with increased numbers of objects and associated counters, there is the potential to get an increasingly insightful look into the performance of systems on your network.

Installing SNMP

In this section let's explore how to get SNMP up and running. These steps apply specifically to Windows 2000 Workstation and Server. For Windows 2000 Professional configuration, use the network icon in the Control Panel to access the network services needed to complete these steps.

1. Double-click on My Computer from the Desktop. My Computer's window opens, showing all storage devices on your system, in addition to icons for Network Connections, Scheduled Tasks, Control Panel, and Printers.
2. Double-click on the Control Panel icon, and the applets included are shown on-screen. Notice there is a short cut for Network Connections within the Control Panel. If a network connection hasn't yet been installed for the system on which you are doing these steps, the shortcut for Network Connections may not be present.
3. If this happens, go back to the Active Desktop (Windows 2000's main desktop) and right-click on the icon My Network Places. The Network Connections dialog box then appears.
4. For adding SNMP support for an existing network connection, right-click on the icon in the Network Connections window to get access to the Properties dialog box. Figure 7.4 shows an example of the Properties page for a given Network Connection. Notice that each connection in Windows 2000 is represented as an icon, very similar to the way that previous versions of NT represented connections as discrete objects.
5. Right-click on the Local Area Connection icon, selecting Properties from the abbreviated menu that appears adjacent to the icon selected. The Local Area Connection Properties dialog box appears, as shown in Figure 7.5.
6. Click once on Add.... The Select Network Component Type dialog box appears, as shown in Figure 7.6.

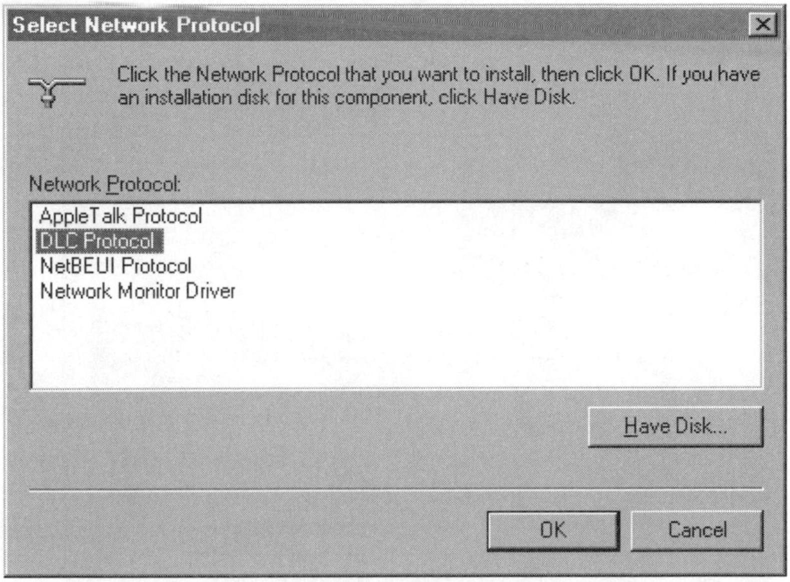

Select Network Protocol

Click the Network Protocol that you want to install, then click OK. If you have an installation disk for this component, click Have Disk.

Network Protocol:

AppleTalk Protocol
DLC Protocol
NetBEUI Protocol
Network Monitor Driver

Have Disk...

OK   Cancel

FIGURE 7.4 Network Connections window provides icons of defined connections and is the home for the Make New Connection wizard.

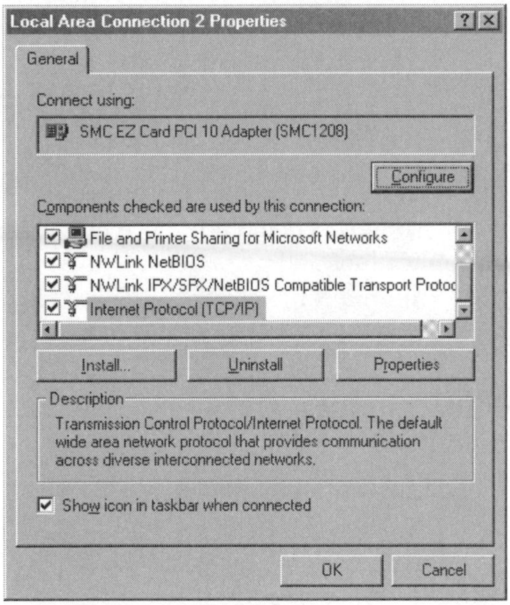

Local Area Connection 2 Properties

General

Connect using:

SMC EZ Card PCI 10 Adapter (SMC1208)

Configure

Components checked are used by this connection:

File and Printer Sharing for Microsoft Networks
NWLink NetBIOS
NWLink IPX/SPX/NetBIOS Compatible Transport Protocol
Internet Protocol (TCP/IP)

Install...   Uninstall   Properties

Description

Transmission Control Protocol/Internet Protocol. The default wide area network protocol that provides communication across diverse interconnected networks.

Show icon in taskbar when connected

OK   Cancel

FIGURE 7.5 Each network connection has extensive properties configurable from the Properties dialog box.

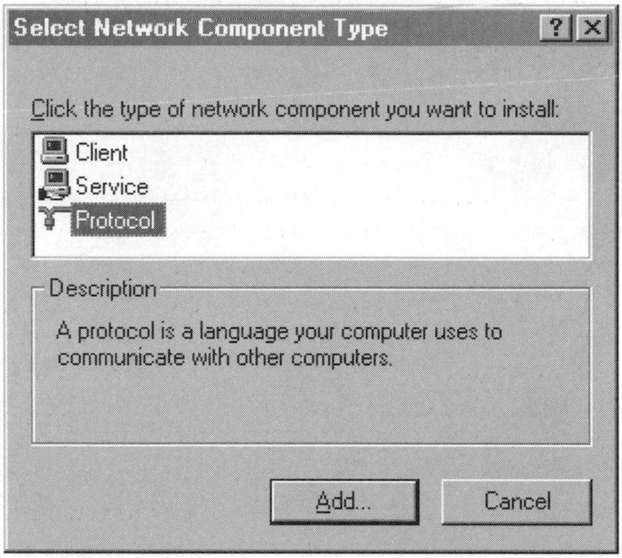

FIGURE 7.6 New to Windows 2000, the Network Component Type dialog box gives options for selecting Client, Service, or Protocol for installation.

7. Click once on Protocol, then click once on Add.... The Select Network Protocol dialog box appears.
8. Select SNMP from the list of protocols, then click OK.
9. Click Cancel to close the Select Network Component Type dialog box.
10. Click OK to close the Local Area Connection Properties dialog box. The SNMP protocol has now been defined.
11. Close all applications and press Control+Alt+Delete.
12. Using the Windows NT Security dialog box, log off and then log on again to place the SNMP selections in the registry.

Once SNMP has been installed, you'll be able to see objects and counters specific to this protocol within the Performance Monitor's options. The Performance Monitor is included in Windows 2000, as it is in previous versions of Windows NT.

ARCHITECTURE OF THE WINDOWS INTERNET NAME SERVICES (WINS)

Each workstation on a network has its own IP address and name when the Windows Internet Naming Services (WINS) is in effect on a network. When a network is configured using WINS, each new workstation on the network broadcasts its name to all other computers on the network; an entry for the computer is inserted into the global name database for the network. Microsoft network products support this system works well on local networks on which all protocols. Any Microsoft operating system can be configured using only TCP/IP protocols, which can be used for communicating via NetBIOS names within the context of a local, non-routed network. Both WINS and DNS are used in networks to assure users that they will be able to use hosts on other networks by easy-to-understand rather than often-truncated and difficult to understand 32-bit names.

A significant limitation of the NetBIOS naming convention is that the names do not propagate across routers. NetBIOS names are disseminated using broadcast datagrams, which IP routers do not forward. The NetBIOS names on one network, therefore, are invisible to computers on networks connected via routers.

Prior to the introduction of Windows NT Server 3.51, the Microsoft LAN Manager product supported internetwork name resolution using static naming tables stored in the LMHOSTS file. This was a direct result of the role of the etc/hosts file predominantly found in the UNIX-based networking routines used throughout sites using both LAN Manager and UNIX in mixed environments. An LMHOSTS file is a text file that contains mapping between NetBIOS names and IP addresses. In many instances when interoperability is needed between different systems, system administrators would need to manually update LMHOSTS files by entering the IP addresses and system names of systems on the internetwork. This approach was a first-line-of-support activity to ensure connectivity between systems, and especially between UNIX-based systems. This labor-intensive approach to handling compatibility would sometimes cause a system administrator to spend hours resolving connectivity issues.

With the introduction of Windows NT Server, Microsoft announced the Windows Internet Name Service (WINS). Like LMHOSTS, WINS maintains a NetBIOS global naming service for TCP/IP-based connections. Unlike LMHOSTS, WINS is dynamic, extending the automatic configuration of the NetBIOS name directory for local networks. The WINS database is updated automatically as NetBIOS systems insert and remove themselves from the network.

If your Windows NT network will be connected to the Internet, using WINS in conjunction with DNS is possible, enabling WINS to give DNS host names for Microsoft-based hosts within your network.

The intent of this section is to define the intricacies of WINS and the steps you can take to ensure the smooth functioning of a network based on this protocol.

Resolving Names on Microsoft Networks

Within WINS, each of the names and its associated IP addresses needs to be resolved. Providing the logic within the protocol for handling name resolution, NetBIOS takes NetBIOS names on TCP/IP environments and references or resolves their relative identities on the network. NetBIOS over TCP/IP Service (NBT) has evolved from a basic, broadcast-based approach to the current name-service approach. Before you look into WINS in detail, it's important to get a handle on the name-resolution modes supported by NBT. The b-node, p-node, and m-node resolution modes are defined in RFCs 1001 and 1002.

> **B-Node.** Name resolution using broadcast messages is the oldest method that has been used for managing NBT name resolution on networks. B-name resolution works well in small, local networks, but it has several disadvantages that become critical as networks grow. As the number of hosts on the network increases, the amount of broadcast traffic can consume significant network bandwidth. Second, IP routers do not forward broadcasts, and the b-node technique cannot propagate names throughout an internetwork. This node is the default name-resolution mode for Microsoft hosts not configured for WINS name resolution. In pure b-node environments, hosts can be configured using LMHOSTS files to resolve names on remote networks.

> **P-Node.** Hosts configured for p-node use WINS for name resolution. P-node computers register themselves with an available WINS server, which functions as a NetBIOS name server. The WINS server maintains a database of the NetBIOS names, ensures that duplicate names do not exist, and makes the database available to WINS clients.

Each WINS client is configured with the address of a WINS server, which may reside on the local network or a remote network. WINS clients and servers communicate via directed messages that can be routed. No broadcast messages are required for p-node name resolution.

Despite the strengths of p-node name resolution, it has two liabilities:

- All computers using p-node must be configured using the address of a WINS server, even when communicating hosts reside on the same network.
- If a WINS server is unavailable, name resolution fails for p-node clients.

Because both b-node and p-node address resolution present disadvantages, two address modes have been developed that form hybrids of b-node and p-node. These hybrid modes are called m-node and h-node.

M-Node. M-node computers attempt to first use b-node (broadcast) name resolution, which succeeds if the desired host resides on the local network. If b-node resolution fails, m-node hosts then attempt to use p-node to resolve the name.

M-node enables name resolution to continue on the local network when WINS servers are down. B-node resolution is attempted first on the assumption that in most environments hosts communicate most often with hosts on their local networks. When this assumption holds, performance of the b-node resolution is superior to that of p-node. Recall, however, that b-node can result in high levels of broadcast traffic.

Microsoft has also warned through its ATEC Centers and the MCSE program that m-node can cause problems when network logons are attempted in a routed environment.

H-Node. Like m-node, h-node is a hybrid of broadcast (b-node) and directed (p-node) name-resolution modes. Nodes configured with m-node, however, first attempt to resolve addresses using WINS. Only after an attempt to resolve the name using a name server fails does an h-node computer attempt to use b-node. M-node computers, therefore, can continue to resolve local addresses when WINS is unavailable. When operating in b-node mode, m-node computers continue to poll the WINS server and revert to h-node when WINS services are restored.

H-node is the default mode for Microsoft TCP/IP clients configured using the addresses of WINS servers. As a fallback, Windows TCP/IP clients can be configured to use LMHOSTS files for name resolution.

Quick Tip: Although networks can be configured using mixtures of b-node and p-node computers, Microsoft recommends this only as an interim measure. P-node hosts ignore b-node broadcast messages, and b-node hosts ignore p-node directed messages. Two hosts, therefore, can conceivably be established using the same NetBIOS name.

Naming on a WINS Network

Once a WINS server has been configured and the all workstations have been registered with the WINS database, attempts to connect with remote systems will be successful. WINS relies on the database to resolve naming and identity throughout the network. The function of the database is to respond with the appropriate address from the WINS database to enable the connection between server and workstation.

MANAGING LMHOSTS FILES

Although a complete name-resolution system can be based on an LMHOSTS file type of architecture, static naming files can be very difficult to administer, particularly when they must be distributed to several hosts on the network. Nevertheless, LMHOSTS files may be necessary if WINS will not be run on a network or if having a backup is desirable in the case of a WINS service stopping unexpectedly or failing.

Although LAN Manager host files supported little more than mappings of NetBIOS names of IP addresses, Windows 2000 offers several options that make LMHOSTS considerably more versatile.

Format of LMHOSTS Files

A sample LMHOSTS file is installed in the directory C:\WINNT\SYSTEM32\DRIVERS\ETC. This file is typically edited to show both IP addresses and system names, which refer to systems throughout the netwlork with which your workstations need to communicate. LMHOSTS can list IP addresses and names that apply to UNIX-based systems as well.

The basic format of an LMHOSTS entry is as follows:

```
ip address     name
```

The IP address must begin in column 1 of the line. Here is an example of a basic LMHOSTS file:

```
129.160.27.35       PASADENA
        DUARTE
        HOLLYWOOD
```

You can also use keywords to make the LMHOSTS file easier to navigate, which is explained in the following section.

LMHOSTS Keywords

Here is an example of an LMHOSTS file augmented using keywords:

```
LAGUNA              #PRE
SANCLEMENTE         #PRE        #DOM:   WESTERNOFFICE


#BEGIN_ALTERNATE
#INCLUDE \\LAGUNA\PUBLIC\LMHOSTS
#INCLUDE \\SANCLEMENTE\PUBLIC\LMHOSTS
#END_ALTERNATE
```

The #PRE keyword specifies that the entry should be preloaded into the name cache. Ordinarily, LMHOSTS is consulted for name resolution only after WINS and b-node broadcasts have failed. Preloading the entry ensures that the mapping will be available at the start of the name-resolution process.

The #DOM keyboard associates an entry with a domain, which might be useful for determining how browsers and login servers behave on a routed TCP/IP network. #DOM entries can be preloaded in cache by including the #PRE keyboard.

The #INCLUDE keyboard makes loading mappings from a remote file possible. One use for #INCLUDE is to support a master LMHOSTS file stored on logon servers and accessed by TCP/IP clients during startup. Entries in the remote LMHOSTS file are examined only when TCP/IP is started. Entries in the remote LMHOSTS file, therefore, must be tagged with the #PRE keyboard to force them to be loaded into cache.

If several copies of the included LMHOSTS file are available on different servers, you can force the computer to search several locations until a file is successfully loaded. This is accomplished by bracketing #INCLUDE keyboards between the keywords #BEGIN_ALTERNATVE and #END_ ALTERNATE, as was done in the example file just presented. Any successful #INCLUDE causes the group to succeed.

Guidelines for Establishing LMHOSTS Name Resolution

B-node computers not configured to use WINS name resolution can use LMHOSTs to resolve names on remote networks. If the majority of name queries are on the local network, preloading mappings in the LMHOSTS file generally is not necessary. Frequently accessed hosts on remote networks can be preloaded using the #PRE keyword.

#DOM keyboards should be used to enable non-WINS clients to locate domain controllers on remote networks. The LMHOSTS file for every workstation in the domain should include #DOM entries for all domain controllers that do not reside on the local network. This ensures that domain activities, such as logon authentication, continue to function.

To browse a domain other than the logon domain, LMHOSTS must include the name and IP address of the primary domain controller of the domain to be browsed. Include backup domain controllers in case the primary fails or a backup domain controller is promoted to primary.

LMHOSTS files on backup domain controllers should include mappings to the primary domain controller name and IP address, in addition to mappings to all other backup domain controllers.

All domain controllers in trusted domains should be included in the local LMHOSTS files of each server being used throughout the network.

CHAPTER SUMMARY

Microsoft continues to base much of its primary naming system on NetBIOS names. Each computer on the network has its own specific name, and all other computers on the local network know what the name is. When the network acquires a new workstation, the global name database is automatically updated. Consequently, system maintenance is relatively effortless when it comes to maintaining computers and keeping track of system names. This system works well on local networks. Microsoft operating systems can use NetBIOS names within the context of a local, nonrouted network.

One of the unfortunate shortcomings of the NetBIOS naming system is that the names do not propagate across routers. NetBIOS names are disseminated using broadcast datagrams, and IP routers do not forward them. Computers on one network cannot read the NetBIOS names on another network when the networks are connected via a router.

Microsoft did offer an alternative to the NetBIOS naming system that allows for recognition of names across networks (LMHOSTS), but the maintenance needed for this file is significant.

WINS (Windows Internet Naming Service) was specifically developed to allow for dynamic updating of a database of IP names and to provide the flexibility of being read on a global level.

8 Learning to Use Microsoft Mail and Faxing in Windows 2000 Professional

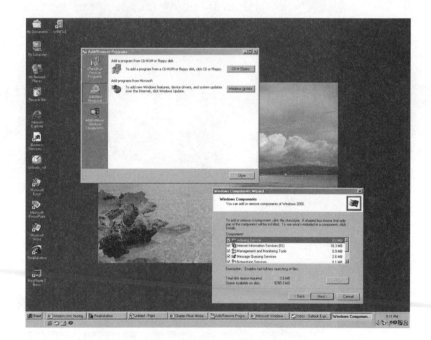

Microsoft's continued fine-tuning of Exchange is exemplified in Windows 2000 Professional and Server. Included in Microsoft Exchange's latest product generations are messaging components including electronic mail and fax. The Exchange client that is so pervasive today in many corporate environments running Windows 2000 Professional, As an administrator you're familiar with the requirements of a client/server application. For Outlook and Outlook Express to work,

Exchange Server will need to be installed on servers throughout your organization. Microsoft's Exchange client Outlook is shown in Figure 8.1.

Initially, Microsoft Exchange clients support three primary messaging services. These are Microsoft Mail, Internet Mail, and direct fax transmission through a fax modem. Microsoft is also working to support CompuServe, and it will undoubtedly add support for other messaging protocols that will enhance Microsoft's ongoing focus on operating system interoperability. Conversely, many companies are planning support for Microsoft Exchange's messaging features, including America Online and MCI Mail. Both of these companies offer support for Microsoft Exchange through add-in modules.

The integration of Microsoft Exchange in third-party messaging products has begun to cause a dilemma for Windows 2000 users. On one hand, Exchange clients are easy to navigate and serve as a central reference point

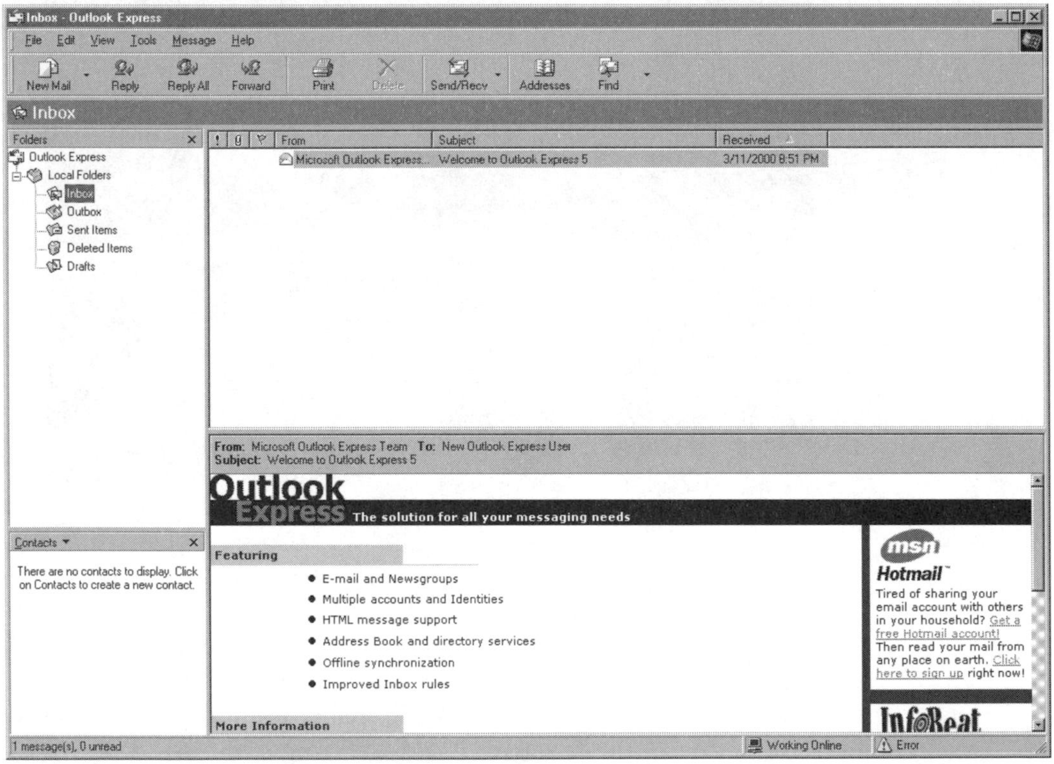

FIGURE 8.1 Outlook is a Microsoft Exchange client pervasively used in conjunction with Windows 2000 Professional.

for the key elements of a communication program that needs to be interacted with to send messages, including e-mail, faxes, and even storage of messages in folders. On the other hand, Windows 2000 users have specific needs that are addressed with the detailed product features of a stand-alone fax or messaging program like Lotus cc:Mail, Lotus Notes, or WinFax. These applications, specifically tailored to a market need, can be customized through a wealth of features to meet the more rigorous requirements of users who need more depth than Outlook provides. For the majority of workstation users, Outlook is sufficient as a messaging application in that it seamlessly supports attachments, has scheduling and out-of-office options that can be configured, and is able to access shared folders from any desktop supported on a common network.

Outlook's acceptance into the Windows 2000 user base continues to gain momentum, primarily due to the fact that many companies using Windows 2000 on the desktop have already committed to Exchange Server on the NT platform in their back-office environments. The client/server combination of Outlook to Exchange uses the same basic model Lotus Notes is employing with Notes clients being able to query and use a Domino server for readily available content and data. Exchange clients, most notably Outlook, have specifically been designed to work seamlessly with Exchange Server.

Just as Sun Microsystems provides basic e-mail tools with Solaris, Microsoft offers Outlook Express that is included with Windows 2000 Professional and Windows 2000 Server. The differences between Outlook and Outlook Express are significant, with the calendaring function in Outlook being widely used throughout many organizations for managing meetings.

The Microsoft Exchange client, or Outlook Express, is the most visible part of Windows 2000's Messaging Subsystem, yet it is only a single element of a more complex equation. The following technologies have served as the foundation for the development of messaging technologies in previous versions of Windows NT 4.0 Workstation and Server, and are included in Windows 2000 Professional as well. They are included here as a brief refresher of technologies underlying Microsoft Exchange clients.

The MAPI (Messaging Applications Program Interface). MAPI is the interface between the Microsoft Exchange or other compatible applications and the drivers that communicate with other information services.

Personal Address Book. Your Personal Address Book contains names, e-mail addresses, telephone and fax numbers, and other information about potential recipients of messages, even if you reach them through different online services. E-mail and fax applications key off the Personal Address book using MAPI, so you can keep a single address book for those with whom you converse.

Personal Folders. Personal Folders is a database that contains messages, forms, and documents. Microsoft Exchange uses Personal Folders to store incoming and outgoing messages. If more than one user shares the same PC, each can use a separate Personal Folder.

Internet Mail Service. This service enables support for sending and receiving e-mail via POP3 Internet mail servers.

Microsoft Mail Postoffice and drivers. Support for Microsoft Mail is provided for legacy purposes. In organizations' migration efforts you might run into a group of users who have for one reason or another not migrated yet. These are the groups Microsoft had in mind when providing this legacy support. The Microsoft Mail Postoffice server is similar to the one in the separate Microsoft Mail product except that it does not provide access to Microsoft Mail gateways. It supports only a single post office, and it also has fewer administration tools. The Microsoft Mail drivers are the interface between Microsoft Mail and MAPI.

Fax drivers. In addition to conventional e-mail messages and files, Exchange can also handle fax pages to and from stand-alone fax machines or fax modems.

When Microsoft developed the architecture for Exchange, it modeled the underlying programming interface to be extensible enough to support front-end programs that the user sees and that are supported by various messaging devices. The Messaging Application Programming Interface (MAPI) in effect acts as the "glue" that holds the various components in the architecture together. MAPI is responsible for coordinating tasks between Microsoft Exchange, Microsoft Mail, CompuServe-based connections, any and all Microsoft network components attached to the system, and the Internet connections used in conjunction with other connectivity-oriented products.

WHAT IS MICROSOFT EXCHANGE?

All the messaging applications delivered as part of Windows 2000 rely on Microsoft Exchange to process inbound and outbound messages. The advantage of this approach is that it provides a single point of access for all communications components. Throughout this section we'll take a tour of how to install Microsoft Exchange, how to work with user profiles within Exchange, how to define properties within an Exchange profile, and much more. Let's get started with a quick tour of how to get Microsoft Exchange installed.

Installing Microsoft Exchange

You can install Microsoft Exchange either during Windows 2000 Setup or directly by itself. If you want to load the Messaging System, which is included in Microsoft Exchange, when you load the rest of Windows 2000, you'll need to select Custom instead of Typical when you are prompted for the type of setup you want to perform. If you took the program's on-screen advice and used the Typical Setup, you'll have to dig out the Windows 2000 Exchange CD-ROM or diskettes before you can load Microsoft Exchange and additional communications programs.

1. To perform a separate installation, click on the Control Panel's Add/Remove Programs icon and choose the Windows Setup tab. If you're installing Windows 2000 from scratch, choose the Custom Setup option. Either way, you'll get a list of components like the one shown in Figure 8.2.

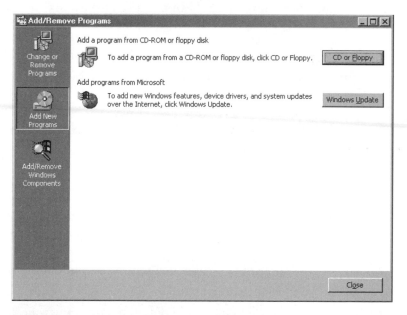

FIGURE 8.2   Choose the applications to install from the Windows 2000 Setup tab of the Add/Remove Program Properties dialog box.

2. Add a check next to each of the services you want to install. If you choose one or more of the messaging-based applications, such as Microsoft Fax, Microsoft Mail, or Internet Mail, the Setup wizard will also automatically include Microsoft Exchange. After you specify the services you want Microsoft Exchange to manage, the Wizard completes the installation for you.

WORKING WITH USER PROFILES

Microsoft Exchange is very property-centric, as much of the Windows NT operating system has become since Windows 2000 and Server. Microsoft Exchange uses configuration files to specify each user's messaging services and other options. If more than one person shares a single workstation, or if multiple systems are used in one location, it's a simple matter in Exchange to have the configuration files transferred to the system where the login was completed. In conjunction with Windows NT Security Model, this approach to having a specific configuration file for each user fits with the direction Microsoft is taking in Windows 2000.

A single profile or configuration file may include one or more messaging services, and each profile manages all the features of the services in that profile. Depending on the way you want to manage your mail, you can create separate profiles for difference services.

The first time you run Exchange clients on a Windows 2000 Professional, the Configuration wizard creates a profile for you. The different sections and steps that follow define the specifics of configuring different properties.

You can modify an existing profile or create a new one by following these steps:

1. Open the Control Panel and double-click on the Mail icon.
2. Click once on the Show Profiles button after the Properties dialog box appears. The Exchange Profiles dialog box next appears. It will display a list of all the profiles that have previously been created on your system.
3. Highlight the name of the profile, and choose one of these options:

    Add. Choose the Add option to create a new profile.

Remove. Choose the Remove option to delete the profile when the name is highlighted.

Properties. Choose the Properties option to define the way Exchange will handle messages in the currently highlighted profile.

Copy. Choose the Copy option to create a new profile with the same properties as the currently highlighted profile. After you copy a profile, you can use the Properties option to personalize the new profile.

4. When you launch Microsoft Exchange, it will start with the default profile active. To change the default profile, click the Show Profiles button, then drop down the menu in the *When starting, use this profile* field and highlight a profile name.

If you create more than one profile, you should assign each profile a name that describes the way you plan to use it, such as the name of the person using it or the name of the service that the profile controls.

Defining Properties in an Exchange Profile

Each of the services, which are also included in any given profile, has a specific set of properties that define the way Exchange uses that service. To view or change the default settings of a given configuration profile in Exchange, follow these steps:

1. Double-click on the Add/Remove Programs icon in the Control Panel.
2. Click once on the Add/Remove Windows Components in the lower left corner of the dialog box.
3. Using the series of selections presented by the Windows Components Wizard, select the options you want for the Exchange profile you're trying to create.

FAX PROPERTIES EXPLAINED

Included in Microsoft Exchange are a series of tools for sending and receiving faxes. The Microsoft Fax Properties dialog box in Figure 8.3 shows both the pages of options and messaging alternatives present in Exchange's implementation of Microsoft Fax.

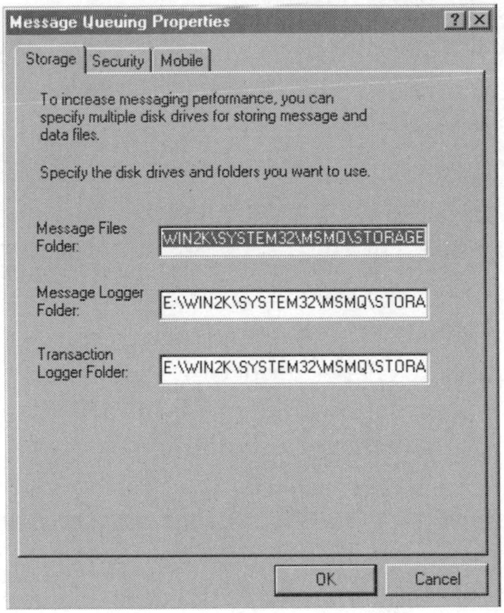

FIGURE 8.3 Use the Fax Properties dialog box to select features of outbound faxes.

EXPLORING MICROSOFT MAIL PROPERTIES

To send and receive messages through a Microsoft Mail post office, you must add Mail to your Microsoft Exchange profile. The Microsoft Mail dialog box has eight tabs, providing a wealth of features for customizing how Mail will work on your workstation. Presented here is a brief synopsis of what is included in each of the eight tabs within the Microsoft Mail Properties dialog box.

Connection. To use Microsoft Mail, you must specify the full path for the office where your messages are being kept. Typically your system administrator will be able to tell you where the post office is located throughout a network. In many cases, the system administrator where you work will have a post office created on a server where others will also be able to connect and use a common subdirectory location for performance expediency. Figure 8.4 shows the Microsoft Mail dialog box with the Connection page selected.

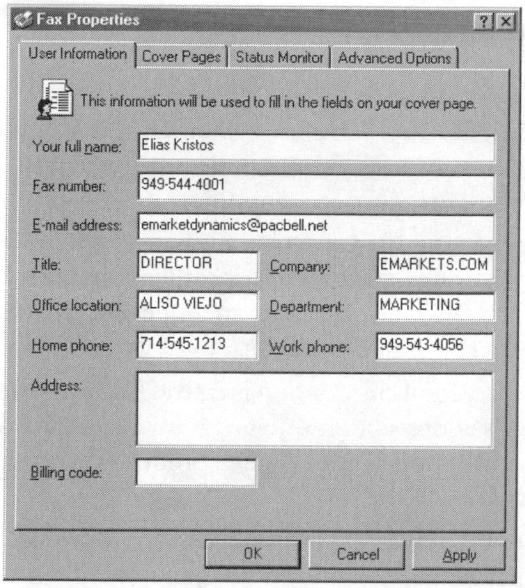

Fax Properties

User Information | Cover Pages | Status Monitor | Advanced Options

This information will be used to fill in the fields on your cover page.

Your full name: Elias Kristos

Fax number: 949-544-4001

E-mail address: emarketdynamics@pacbell.net

Title: DIRECTOR   Company: EMARKETS.COM

Office location: ALISO VIEJO   Department: MARKETING

Home phone: 714-545-1213   Work phone: 949-543-4056

Address:

Billing code:

OK   Cancel   Apply

FIGURE 8.4   The Connection tab defines the path to your post office.

If you are a system administrator or the person responsible for the network and getting Microsoft Mail up and running and you do not yet have a post office created, be sure to read the section "Creating a Microsoft Mail Post Office" later in this chapter.

NOTE

Logon. Use this tab to specify the name and password that your post office uses for your account.

Delivery. Use this tab to define the way messages move between Exchange clients and the post office.

In the context of Microsoft Mail options within the Mail Properties dialog box, it's important to have the Enable incoming option selected to receive e-mail. The Enable outgoing option must be active to send mail.

Many Microsoft Mail installations have gateways to other messaging services, including fax and MCI Mail, among many others. If you don't want to use Mail to send messages to one or more of these services, click on the Address Types button and remote the check mark from the name of the service. For example, if you want to send faxes through your fax modem rather than through your LAN, remove the check mark next to Fax.

LAN Configuration. The LAN Configuration options all require a connection to a LAN. If you connect to your post office through a modem, you can ignore these options.

When Use remote mail is active, Exchange automatically transfer the headers of new incoming messages rather than the full text of the messages. When you want to read a message, use the Remote Mail command in Exchange's Tools menu.

When the Use local copy option is active, Exchange stores a copy of the post office's address book on your local hard drive and uses that list rather than connect you to the post office every time you request an address.

If your connection to the post office is very slow, the external delivery agenda may reduce the amount of time needed for mail delivery.

Log. Use the options on the Log tab to create a history file for Microsoft Mail activity.

Remote Configuration. The Remote Configuration tab controls options that apply when you're using a modem connection to your post office. These options are similar to the ones on the LAN Configuration tab.

Remote Session. Use the Remote Session tab to specify when and for how long you want Exchange to connect to the post office through a modem or to schedule remote mail delivery at specified times.

When the *Automatically start* option is active, Exchange will dial out to your post office as soon as you start Exchange. The *Automatically end* options specify when Exchange should break the connection.

Dial-Up Networking. The options in the Dial-Up Networking dialog box control the way Exchange connects to a remote network through a modem. These are the same options you can reach through the Dial-Up Networking options in the Network Connections icon in the My Computer folder, accessible from the Desktop.

CREATING A MICROSOFT MAIL POST OFFICE

Be sure to check if there's a Microsoft Mail post office located somewhere on your network; if not, you can create one quite easily. Follow these steps using the Microsoft Mail Post Office applet found in the Control Panel.

1. Select the Create a new Workgroup Postoffice option.
2. Click once on Next.
3. Enter a location for the post office. This needs to be a directory located on one of your network drives, or where everyone who will be using this post office can access it. Be sure to share the directory, giving everyone who will use it Full access. Click once on Next.
4. Enter your Administrator account information. Minimally this needs to include your Name, Mailbox name, and a password.
5. Click Finish. Remember to share the Post Office subdirectory so that all users can access it to get their mail.

DEFINING INTERNET MAIL PROPERTIES

To send and receive messages across the Internet, you must add Internet Mail to your Microsoft Exchange profile. You will need a valid POP3 mail account and a working Internet connection to send and retrieve mail. As shown in Figure 8.5, the Internet Mail dialog box has two tabs, General and Connection.

General

Here you must enter the various settings supplied either by your network administrator or by your Internet Service Provider (ISP). You should include your full name and e-mail address under Personal Information, and you'll need to fill in the three fields under Mailbox Information. For *Internet Mail server*, usually the name is "mail" or "pop". Be sure to include the full domain name also, unless both you and the mail server happen to be on the same Internet domain. Be careful to ensure that the mail account name and password are distinct from the account you use to access the Internet. Figure 8.6 shows the General page of the Internet Mail dialog box.

FIGURE 8.5 Using the options for configuring Internet Mail.

FIGURE 8.6 Setting the properties on the Internet Mail Service.

Clicking on Message Format brings up a small dialog box that lets you specify whether to use MIME and which character set to use for Internet Mail. Figure 8.7 shows an example of the Message Format dialog box.

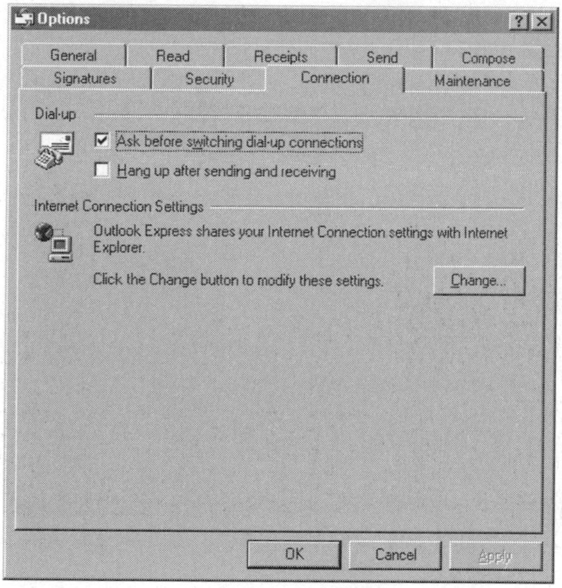

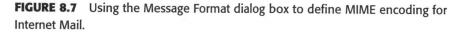

FIGURE 8.7   Using the Message Format dialog box to define MIME encoding for Internet Mail.

You will usually want to enable MIME, as this option allows you to send and receive files and other information as enclosures or attachments. Note that enabling/disabling MIME changes the default font selected for your Internet mail. With MIME enabled, the ISO 8859-1 character sets become the default. This is usually preferable to US-ASCII, however, if you want to your messages to be compatible with the widest variety of other platforms.

The Advanced Options button simply allows you to forward outbound e-mail to a different mail server. This can be useful if one server handles your incoming mail and another handles outgoing. If you need this option, simply enter the mail server name in the dialog box and click OK.

Connection

The Connection screen lets you define how to connect to the mail server. Selections for this option are the same for both Windows 2000 Professional and Windows 2000 Server, which are either via the network or by modem. If you're using a dial-up account, select the desired connection settings. Also be sure to note the Login As button, as it is useful if you do not have "remember password" enabled normally for dial-out. Obviously, you may still want to supply it in this case, so Exchange can dial in unattended to retrieve your e-mail.

You can also choose to work offline and use Remote Mail, by enabling the checkbox with this name. Log File is useful for troubleshooting problems, but normally it is not used. Using the Schedule button you specify how frequently Exchange checks your Internet mail server for incoming mail.

Logon

Using the Logon tab specifies the name and password that the post office you are a member of uses for your account.

Delivery

The Delivery tab controls the way messages move between your post office and the Exchange client. Specifically the Enable incoming option must be active to receive mail. The Enable outgoing option must be active to send mail.

Many of the Microsoft Mail installations have gateways to other messaging services, including fax and MCI Mail, among many others. If you don't want to use Mail to send messages to one or more of these services, click on the Address Types button and remove the check mark from the name of the service. For example, if you want to send faxes through your fax modem rather than through your LAN, remove the check mark next to Fax.

LAN Configuration

The LAN Configuration options all require a connection to a LAN. If you connect to your post office through a modem, you can ignore these options.

When Use remote mail is active, Exchange automatically transfers the headers of new incoming messages rather than the full text of the messages. When you want to read a message, use the Remote Mail command in Exchange's Tools menu.

CHAPTER SUMMARY

One of the most pervasive approaches to communicating is the use of electronic mail and electronically enabled faxing. This chapter focuses on how to get these two tools up and running in Windows 2000. Using the steps included in this chapter, you'll be able to use the Fax and Mail options within Microsoft Exchange.

9 Connecting with Windows 2000 Professional

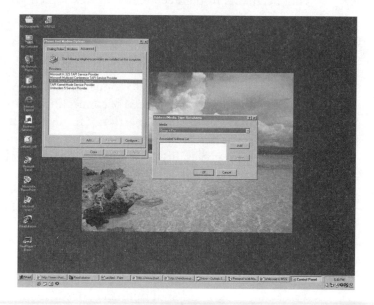

Both client/server applications and the Internet continue to redefine the very core of businesses, taking products that would be sold regionally to national distribution, depending on the strength of a company's distribution channel. Today the Internet can take a company that has limited physical distribution and literally make it globally focused by getting their website quickly published.

The Internet is the most significant change agent affecting technology; it's also the area of technology that is changing most rapidly. This same technology is making it possible for organizations, companies, and universities to create their own their own private networks, or intranets.

WINDOWS 2000 AND THE INTERNET

Support for multithreaded applications, enhanced TCP/IP commands, and time-tested subsystems for running Windows 16-bit, Windows 32-bit, OS/2, and POSIX applications makes Windows 2000 one of the best choices for hosting Internet content. Various types of Internet connections are possible given the latest generation of Windows 2000, and we'll take a look at the procedures you can use for connecting your workstation to the Internet. Microsoft's increasing emphasis on the Internet is also leading to enhancements in the Peer Web Services, which give you the flexibility of publishing Internet content on your own.

Exploring Internet Connections with Windows 2000

There are several approaches to connecting Windows 2000 workstations to the Internet. These vary in speed of connection, relative level of automation, and level of reliability of connection. Throughout this section we'll look at the various approaches you can use for getting a connection to the Internet. Starting with availability in Windows 3.1, HyperTerminal is by far the longest-running solution in the Windows family, yet it does not typically provide the throughput users have come to expect with the widespread availability of high-speed TCP/IP connections throughout an organization. At the opposite end of the spectrum is accessing the Internet through a Network Interface Card (NIC), which is always installed onto the system bus of the workstation you are working with. The NIC Card in a workstation is connected to a local area network that is, in turn, routed to the Internet. The most common variation of Internet connectivity is to use a dial-up connection for connecting to an ISP. Outside of using an ISP, many users are using America Online or CompuServe to connect to the Internet.

You can explore each of these types of configurations in a little more detail. In order to better understand the similarities and differences among connection methods, it's helpful to know that the Internet uses a common set of rules for communicating between systems on a network. This approach to allowing for system communication is enabled by using the TCP/IP protocol, which allows many different types of systems to communicate with one another. Sending and receiving data through the Internet require the use of TCP/IP, as it is pervasively used for linking disparate systems throughout the many networks make up the Internet.

Beginning at the most basic type of connection, Figure 9.1 shows an example of a remote-terminal connection. As the most fundamental of

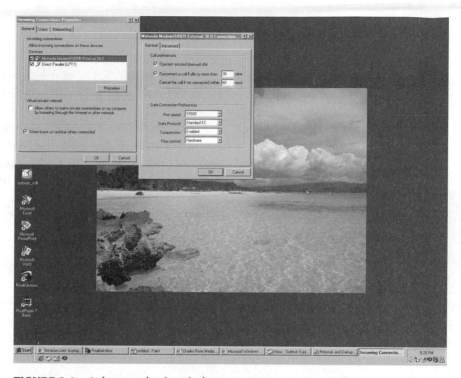

FIGURE 9.1   At its most basic, Windows 2000 can connect to the Internet via dial-up modem.

communications approaches for ensuring connections between remote systems and Windows 2000 workstations, this approach relies on creating a serially based link from one system to another. This approach, described in Figure 9.1, is based on using terminal emulation features within communications programs to enable the connection. Unless the host system is located in the same building, you'll use a modem to dial up and connect with another system through a phone line. The same type of connection appears on a workstation when a shell account is running within a TCP/IP session.

As with any connection method, there are strengths and weaknesses to this approach of making a connecting with the Internet. You'll still need to have a modem, yet it is a peripheral that can be installed in a progression of systems as you migrate hardware platforms (assuming they each have PCI slots). Using a remote-terminal connection also puts very little strain on a host computer, allowing it to support a large number of simultaneous remote terminal sessions. Due to the TCP/IP gateway to the Internet, data caches, and very often data files located on the Internet, remote-terminal connections are best used when the relative horsepower of a client system is significantly lower than that of a host system. That's because the most demanding tasks can be completed on the host, and the data or information required on the client system can be quickly transferred via modem.

A second approach to connecting a Windows 2000 workstation to the Internet is via a local-area network (LAN). Figure 9.2 shows an example of a direct TCP/IP link through both a SLIP (serial line Internet protocol) and PPP (point-to-point protocol) connection. With Microsoft's increasing focus on adding Internet connectivity and enabling electronic commerce through the BackOffice suite of products, constructing a LAN-based networking solution is the best alternative in terms of speed and consistency of connection to the Internet.

A TCP/IP connection is by far the fastest and most flexible approach to connecting to the Internet. Many NT-based programs provide graphical displays, graphically oriented file transfer, and other features that are not available or are excruciatingly slow on remote-terminal connections. Yet for all its speed, TCP/IP is by far the most expensive option and relies on your enterprise or organization being connected to the Internet via a point of presence through an ISP.

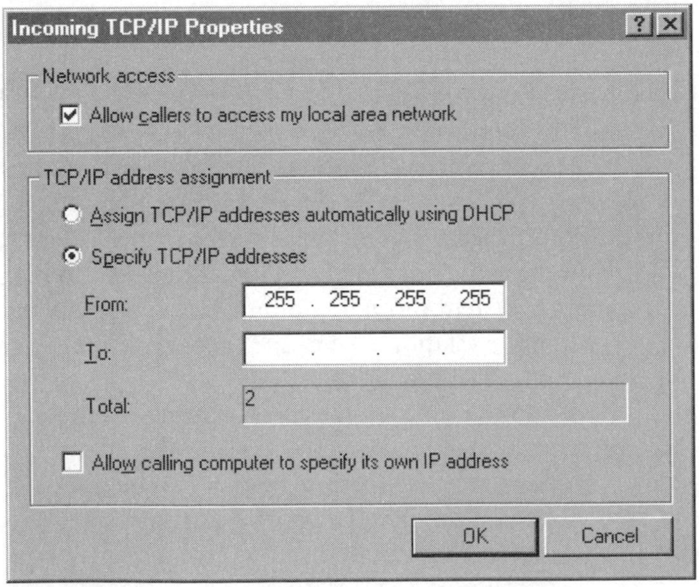

FIGURE 9.2 Connecting to the Internet via a LAN.

Understanding Internet Addresses

Every system must have an identity to communicate on your network. The majority of TCP/IP-based networks today have a static IP address, or one that doesn't change. Increasingly as notebook and laptop computers have become capable of running Windows NT 4.0 and Windows 2000, and with the rapid increase in the number of Internet Service Providers (ISPs), the Dynamic Host Configuration Protocol (DHCP) has become more and more widespread. In the case of an organization that has a highly mobile workforce, the idea of having DHCP enabled as the method by which IP addresses are defined makes perfect sense. At a functional level, IP addresses are said to be either static or dynamic. In the latter case, a dynamic IP address mechanism is one of the most popular today due to the flexibility it provides administrators and users alike.

But what is an IP address? And how does it work? Let's take a look at what makes up an IP address and how its structure allows each system on a network to have a unique identity. First, an IP address is composed of a 32-bit (4-byte) number, and for ease of communication, it is usually defined as a four-part numeric address. A typical address might be 129.135.160.27, with each of the components of the address responsible for a specific function in the subnet mask architecture, as defined by the system administrator for a given network. Figure 9.3 shows an example of how the IP address is defined within the Microsoft TCP/IP Properties dialog box. Notice that this entry also provides for definition of the subnet mask values and the Default Gateway for the specific network on which the individual workstation is being installed.

If you've spent time managing a UNIX-based network, you know that using the etc/hosts file for arbitrating routing glitches between one system and another is one of the most helpful tools for getting one system to recognize another on the same network. Many UNIX systems also support Domain Name Services, where an entire range of IP addresses has a unique system name that gives each workstation a unique identity. In nearly all the cases where a name also exists, a domain exists that shows the type of business or other institution that uses the address or defines the domain from which the system is originating. For example, look at *gateway.com*; the .com designates the Gateway Web site as being commercially focused (hence the .com). Many religious organizations, such as *salvationarmy.org* end with the domain .org as this designates a domain that is predominantly used by nonprofit organizations. Likewise with the .edu extension, for example, *stanford.edu*, which defines an educational institution. Many

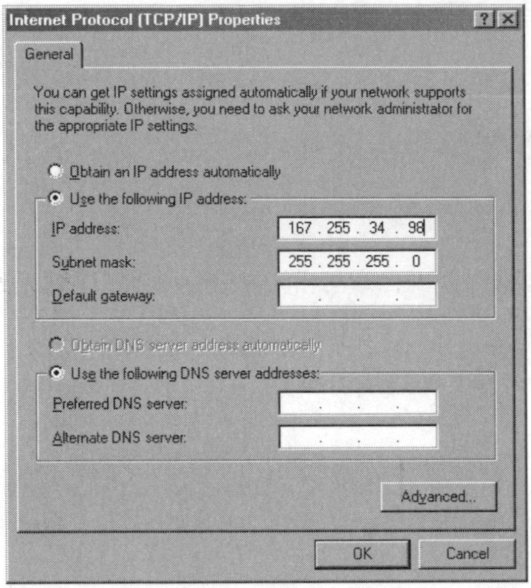

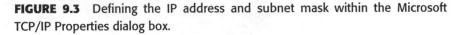

FIGURE 9.3 Defining the IP address and subnet mask within the Microsoft TCP/IP Properties dialog box.

organizations have servers that include a Domain Name Server, which converts these domain names to numeric addresses. This is also the function, at a much more bounded and smaller level, for the etc/hosts file that many system administrators use for resolving conflicts between systems on a network.

Throughout this chapter we have discussed several types of access methods, each translating the IP address and the accompanying name to the domain server, where a Domain Server translates the name into numeric values.

So what is an IP address? It is the numeric definition of routing for a specific system to be able to be recognized and communicate with other systems throughout a network. IP addresses can be dynamically assigned, as is the case with the DHCP features within the TCP/IP command set, or they can and often are defined from a static IP address standpoint, where a system stays in one location and does not change locations or users.

Quick tip: System Administrators are well aware of your IP address, and before changing it or modifying any values of it or the Subnet Mask values, be sure to check with them first. Many system administrators have specific iden-

TIP

tities for each system on a network and can assist you in optimizing your Windows 2000 workstation for access to the Internet. If you're a system administrator you'll find this chapter informative from the standpoint of configuring workstations for access to LANs and the Internet.

AT THE FOUNDATION: REMOTE TERMINAL EMULATION IN WINDOWS 2000

At the lowest level of performance and connectivity available, using Remote Terminal Emulation is synonymous with having your system act like a remote terminal to a host mainframe. The level of functionality and performance associated with terminal emulation is limited to what the host system provides during the communications session. When workstation users opt for this approach, the applications being accessed are nearly all alphanumeric from the standpoint of their graphical interfaces. It's possible to connect to the Internet using remote terminal emulation programs, yet the ISP you are working with needs to provide the account and setup parameters compatible with this approach of configuring a Windows 2000 Professional workstation to the Internet via browsers.

The most common and most pervasively used remote-terminal access programs are on the UNIX platforms and between UNIX and Windows NT systems. In the context of a UNIX command session, the UNIX-like commands such as Telnet, FTP, Ping, and Grep are all predominantly used within these sessions.

When working with a UNIX workstation, the shell account replicates exactly what the Command Prompt window shows on a Windows 2000 Professional-based workstation. Alphanumeric characters are displayed on the screen, enabling binary transfer of files through a modem from one location to another. All the TCP/IP commands are being interpreted and completed on the host. As commands are completed on the client workstation, the representation is displayed locally yet the command is completed on the host system. This is increasingly the case as applications are ported to the Internet for use through only a browser.

If you choose to configure workstations to have a remote-terminal connection to the Internet, you don't have to worry about network configuration because the person running the host system has taken care of these issues. You merely need to know your own account name and password, and the host's Internet address (name), and you are ready to configure a Windows 2000 Professional to work in conjunction with the Internet.

As mentioned earlier, it is assumed that you want to connect via LAN or modem to the Internet. Connecting via LAN simply requires that you have your networking hardware installed and connected to the network, and that you have the necessary NTT software installed to get on the network. In other words, once you have a network connection, you can get to the Internet.

USING MODEM CONNECTIONS VIA DIAL-UP NETWORKING

To begin, you'll need to have an Internet Service Provider (ISP) selected if you choose to use this as your only approach to getting to the Internet. Here's a checklist of items you need to get from your ISP to enable Dial-Up Networking with Windows 2000 Professional.

Get the connection information from your ISP.

Install the modem into your system if not already done.

Install NT's modem support for your modem. Many of the installation scripts from ISPs will search for and install modems. It's best to go through the process of using the Dial-Up Networking installation steps to get a modem defined first, as the ISP installation scripts vary in their effectiveness under Windows 2000.

Install Dial-Up Networking by going through Explorer. Using the More… button from within the Dial Up Networking dialog box, be sure to define the baud rate, data, and the stop and parity bits.

Check to make sure Remote Access Services has TCP/IP installed. If it doesn't, you'll experience significantly slower and possibly sporadic access times through your modem. Make sure you have TCP/IP enabled for Remote Access Services.

Check to make sure the BIOS Setup for the workstation you are configuring also has the COM port turned on and configured with an IRQ value that doesn't conflict with any others. Typically if the COM port for the modem you've installed is the last item configured through the BIOS, it will be assigned a noncompeting IRQ value relative to every other previous device.

Customize the Dial-Up Connection, ensuring that the options reflect what the ISP has told you.

Even though this list looks formidable, it's actually interesting to go through these steps and see in detail how Windows 2000 Workstation's options vary from those of NT 4.0 Workstation and how to get to the specific actions that a workstation will take to get a connection established. While the preceding list may look daunting to some, it isn't, and it need only be done once, provided all the necessary information from the ISP is collected in advance and that the modem is working, has the right drivers, and so on. Let's take a look at each of these steps in greater detail.

Getting Connected with an ISP

With millions of individuals, organizations, and businesses connected to the Internet, ISPs have generated a significant base or knowledge about how Windows NT workstations connect with the Internet. As a result, most ISPs are very good about giving you exactly the pieces of information you'll need. Many ISPs now provide the information on the attributes of many documents provided during setup.

The DHCP Protocol is a favorite among ISPs as it frees their clients from having to store and identify their specific IP addresses. When an ISP has the majority of their client dialing up via modem, the DHCP protocol makes the most sense as it greatly simplifies configuration.

Here are the items you need to get from your ISP:

- Primary and secondary (if available) Domain Name Server (DNS)
- User ID
- Password
- Dial-In phone number

If the service provider does not use dynamic address assignments, you'll need to obtain your IP address and gateway address from the service provider. If your ISP is like many, it has adopted the Dynamic Host Configuration Protocol (DHCP) to ensure that the allocation of IP addresses can be done seamlessly, every time you dial up and get logged onto the Internet via its servers.

You'll also find out whether the account you'll use to connect to the Internet is a SLIP or PPP account. While many ISPs will not specifically mention this, many of them use PPP, or Point-to-Point protocol, for handling this connection. PPP is more reliable than SLIP and provides encrypted passwords, data compression and error checking.

After getting the configurations defined for the ISP you'll be using, the next step is to install your modem. This involves getting the NT modem driver software installed and then testing the modem with the phone line you have available for dialing up.

Installing a Modem in Windows 2000

First, install modem physically in your system, if it isn't already, in a spare slot; if it is an external modem, connect it to the serial port. The following steps apply to either internal or external modems used in Windows 2000.

1. Open the Modems applet in the Control Panel. If there are already modems installed, click once on the Add button to install a second or third modem.
2. Click once on the Modems page of this dialog box. You'll see all the modems currently installed.
3. To add another modem, click once on Add... at the bottom of this specific page.
4. The first screen of the Install New Modem wizard shows the name of the modem you're installing. If the name of the modem is correct, click Next.

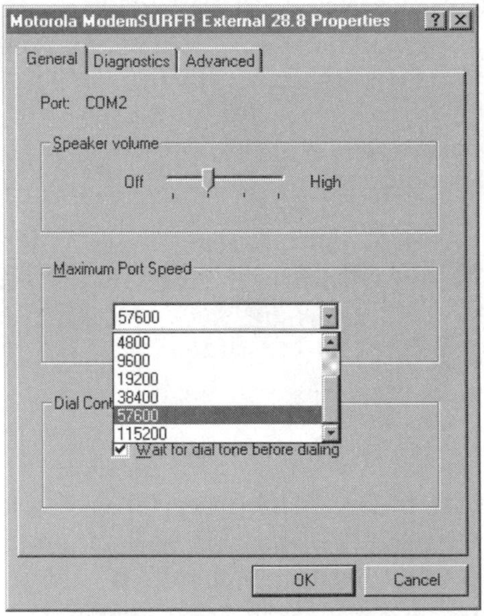

FIGURE 9.4 Using the Install Model Wizard to get a new modem created in Windows 2000 Professional.

5. You're now asked to enter the Location Information. Change the settings as appropriate. If you live in an area that does not have touch-tone dialing, be sure to change this setting to Pulse.
6. Click Finished. You're now returned to the main Modem Properties dialog box.
7. Click Close. Your modem is now ready to use.

EXPLORING REMOTE ACCESS SERVICES IN WINDOWS 2000

Using the Remote Access Service (RAS) in Windows 2000, you can also connect with the Internet. Throughout this section you'll get a thorough understanding of how to use RAS for connecting to the Internet and between client systems located throughout your organization. When RAS is installed on computers running Windows NT, your clients can connect over telephone lines through RAS to a remote network. The RAS server acts as a gateway between the remote client and the network. After a user has made a connection, the telephone lines become transparent to the user, and the user can access all network resources as if sitting at a computer that is directly attached to the network. For example, RAS makes a modem act like a network adapter card, projecting a remote computer onto a LAN.

In Windows NT version 4.0, RAS on the client side is called *Dial-Up Networking* and has a user interface that is consistent with Microsoft Windows95. Supporting Windows NT requires knowledge of how a remote client can access resources and services through RAS.

The goals of this section are to define how RAS works in conjunction with Dial-Up Networking and to explain how the telephony API (TAPI) works. You'll also get an overview of how to install and configure Remote Access Services and Dial-Up Networking and how to troubleshooting Remote Access Services.

Exploring RAS and Dial-Up Networking

RAS and Dial-Up Networking enable the extension of a network beyond a single location. RAS enables incoming connections from users at remote clients that are using Dial-Up Networking or other Point-to-Point Protocol (PPP) or Serial Line Internet Protocol (SLIP) dial-up software. Dial-Up Networking provides low-speed connections and is used by clients that connect to an RAS server or an Internet service provider (ISP).

Using RAS and Dial-Up Networking, clients can be connected to remote networks. After a connection is made, the remote links become transparent, and a client can gain access to network resources as if the client is directly attached to the network. After Dial-Up Networking is installed, the phone book feature can be used to record telephone numbers that are needed to connect to remote networks.

Remote clients can connect to an RAS server through a Public Switched Telephone Network (PSTN), an X.25 network, or an Integrated Services Digital Network (ISDN). They can also connect remotely over a TCP/IP network, such as the Internet, by using the Point-to-Point Protocol (PPTP).

Compatibility between PSTNs and Modems

Windows NT RAS uses standard modem connections over Public Switched Telephone Networks (PSTN). Most modems that comply with industry standards can interoperate with other modems. However, many difficult-to-diagnose problems can result from incompatible modems. Previous versions of Windows NT can automatically detect modems, and tests with Windows 2000 also proved to be compatible with modem connections over PSTN-based connections. This is especially useful when the user is not sure which modem is installed on the remote clients (for example, if his or her computer has an internal modem installed). If there is a problem detecting a modem automatically, it is possible to install a modem manually through the Modems program in Control Panel.

Integrating ISDN

Integrated Services Digital Network (ISDN) is a digital system that offers much faster communication than PSTN, communicating at speeds of 64 Kbps or faster. ISDN lines must be installed at both the server and the remote site. Additionally, an ISDN adapter must be installed in both the server and the remote client. The ISDN adapter and the X.25 adapter are treated as network adapter cards, thereby giving remote computers a direct data feed across a WAN to the LAN.

POINT-TO-POINT TUNNELING PROTOCOL

RAS servers are usually accessed directly through a modem, an ISDN card, or an X.25 PAD. They can also be accessed indirectly via the Internet with the Point-to-Point Tunneling Protocol (PPTP). PPTP is a networking technology that supports multiprotocol virtual private networks (VPNs). This support enables remote users to gain secure access to corporate networks across the Internet. Using PPTP, first a connection to the Internet is established, and then a connection to the RAS server on the Internet is established.

REMOTE ACCESS PROTOCOLS

RAS supports two kinds of protocols: those that transmit data over LANs and those that transmit data over WANs. Windows NT supports LAN protocols such as TCP/IP, NWLink IPX/SPX-compatible transport protocol, and NetBEUI, and remote access protocols such as SLIP, PPP, and the Microsoft RAS protocol.

Windows NT RAS supports NetBEUI, TCP/IP, and IPX. For this reason, Windows NT RAS can be integrated into existing Microsoft-based, UNIX, or Novell NetWare networks using the PPP remote access standard. Clients running Windows NT RAS can also connect to existing SLIP-based remote access servers (primarily UNIX servers). When RAS is installed and configured, any supported protocols already installed on the computer are automatically enabled for RAS. RAS connections can be established through SLIP or PPP.

Serial Line Internet Protocol

SLIP is an industry standard that addresses TCP/IP connections made over serial lines. Windows NT Dial-Up Networking gives clients running Windows NT access to Internet services supporting SLIP. SLIP has several limitations that include the requirement of static IP addresses, so SLIP servers cannot utilize DHCP or the Windows Internet Name Service (WINS). SLIP also typically relies on text-based logon sessions, and it usually requires a scripting system to automate the logon process. SLIP also supports TCP/IP, but it does not support IPX/SPX or NetBEUI, and it transmits authentication passwords as clear text.

Point-to-Point Protocol

PPP was designed as an enhancement to the original SLIP specification. PPP is a set of industry-standard framing and authentication protocols that enable RAS clients and servers to interoperate in a multivendor network.

PPP provides a standard method of sending network data over a point-to-point link. PPP supports several protocols, including Macintosh AppleTalk, DEC DECnet, Open Systems Interconnection (OSI), TCP/IP, and IPX. Windows NT supports NetBEUI, TCP/IP, and IPX.

Windows NT Protocol Support over PPP

PPP support enables computers running Windows NT to dial in to remote networks through any server that complies with the PPP standard. PPP compliance also enables a computer running Windows NT Server to receive calls from, and provide access to, other vendors' remote access software.

The PPP architecture enables clients to load any combination of Net-BEUI, TCP/IP, and IPX. Applications written to the Windows Sockets (WinSock), NetBIOS, or IPX interface can be run on a remote computer running Windows NT. Supporting TCP/IP makes Windows NT "Internet ready" and allows remote clients to access the Internet through WinSock applications. Dial-up Networking clients that have both the IPX interface and Client Service for NetWare (CSNW) installed can access NetWare servers. Dial-up Networking clients that do *not* have CSNW installed can still access a NetWare server if Gateway Service for NetWare (GSNW) is installed on an RAS server. The RAS server then functions as a gateway to a NetWare server. In this case, IPX is not required on the client. RAS Setup automatically binds to NetBEUI, TCP/IP, and IPX if they are installed on the computer when RAS is installed. After RAS is installed, each protocol can be configured separately for use with RAS.

PPP Multilink Protocol

The PPP multilink protocol provides a means to increase data transmission rates by combining multiple physical links into a logical bundle that increases bandwidth. RAS with PPP multilink protocol can be used to combine analog modem paths, ISDN paths, and even mixed analog and digital communications links on both clients and servers. For example, a client with two 28.8 Kbps modems and two PSTN lines can use the PPP multilink protocol to establish a single 57.6 Kbps connection to a PPP multilink protocol server. This will speed up access to the Internet or to an intranet and reduce the time required for remote connection, thus reducing the cost of remote access. Both the Dial-Up Networking client and the RAS server need to have the PPP multilink protocol enabled for this protocol to be used.

GATEWAYS AND ROUTERS

Windows NT RAS includes a NetBIOS gateway by which remote clients can gain access to NetBIOS resources, such as file and print services, on a network. This enables clients running NetBEUI to gain access to RAS servers regardless of which protocol is installed on the server. The NetBIOS gateway does this by translating the NetBEUI packets into IPX or TCP/IP formats that can be understood by remote servers.

IP and IPX Routers

Windows NT enhances the RAS architecture by adding IP and IPX router capabilities. A RAS server that has IP and IPX routers installed can act as a router to link LANs and WANS and connect LANs that have different network topologies, such as Ethernet and Token Ring.

In addition, an RAS server can be an IPX router and an SAP agent for Dial-Up Networking clients. SAP is similar in functionality to the Windows NT Browser service. After it is configured, an RAS server enables remote clients to access NetWare file and print services and to take advantage of WinSock applications.

EXPLORING THE TELEPHONY API

The Windows NT Telephony API (TAPI) provides a standard way for communications applications to control telephony functions for data, fax, and voice calls. TAPI virtualizes the telephone system by acting as a device driver for a telephone network. TAPI manages all signaling between a computer and a telephone network, including such functions as establishing, answering, and terminating calls. TAPI can also include supplementary functions, such as hold, transfer, conference, and call park, found in PBXs, ISDN, and other telephone systems.

TAPI Settings

TAPI allows users to centrally configure a computer for local dialing parameters. The basic TAPI settings for a system are set up when a TAPI-aware program is run for the first time. Dial-Up Networking is a TAPI-aware application. If a TAPI-aware application has not been run, the TAPI configuration will be automatically installed when Dial-Up Networking is installed. Three TAPI settings that can be configured are locations, calling cards, and drivers.

Locations

A *location* in Windows NT Dial-Up Networking is a set of information that TAPI uses to analyze telephone numbers in international number format and to determine the correct sequence of numbers to be dialed. A location does not need to correspond to a particular geographical location, although it usually does. A location could include the special numbers needed to dial out from an office or hotel room. Locations can be named anything that can help the user remember them.

Location information includes the following:

- Area (or city) code
- Country code
- Outside line access codes for both local and long-distance calls
- Preferred calling card

Calling Cards

TAPI uses calling cards to create the sequence of numbers to be dialed for a particular calling card. The number is stored in scrambled form and will not be displayed after it is entered. This is a security feature that is used to avoid unauthorized access to the number. Multiple calling cards can be defined.

Drivers

TAPI drivers, also known as TAPI Service Providers (TSPs), are software components that control TAPI hardware (for example, a PBX, voice mail card, phone system, or other equipment). Usually, TAPI drivers are installed with the TAPI hardware. However, the TAPI driver for modems (Unimodem.tsp) is automatically installed with the operating system.

Configuring a TAPI Location

Preparing a computer running Windows NT to use TAPI involves configuring a TAPI location. Configure TAPI locations through the Dialing Properties dialog box, which is accessible through the Telephony program in Control Panel. The Dialing Properties dialog box contains tabs through which various TAPI options can be configured.

TABLE 9.1   Configuration Options for the Locations Tab

Option	Use This Option to
I am dialing from list and the New button	List the locations that are currently set up. To set up an additional location, click New.
The area code is	Enter the area code for the TAPI location. If the location is in a country other than the United States, type the city code without leading 0s. For example, if the city code is 071, type 71.
I am in	Display the current country name.
To access an outside line	Type the number(s) required to access an outside line for local and long-distance calls. In many cases, these numbers will be the same. If no number is required to access an outside line, leave both spaces blank.
Dial using Calling Card	Specify that the displayed calling card will be used when calling from this location.
Change button	Change the calling card to be used for this location.
This location has call waiting. To disable it, dial	Specify whether this location uses call waiting. Call waiting should be turned off when dialing from a computer. Contact the local telephone company for information about disabling call waiting.
The phone system at this location uses	Specify either tone or pulse dialing.

Table 9.1 lists the configuration options available on the Locations tab in the Dialing Properties dialog box.

INSTALLING AND CONFIGURING RAS

Configuring RAS differs from configuring Dial-Up Networking clients. Although Dial-Up Networking clients are configured primarily to dial in to remote networks, RAS servers are configured to provide access to network services for those clients. RAS server configuration involves configuring communication ports, network protocols (such as NetBEUI, TCP/IP, and IPX), and encryption settings.

Installing RAS

RAS can be installed either during or after the installation of Windows NT 4.0. If Remote access to the network is selected during setup, both RAS and Dial-Up Networking will be automatically installed. One or both services can be installed manually after installation of Windows NT.

Whether RAS is installed during Windows NT installation or through the Network program in Control Panel, the following information is required:

- The model of the modem that will be used
- The type of communication port to use for the RAS connection
- Whether this computer will be used to dial in, dial out, or both
- The protocols to be used
- Any modem settings such as baud rate or Kbps
- Security settings, including callback

After Windows 2000 is installed, it is possible to install Dial-Up Networking manually. It can be installed through the Dial-Up Networking icon located in My Computer or the Dial-Up Networking icon located on the Accessories menu.

Configuring an RAS Server

The first step in configuring an RAS server is to specify the hardware that RAS will use, including the type of modem and the port to which the modem will be connected.

The drivers and ports used by RAS servers are configured through the Remote Access Setup dialog box in the Network program of Control Panel. Click the Services tab, click Remote Access Service, and then click Properties. The Remote Access Setup dialog box appears. Table 9.2 lists the configuration options available through this dialog box.

RAS Server Port Configuration Options

To configure the RAS server ports, in the Remote Access Service dialog box, click Configure. Table 9.3 explains the options listed in the Configure Port Usage dialog box.

Port configuration options affect only the specified port. For example, if the COM1 port for the server is configured to receive calls and the COM2 port is configured to dial out and receive calls, a user at a remote client can

TABLE 9.2   Configuration Options for the Locations Tab.

Option	Use This Option to
Add	Make a port available to RAS and install a modem, X.25 PAD, or a VPN for PPTP.
Remove	Make a port unavailable to RAS.
Configure	Change the RAS settings for the port, such as the attached device or the intended usage (dialing out only, receiving calls only, or both).
Clone	Copy the same modem setup from one port to another.
Network	Configure the network protocol and the multilink and encryption settings.

TABLE 9.3   Configuration Options of the Configure Port Usage Dialog Box

Option	Use This Option to Enable
Dial out only	Dial-Up Networking clients to use the port to initiate calls.
Receive call only	RAS servers to receive calls from Dial-Up Networking clients on the port.
Dial out and Receive calls	RAS servers to use the port for either Dial-Up Networking client or a server functions.

call in on either COM port, but a local user could use only COM2 for outbound Dial-up Networking calls. After selecting the appropriate Port Usage option, click OK. The Remote Access Setup dialog box reappears.

Configuring Protocols on the Server

The RAS server enables users at a variety of remote clients to connect to the server through different protocols. In general, the RAS server and the LAN should run the same protocols. This allows RAS clients to use any combination of supported protocols to gain access to remote resources. Protocols can be installed through the Protocols tab in the Network program in Control Panel.

In the Remote Access Service dialog box, click Network to use the Network Configuration dialog box to select and configure the LAN protocols. Network protocol configuration applies to RAS operations on all RAS-enabled ports. Table 9.4 describes the protocol configuration options available in the Network Configuration dialog box.

TABLE 9.4   Protocol Options in the Network Configuration Dialog Box

Options	*Use This Option to*
Dial-out protocols	Select the dial-out protocols.
Server settings	Select and configure the protocols that the RAS server can use for servicing remote clients.
Encryption settings	Select an authentication level ranging from clear text for down-level clients to Microsoft encrypted authentication for clients running Windows NT or Windows95.
	If Require Microsoft encrypted authentication is selected, the Require data encryption check box can also be selected.
Enable multilink	Enable the Dial-Up Networking PPP multilink protocol. To use the PPP multilink protocol, both the client and the server must have the PPP multilink protocol enabled.

Configuring an RAS Server to Use NetBEUI

If the NetBEUI protocol has been installed, the RAS Setup program enables NetBEUI and the NetBIOS gateway by default. RAS servers use NetBEUI to provide remote clients with access to small workgroups or department-sized LANs. NetBEUI is the smallest, and often the fastest, protocol used over RAS.

To configure a RAS server to use NetBEUI, in the Network Configuration dialog box, select the NetBEUI check box, and then click Configure next to NetBEUI. The RAS Server NetBEUI Configuration dialog box appears. Use this dialog box to enable remote NetBEUI clients to gain access to the following:

Entire network. This option grants remote clients permission to gain access to resources on the network.

This computer only. This option grants remote clients permission to gain access only to the resources on the RAS server.

Recall that the NetBIOS gateway translates NetBEUI packets to IPX or TCP/IP as needed.

Configuring an RAS Server to Use TCP/IP

To configure an RAS server to use TCP/IP, in the Network Configuration dialog box, select the TCP/IP check box, then click Configure. The RAS Server TCP/IP Configuration dialog box appears. Use this dialog box to grant network access permissions and IP addresses to Dial-Up Networking clients. Table 9.5 outlines the available configuration options.

TABLE 9.5 Options for the RAS Server TCP/IP Configuration Dialog Box.

Option	Use This Option to
Allow remote TCP/IP clients to access	Allow Dial-Up Networking clients to gain access to the entire network or only the resources on the RAS server.
Use DHCP to assign remote TCP/IP client addresses	Use a DHCP server to dynamically assign an IP address to a Dial-Up Networking client. Dial-Up Networking clients require an IP address to communicate on TCP/IP networks.
Use static address pool	Configure the IP address range; designate beginning and ending values for the IP address range. Use the Add and Remove buttons to exclude any IP addresses that are not to be used.
Allow remote clients to request a predetermined IP address	Enable Dial-Up Networking clients to request a predetermined IP address.

Configuring an RAS Server to Use IPX

Use the RAS Server IPX Configuration dialog box to grant remote IPX clients access to the network and to allocate network numbers.

To configure a RAS server to use IPX, in the Network Configuration dialog box, select the IPX check box, and then click Configure. The RAS Server IPX Configuration dialog box appears.

Dial-Up Networking clients can gain access to NetWare server file and print sharing resources through RAS servers that support IPX.

Use the RAS Server IPX Configuration dialog box to grant network access permissions and to allocate NetWare network numbers to Dial-Up Networking clients. Table 9.6 outlines the configuration options.

TABLE 9.6   Options for the RAS Server IPX Configuration Dialog Box

Option	Use This Option to
Allow remote IPX clients clients to access	Allow Dial-Up Networking clients to gain access to the entire network or this computer only.
Allocate network numbers automatically	Assign network numbers automatically to Dial-Up Networking clients. The same network number can be assigned to all IPX clients.
Allocate network numbers	Assign network numbers manually to Dial-Up Networking clients.
Assign same network number to all IPX clients	Assign a single network number to all IPX clients. Only one network number will be added to the routing table for all active Dial-Up Networking clients.
Allow remote clients to request IPX node number	Enable Dial-Up Networking clients to request an IPX node number rather than use the node number assigned by the RAS server.

INSTALLING AND CONFIGURING DIAL-UP NETWORKING

Dial-Up Networking enables users at remote clients to connect to a network from a remote site, such as home or a hotel. Dial-Up Networking is used to call the dial-up server and establish a telephone connection with the network. After the connection has been made, a Dial-Up Networking client can be used as if it were connected directly to the network. A number of options can be set in Dial-Up Networking, including phonebook entries, logging on using a dial-in entry, and the AutoDial feature.

Installing Dial-Up Networking

Dial-Up Networking is automatically installed during Windows NT installation if Remote access to the network is selected during Setup. Dial-Up Networking is also automatically installed on computers running Windows NT Server or Windows NT Workstation when RAS is installed, if RAS is configured to dial out and receive calls, or to dial out only. Dial-Up Networking can also be manually installed by double-clicking the Dial-Up Networking icon in My Computer.

Configuring Phonebook Entries

Using a modem, ISDN or another WAN adapter uses Dial-Up Networking to connect a client to remote networks. A *phonebook* entry stores all the settings needed to connect to a particular remote network. The Dial-Up Networking client stores all of its configuration data for a single connection in a phonebook file. A phonebook can be specific to an individual user or shared among all users on the computer. A phonebook shared in this way is called a *system phonebook*.

1. To create or edit phonebook entries, access Dial-Up Networking through either My Computer or the Accessories menu.
2. To use the Accessories menu, click the Start button, and then point to Programs.
3. Use the New Phonebook Entry wizard to create the first phonebook entry. After gaining experience with phonebook entries, it may be more efficient to turn off the wizard by selecting the I know all about phonebook entries and would rather edit the properties directly check box.

TIP

Quick tip: To use the New Phonebook Entry wizard again, in My Computer, double-click the Dial-Up Networking icon, click More, and then click User Preferences. Then, click the Appearance tab, click Use wizard to create new phonebook entries, and then click OK. The next time a new phonebook entry is created, the wizard will automatically start.

New Phonebook Entry Configuration

To create or configure a phonebook entry, in My Computer, double-click the Dial-Up Networking icon, and then click New. If the New Phonebook Entry wizard is disabled, when you click New, the New Phonebook Entry dialog box appears. Use the tabs in the New Phonebook Entry dialog box to configure the parameters described in Table 9.7.

In addition, the following TCP/IP settings (available on the Server tab) may need to be configured based on the dial-up server type that is selected. The TCP/IP settings are only available for PPP and SLIP servers; they are listed in Table 9.8.

TABLE 9.7   Parameters for New Phonebook Entry

Tab	Use This Tab
Basic	To configure a name for the phonebook entry.
	To enter the telephone number and any alternate telephone numbers and to use Telephony dialing properties, such as when calling long distance or using a credit card.
	To specify and configure the device used by the phonebook entry.
	To enable the PPP multilink protocol, in the Dial Using list, click Multiple Lines, then click Configure. In order to use the PPP multilink protocol, multiple devices, such as modems, must be installed.
Server	To select the dial-up server type, choose PPP, SLIP, or an earlier RAS protocol. The other options available depend on the server type selected, but they include selecting a network protocol, such as NetBEUI, TCP/IP, or IPX/SPX compatible transport, and selecting software data compression.
Script	To specify a terminal window or script file if manual intervention is required before or after dialing to establish a remote access session.
Security	To select a level of authentication and encryption.
X.25	To select an X.25 network provider and to configure connectivity information required by the X.25 network provider.

TABLE 9.8   TCP/IP Settings on the Server TAb

Option	Description
IP address	Automatically assigned by the dial-up server or manually configured on clients.
Name Server addresses	Assign DNS and WINS server addresses. These can be assigned by a DHCP server or manually configured at the client.
Use IP header compression	Enables header compression for low-speed serial links.
Use default gateway on remote network	Select this check box if the Dial-Up Networking client is using a network card to connect simultaneously to a LAN. When this check box is selected, packets that cannot be routed on the local network are forwarded to the default gateway on the remote network. In addition, address conflicts between the remote and local networks are resolved in favor of the remote network.

Logging on through Dial-Up Networking

When Dial-Up Networking is installed, Windows NT includes a logon option that enables users to log on to a domain using Dial-Up Networking. With this option, users can select a Dial-Up Networking phonebook entry that they will use to log on. Dial-Up Networking then establishes a connection to the RAS server so that a domain controller for the specified domain can validate the logon request.

Dial-Up Settings

The dial-up settings for establishing a connection for logging on are configured using the Logon Preferences dialog box on the Dial-Up Networking client. To access the Logon Preferences dialog box, click More in the Dial-Up Networking dialog box, and then on the More menu, click Logon Preferences.

Table 9.9 describes the logon options that can be configured in the Logon Preferences dialog box.

TABLE 9.9 Options in the Logon Preferences Dialog Box

Tab	Use This Tab to
Dialing	Specify the number of and interval between redial attempts. It can also be used to set an idle connection timeout period.
Callback	Configure the server to disconnect and to call the client back following authentication. This reduces telephone charges and increases security.
Appearance	Configure the Dial-Up Networking interface that appears during logon, including options to allow number preview before dialing, to show the location setting before dialing, to allow location edits during the logon process, to show connection progress while dialing, to close on dial, to allow phonebook edits during the logon process, and to use the wizard to create new phonebook entries.
Phonebook	Specify the system phonebook or an alternate phonebook to be used during logon.

User Profiles with Dial-Up Networking

Windows NT uses the same logon process for logging on to a LAN directly or through Dial-Up Networking. This process is identical for direct and remote logon because a copy of a user's profile is cached on the client each time the user logs off. Consider using the locally cached user profile rather than the server-based profile when logging on through Dial-Up Networking. For example, if the server containing a server-based profile is unavailable, any customization of the desktop that is stored in that profile will not occur. If there is a locally cached user profile, these customizations will occur.

Configure Windows NT to use the locally cached user profile through the User Profiles tab, which is accessible through the System program in Control Panel.

AutoDial

Windows NT 4.0 Dial-Up Networking supports *AutoDial*. AutoDial maintains network addresses and maps them to phonebook entries. This mapping allows automatic dialing when a user references the network address from an application or from the command line.

AutoDial Mapping Database

The AutoDial database can include IP addresses (for example, 127.95.1.4), Internet host names (for example, www.microsoft.com), or NetBIOS names (for example, PRODUCTS1). Each address in the database is associated with a set of entries. These are entries that RAS can use to dial from a particular TAPI dialing location.

Automatic Reconnection

AutoDial tracks all Dial-Up Networking connections so those clients can be automatically reconnected.

AutoDial attempts to make a reconnection in the following situations:

If a client is disconnected from the network, AutoDial will attempt to establish a connection whenever an application is used that references a network connection.

If a client is connected to a network, AutoDial attempts to create a network connection for addresses that it has previously learned.

Enabling and Disabling AutoDial

A user can enable and disable AutoDial in the User Preferences dialog box for a phonebook entry. To enable AutoDial, in the Dial-Up Networking dialog box, and then in the Phonebook entry to dial list, select an entry. Click More, and then click User Preferences. Click the Dialing tab, and then in the Enable auto-dial by location list, select each location listed. To disable AutoDial, on the Dialing tab, click to clear each location listed in the Enable auto-dial by location list.

AutoDial works only when the Remote Access Autodial Manager is running. To determine if the Remote Access AutoDial Manager is running, double-click the Services icon in Control Panel. If the Remote Access Autodial Manager is started, then AutoDial is able to function. If the Remote Access Autodial Manager is not running, start it by selecting it, clicking Startup, setting the Startup Type to either Automatic or Manual, and then clicking Start.

Quick tip: Windows95 and Windows NT versions earlier than 4.0 do not support AutoDial. AutoDial does not support IPX connections. AutoDial works only with the TCP/IP and NetBEUI protocols. For more information about AutoDial, see the Dial-Up Networking (RAS) Help.

TIP

TROUBLESHOOTING REMOTE ACCESS SERVICES

This topic describes some of the common errors that can occur when using RAS, along with guidelines and tools for solving these problems.

Event Viewer

Event Viewer is used to view the system log, which contains events for all Windows NT internal services and drivers. Event Viewer is useful in diagnosing RAS problems because many RAS events are entered in the system log. For example, if the Dial-Up Networking client fails to connect or if the RAS server fails to start, check the system log.

Problems with PPP Connections

If a user has problems being authenticated over PPP, a Ppp.log file can be created to provide debugging information to troubleshoot the problem. The Ppp.log file is stored in the *systemroot*\System32\Ras folder, and it is enabled by changing the following registry parameter value to **1**:

Authentication Problems

If a Dial-Up Networking client is having problems being authenticated over RAS, try to change the authentication settings for that client. Try the lowest authentication option on each side, and if successful, start increasing the authentication options to determine the highest level of authentication that can be used between the two systems.

Dial-Up Networking Monitor

The Dial-Up Networking monitor, which can be accessed through the Dial-Up Monitor program in Control Panel, shows the status of a session that is in progress. It shows the duration of the call, the amount of data being transmitted and received, and the number of errors. In addition, it can show which lines are being used for multilink sessions.

Multilink and Callback

If a user at a client uses a multilink-enabled phonebook entry to call a server that is configured to call the user back, when the callback is made it will be to one of the multilink devices. The reason for this is that the RAS Admin utility allows only one number to be stored for callback purposes for each user account. Therefore, the RAS server calls only one of the devices, and the multilink functionality is lost.

If using ISDN with two channels that have the same telephone number makes the link between the Dial-Up Networking client and the RAS server, then multilink will work with callback.

AutoDial Occurs during Logon

During the logon process, when Windows NT Explorer initializes, any persistent network connections or desktop shortcuts that reference network locations will cause AutoDial to attempt to make a connection. To avoid this, disable AutoDial or remove the persistent connections and shortcuts.

CHAPTER SUMMARY

The intent of this chapter is to provide a roadmap and assistance with the task of getting Windows 2000 workstations connected with client/server-based applications and the Internet. Focusing on the Dial-Up Networking and Remote Access Services that have quickly generated a significant installed base of clients, Microsoft's focus is on being able to share data over a variety of networking protocols with servers located behind the firewalls of your organization. This chapter has focused on the flexibility that RAS provides in terms of support for NetBEUI and IPX, two protocols that are prevalent in many smaller businesses and organizations throughout the world.

10 Using Windows 2000 Professional with Novell NetWare

The need to share printers and files has continually driven Microsoft and Novell to work together, on behalf of their shared customers, for a connectivity solution. Windows 2000 includes many useful tools for ensuring connectivity in those networking environments where Novell is also being used. This chapter provides a series of guidelines for organizations to getting the most out of their investment in Windows 2000 Professional and Server.

WINDOWS 2000 SERVICES FOR NETWARE

To facilitate network interoperability in a mixed Windows, Windows NT, and NetWare environment, Microsoft developed a complementary set of services to enable each platform to integrate with the others and to migrate from NetWare-based services. Two services are shipped together in one package, known as the Services for NetWare. The two services are File and Print Services for NetWare (FPNW) and the Directory Service Manager for NetWare (DSMN). The enhancements to these two products include the following:

- Support for the Windows 95 user-interface
- Support for the Administration Tools for Windows 95, including the ability to administer FPNW/DSMN servers from Windows 95 clients
- Enhanced multiprocessor support
- Web administration tool for Windows 2000 Server supporting FPNW
- The ability to administer an FPNW server through a Web browser
- Significant increase in performance over FPNW for Windows NT 3.51

File and Print Services for NetWare (FPNW)

File and Print Services for NetWare allows users of NetWare 2.x/3.x (and 4.x in bindery emulation mode only) to utilize the multipurpose features of Windows 2000 Server by enabling the seamless integration of Windows 2000 Server into an existing NetWare network.

Most of the Windows 2000 Server software needed for the migration is included in the packaged product. The File and Print Services for Net-Ware, however, is an add-on product and must be licensed separately. File and Print Services for NetWare allows a standard NetWare client to access file and print resources on a Windows 2000 Server-based computer. The server to which the users connect will actually be a Windows 2000 Server, but to the user it will appear to function as a NetWare 3.x server.

FPNW is not a requirement for the migration process, but it will enhance the migration because it includes an updated Migration Tool NWCONV.EXE utility that migrates an entire bindery of NetWare user and group accounts. FPNW allows Windows 2000 Server accounts to become NetWare enabled and used on Windows 2000 Server. This migrates NetWare-specific information including grace logons, limited concurrent connections, and station restrictions. Volume restrictions, though, are not supported.

Directory Service Manager for NetWare

Directory Service Manager for NetWare (DSMN) has features that will ease administrators' tasks and give NetWare clients robust and seamless access to the servers. DSMN takes advantage of the Windows 2000 global directory services capabilities by allowing the administration and management of a mixed Windows 2000 Server and NetWare network to take place on a global level, with a central point of management and single logons to services and applications. DSMN allows administrators to import accounts from one or many NetWare bindery servers and manage them from Windows 2000 Servers. DSMN gives administrators the flexibility of synchronizing passwords and system-level functions between Windows 2000 and NetWare servers. DSMN specifically includes the following: a simple interface for propagating users from NetWare to Windows 2000 Server management of the Windows 2000 Server-based system from anywhere, including remote dial-up, Windows 95 desktops, and Windows 2000 Workstation. DSMN also simplifies the end user using the NetWare client by doing the following:

- Allowing the user to RAS into Windows 2000 Server using the same account information
- Needing only one logon for the user to access applications and file and print services
- Allowing new users to be up and running because the administrator can create the account on the Windows 2000 Server and Windows 2000 Server will propagate the account information to the NetWare servers
- Letting the user have the same name and password on all the servers on the network, Windows 2000 and NetWare alike

Windows 2000 Server can use three other utilities to maximize efficiency when integrating Windows 2000 Server and NetWare. These utilities include Gateway Services, NWLink, and the Migration Tool for NetWare.

Gateway Services for NetWare

Gateway Services for NetWare allows a Microsoft networking client (LAN Manager, MS-DOS, Windows for Workgroups, Windows95, or Windows 2000 Workstation, including remote clients) to access NetWare server services through the Windows 2000 Server-based machine. The only protocol required for a remote user is the protocol used for the Remote Access Server connection. This means that you could run IPX, TCP/IP, or NetBEUI. Additionally there is no technical limit to the number of remote users that can dial in serially to take advantage of the gateway. Licensing requirements should be addressed by careful review of the appropriate license agreements.

COMPARING NOVELL NETWARE AND WINDOWS 2000 SERVER

The choice of a network operating system is a strategic decision driven by organizations' demands for a platform on which to build business solutions. Looking at the spectrum of servers and their associated applications today, it's important to note the differences between Windows 2000 Server and NetWare. The intent of this section is to define the differences between the network operating systems.

Exploring Windows 2000 Server

Windows 2000 Server provides an integrated solution for application, file, and print sharing. Windows 2000 Server has been designed from the ground up as an integrated, multipurpose operating system. As opposed to combining non-integrated services, Windows 2000 Server provides complete integration between its services, resulting in easier management and lower total cost of ownership (TCO). For instance, once authenticated to the directory, users don't need to authenticate themselves again to access other applications and services.

File and print sharing. The file and printer sharing implementation in Windows 2000 Server gives customers an advanced solution, offering a distributed file system, Internet printing, content indexing, dynamic volume management, and plug-and-play support.

Networking and communications. The networking infrastructure is complete and manageable—it offers true dynamic configuration, integrated dial-up, and Virtual Private Networking (VPN) with support for the latest IETF VPN protocol suite, telephony and Quality of Service (QoS).

Application services. Windows 2000 Server provides scalability in terms of CPU and memory support. The combination of Cluster ing Services, component load balancing, and the Windows Load Balancing Service gives customers a comprehensive availability/load-balancing solution to further increase system scalability and reliability. Windows 2000 Server and COM+ provide a flexible and robust platform on which to build distributed applications. Finally, integrated Terminal Services gives customers a comprehensive thin-client solution.

Internet services. The Windows 2000 Internet services feature-set are also a complete solution, offering numerous unrivaled Internet services: management, publishing, streaming media, and performance enhancement capabilities.

Management services. The Active Directory services in Windows 2000 Server are built completely around Internet-standards and offer extensibility and scalability. This makes it a solution on which to build enterprise-level, directory-enabled applications. Microsoft Management Console (MMC) gives customers a single customizable interface for managing networking services and applications. The combination of IntelliMirror™ management technologies, Windows Installer, and Group Policy Services easily presents a comprehensive solution for software distribution and desktop management. Security support in Windows 2000 Server provides support for Kerberos, smart card authentication, fully integrated public key infrastructure, and file system encryption services.

Windows 2000 Server does lack an extensible, hierarchical directory. Although the directory in Windows 2000 Server provides organizations with a centralized directory for managing users and groups and single logon services, it simply cannot compete with the feature-set in either Active Directory services or Novell Directory Services (NDS).

File and print sharing. File and printer-sharing support is robust, although Windows NT's feature-set cannot compare favorably with either Windows 2000 Server or NetWare 5.0. On the other hand, Windows 2000 Server surpasses NetWare 5.0 running on the NetWare File System (NWFS) in several capacities—such as offering fewer file system limitations, integrated namespace support, data striping, and data striping with parity.

Networking and communications. Although not as comprehensive as Windows 2000 Server, Windows NT networking support is still far better than that of NetWare 5.0—offering several unmatched capabilities such as integrated dial-up access and VPN support.

Application services. Application support is outstanding, providing numerous capabilities such as message queuing, clustering and load balancing, and a thin-client solution in the form of the Terminal Server Edition of Windows 2000 Server. Distributed component support is also superior—the combination of COM and Transaction Server offers many capabilities not found in NetWare 5. Furthermore, with the addition of Active Server Pages (ASP), the power of COM and Transaction Server-based applications can be extended to the Web.

Internet services. The Internet services found in Windows 2000 Server are far superior to those of NetWare 5.0, offering load balancing, content management, and protocol support including SMTP and NNTP that is simply unrivaled.

NetWare 5.0

Finally, Novell NetWare 5.0 provides the services in the major customer deployment scenarios; however, as this chapter describes in detail, NetWare 5.0 falls short in providing customers with an integrated solution. Many of the services are simply provided as add-ons and lack common installation, management interfaces, and security infrastructure. Because of the lack of an integrated architecture, NetWare 5.0 at times is very difficult to use and administer. Furthermore, NetWare 5.0 lacks many of the features, such as clustering, load balancing, VPN support, distributed file system, dynamic volume management, and others that provide customers with better availability and lower TCO. Even Novell's strongest feature—NDS—fails to deliver the infrastructure to provide support beyond basic user management needed for directory-enabled applications.

File and print sharing. File and print services are extremely robust, but functionality limitations in Novell Storage Services and client and device compatibility issues with Novell Distributed Print Services negate their usefulness and all of the benefits for many customers.

Networking and communications. Native TCP/IP support is provided, but client compatibility will be an issue for many existing environments. Hot Plug PCI support is innovative, but its lack of hardware support puts it out of the reach of many customers. Although a routing and remote access solution is provided, it simply cannot match the advanced feature-sets found in the Microsoft solutions such as integrated VPN support or connection sharing.

Application services. NetWare 5.0 provides services to develop Java/CORBA applications. Its lack of availability/load balancing services, message queuing services, support for other languages for CORBA applications, comprehensive distributed component functionality, such as integrated transactions, and no solution for thin-clients all make it a relatively limited choice.

Internet services. The Internet services implementation found in NetWare 5.0 provides organizations with the basic HTTP and FTP services required to host Internet and intranet sites. It lacks true operating system integration, a comprehensive Web application framework, and support for many key features and standard protocols found in the Microsoft products such as comprehensive multisite hosting, Web-DAV, SMTP, and NNTP.

Management services. Although it boasts an extremely impressive feature-set, NDS is beset by scalability or latency issues and a general lack of support for Internet-standards, making it not as appropriate a choice for the enterprise as Windows 2000 Server. The GUI administration tools in NetWare 5.0 provide are adequate, offering in some ways better integration than Windows 2000 Server. It still cannot compare, though, to the MMC in Windows 2000 Server. The included Z.E.N.works Starter Pack provides an excellent software distribution solution—offering some capabilities not found in Windows 2000 4.0—but it lacks many of the desktop management features, such as user data management, found in Windows 2000 Server.

WHAT IS NWLINK?

The NWLink IPX/SPX/NetBIOS compatible transport protocol (referred to as NWLink) is Microsoft's implementation of Novell's NetWare Internetwork Packet Exchange/Sequenced Packet Exchange (IPX/SPX) protocol. NWLink is most commonly used in environments where clients running NetWare are used to access resources on computers running Microsoft operating systems.

Understanding NWLink Features

NWLink allows workstations running Windows 2000 to communicate with other network devices that are using the IPX/SPX protocol. NWLink can be used in small network environments that use only clients running Windows 2000 and other Microsoft operating systems.

NWLink supports the networking application programming interfaces (APIs) that provide the interprocess communications (IPC) services described in Table 10.1.

TABLE 10.1 NWLink's APIs

Networking API	Description
WinSock	Supports existing NetWare applications written to comply with the NetWare IPX/SPX packets interface.
NetBIOS over IPX	Implements as NWLink NetBIOS; supports communication between a NetWare client running NetBIOS and a computer running Windows 2000 and NWLink NetBIOS.

NWLink also provides NetWare clients with access to applications designed for Windows 2000 Server, such as Microsoft SQL Server and Microsoft SNA Server. To give NetWare clients access to file and print resources on a computer running Windows 2000 Server, you should install File and Print Services for NetWare (FPNW).

The 32-bit Windows 2000 implementation of NWLink has the following features:

- Supports communications with NetWare networks
- Supports sockets and NetBIOS over IPX
- Provides NetWare clients with access to Windows 2000 servers

Getting NWLink Up and Running

Following the steps here, you'll be able to install NWLink. These steps also apply to any protocol being installed in Windows 2000.

1. Right-click on My Network Places, and then click once on Properties.

2. In the Network and Dial Up Connections window, right-click Local Area Connection, and then click Properties. The Local Area Connection Properties dialog box appears, displaying the network adapter in use and network components configured for this adapter.
3. Click Install.
4. In the Select Network Component Type dialog box, click Protocol, and then click Add.
5. In the Select Network Protocol dialog box, in the Network Protocol list, click NWLink IPX/SPX/NetBIOS Compatible Transport Protocol, and then click OK.

TIP

Quick tips to keep in mind:

- *You must be a member of the Administrators group to install NWLink.*
- *When you install NWLink, it is installed for all your connections. If you do not want NWLink installed for a certain connection, right-click that connection, click Properties, and on either the General or Networking tab, clear the NWLink IPX/SPX/NetBIOS Compatible Transport Protocol check box.*
- *To confirm that NWLink has been initialized properly, at a command prompt, type* **ipxroute config**. *You should see a table with information about the bindings for which NWLink is configured.*

Configuring NWLink

NWLink configurations are actually made up of three different components. They are the frame types, network number, and internal network number. By default, Windows 2000 detects the frame type and a network number automatically when you install NWLink. Windows 2000 also provides a generic internal network number. You must manually specify an internal network number if you plan to run FPNW or IPX routing. Figure 10.1 shows the NWLink Configuration dialog box.

Let's take a look at each of the configuration components of NWLink:

Frame type. This defines the way that the network adapter cards format data. To ensure proper communication between a computer running Windows 2000 and a NetWare server, you'll need to configure the NWLink frame type to match the frame type on the NetWare server.

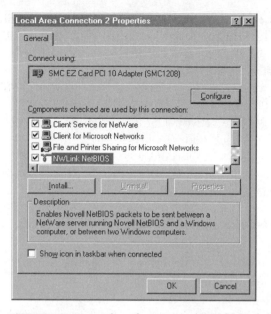

FIGURE 10.1. Using the Networking options in Windows 2000 to configure NWLink.

Network number. Each frame type configuration on a network adapter card requires a special network number, which must be unique for each network segment. All computers on a segment using the same frame type must use the same network number to communicate with one another.

Although Windows 2000 automatically detects a network number during NWLink installation by default, you can also manually specify a network number using the Registry Editor.

Setting a network number in the registry for a given frame type requires entering two corresponding entries, NetworkNumber and PktType, in HKEY_LOCAL_MACHINE\SYSTEM\CurrentCOntrolSet\Services\ NwInkipx\Parameters\Adapters\Adapter subkey in the registry.

NetworkNumber specifies the network number (in hexadecimal) for the adapter. If the value for this entry is 0, NWLink gets the network number from the network while it is running. Network numbers are four bytes (four hexadecimal characters). The NetworkNumber entry takes an REG_MULTI-SZ data type. PktType specifies the packet form to use. The PktType entry takes a REG_MULTI_SZ data type.

Internal network number. This defines the computer on the network for internal routing. This eight-digit hexadecimal number, or virtual network number, is by default set to 00000000.

The internal network identifies a virtual network segment inside the computer. That is, the internal network identifies another (virtual) segment on the network. So, if an internal network number is configured for a computer running Windows 2000, a NetWare server or a router adds an extra hop to its route to the computer.

You must manually assign a unique, internal network number in the following scenarios:

- When FPNW is installed and there are multiple frame types on a single adapter
- When FPNW is installed and NWLink is bound to multiple adapters in the computer

An application is using NetWare Service Advertising Protocol (SAP). SQL Server and SNA Server are examples of applications that can use SAP. This acronym has no relation to the German software company whose name is SAP by the way.

Quick tips to keep in mind:
If a computer has multiple network adapter cards bound to NWLink, and if you want each one to use a different frame type, configure each network adapter card to use the Manual Frame Type Detection option. You'll need to specify the frame type, network number, and internal network number for each network adapter card.

TIP

A connection between two computers that use different frame types is possible if the NetWare server is acting as a router. This is inefficient, however, and could result in a slow connection.

To open Network and Dial-up Connections, click Start, point to Settings, click Control Panel, and then double-click Network and Dial-up Connections.

By default, NWLink automatically detects the frame type used by the network adapter to which it is bound. If NWLink detects no network traffic or if multiple frame types are detected in addition to the 802.2 frame type, NWLink sets the frame type to 802.2.

You can determine which you are using by typing ipxroute config at a command prompt.

NWLink's Role in the Microsoft Connectivity Strategy

NWLink is Microsoft's implementation of Novell's NetWare Internetwork Packet Exchange/Sequenced Packet Exchange (IPX/SPX) protocol. NWLink is most commonly used in environments where clients running Microsoft operating systems are used to access resources on NetWare servers or where clients running NetWare are used to access resources on computers running Microsoft operating systems. NWLink supports WinSock and NetBIOS over IPX networking APIs. WinSock supports existing NetWare applications written to comply with the NetWare IPX/SPX Sockets. NetBIOS over IPX is implemented as NWLink NetBIOS and supports communication between a NetWare client running NetBIOS and a computer running Windows 2000 and NWLink NetBIOS.

INSTALLING GSNW

Follow these steps to install GSNW:

1. From the Start menu, choose Settings, and then open Control Panel. Select the Network applet icon.
2. Choose Add Software, select Gateway Service for NetWare, and then choose Continue.
3. Be sure that the path to the Windows 2000 Server distribution files is correct and then select Continue. Click OK to complete the network reconfiguration
4. Restart the Windows 2000 Server.
5. Log back on with the Administrator account. You will be prompted to select a "Preferred Server for NetWare." You should be provided with a list of all known Novell servers on the network from which you can choose. Now, as a NetWare client, when you log on to this system your account will automatically be validated by the Windows 2000 Server domain and will also be authenticated by the preferred NetWare server that you select. If your password on the Windows 2000 Server domain is different from that on the NetWare server, you will be prompted to enter the NetWare password.

6. To verify that the Gateway Service for NetWare has been installed, from the Start menu, choose Settings, and then open Control Panel. Select the Network applet icon. You should have a new applet icon titled "GSNW." If you start the GSNW applet, a dialog box will appear in which you can change your preferred server and print options.

If you do not need to install the gateway feature of GSNW, then skip the steps that follow and continue with installation of FPNW in the following section. Be sure that the account, which you are logged on to, on the Windows 2000 Server has administrative rights and is also a NetWare account. Make any necessary changes through User Manager for Domains.

Complete the following steps on the NetWare server:

1. Create (or identify) a NetWare Gateway User Account. This is an account that the GSNW service uses to connect to the NetWare Server.
2. Create a group called NTGATEWAY.
3. Make the Gateway User Account a member of the NTGATEWAY group.

Complete the following step on the Windows 2000 Server:

1. Go to the GSNW applet in Control Panel and select the Gateway button. Check Enable Gateway and provide the Gateway User Account and password.
2. The next step is to add connections to a NetWare resource and share the redirection. Use the Universal Naming Convention (UNC) syntax to specify the Network Path.

INSTALLATION OF FILE AND PRINT SERVICES FOR NETWARE (FPNW)

The File and Print Services for NetWare (FPNW) is a separate, add-on product for Windows 2000 Server. It provides a Windows 2000 Server with a NetWare Core Protocol-compatible (NCP-compatible) server service. In layman's terms, this allows the Windows 2000 Server to act as a NetWare

Server to all NetWare clients. A NetWare client sees the Windows 2000 Server when an SLIST command is executed.

The Windows 2000 Server will appear in the client's File Manager or Explorer list of NetWare or NetWare-compatible servers. A client can "MAP" to a shared volume and directory on an FPNW-enabled Windows 2000 Server just as if it were a NetWare server. A NetWare client can likewise connect to a printer on the Windows 2000 Server.

Finally, a NetWare client will be able to log on to the Windows 2000 Server and have configured system and personal logon scripts execute. FPNW installation files are on a compact disc. If you need to install them from a floppy, copy the appropriate disk directories (there are two of them) to floppies.

TO INSTALL FILE AND PRINT SERVICES FOR NETWARE

Follow these steps:

1. From the Start menu, choose Settings, and then open Control Panel. Start the Network Applet and choose Add Software. Select Have Disk and click Continue.
2. In the Insert Disk dialog box, enter the path to the FPNW installation files and click OK. Select File and Print Services for NetWare and click OK. You must complete the Install File and Print Services for NetWare dialog box.
3. Specify the Windows 2000 Server directory that will be used as the SYS volume. The default is C:\SYSVOL. Note that for file-level security, you will want the SYS volume to be on an NTFS-formatted volume.
4. The server will also be given a new name that will identify it on the NetWare network. The default name is the Windows 2000 Server Computername followed by an underscore and FPNW. You can change this name. Important: In many cases, you will want to rename your NetWare server—for example, (OLD_NW312)—and assign the previous NetWare server name to your Windows 2000 Server configured with FPNW, thus allowing your existing clients to run untouched.
5. A Windows 2000 Server Supervisor account will be created. You must supply and confirm the password for the Supervisor account.

6. In the Tuning section, choose Minimize Memory Usage if the system is primarily used for applications, Balance between Memory Usage and Performance if the server is both an applications server and file and print server, or Maximize Performance to provide the best file and print sharing performance (this will use additional system memory).

7. Click OK when complete. If you chose a drive for the SYSVOL that wasn't NTFS, you will receive a warning indicating that security cannot be enforced. You have the option to change the SYS location.

8. A special Windows 2000 Server account called FPNW Service Account will be created for running the FPNW services. You must supply and confirm a password for that account. Click OK. Click OK from the Network Settings dialog box and restart the computer.

9. Log on to the system as Administrator. If you open the Control Panel from the Start menu you will see a new FPNW icon. A successful installation of FPNW will also create an FPNW menu in Exploring c:\WINNT\system32 and a modified user accounts dialog box, adding a button for NetWare client options.

Upon successful completion of this last step (as indicated by the absence of errors in the Event Viewer/System Log); you have achieved Windows 2000 Server coexistence in a NetWare environment. Migration can follow readily as you replace the functionality of the NetWare servers and decommission them, if your deployment calls for it. First, be sure that you have documented the actual time taken to complete the preceding steps for each server. Also, you should document any problems that were encountered along with their resolutions.

CHAPTER SUMMARY

Focusing on organizations that need to integrate both Novell NetWare and Windows 2000 Workstation and Server, this chapter explores the most commonly used tools for making these two diverse operating systems worktogether. This is especially critical for those of you who are migrating both operating systems within the coming years, as Windows 2000 with Active Directories poses a comparable set of functional capabilities to the NDS capabilities included in NetWare 5.0. The topics covered in this chapter also describe how to make file and print services work across both platforms, ensuring members of the teams in your organizations can continue to use files, printers, and in the case of using NWLink, even applications.

11 Using Microsoft Peer Web Services in Windows 2000 Professional

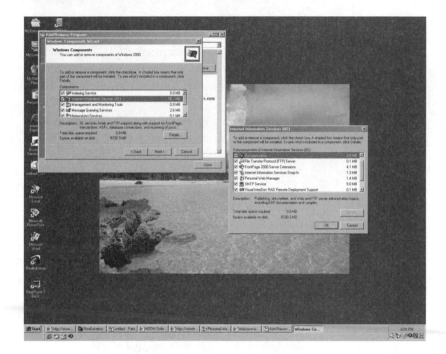

Increasingly as Windows 2000 workstations and servers are developed specifically for creating Web content, the role of your workstations in the Web content generation, posting, and maintenance processes becomes more critical. The Peer Web Services (PWS) suite of applications was first available on Windows 2000 Professional and was developed to give workstation users the flexibility of completing low-end Web publishing in conjunction with business applications. Many companies use the features of PWS for creating and maintaining individual department Web sites. PWS is actually a subset of the Internet Information Server components, which are included in Windows 2000 Server.

Using Peer Web Services, you can complete the following tasks:

- Publish a department home page on your corporate intranet that lists sales records, reports, members of your department, or other departmental documents
- Publish personal home pages that describe your interests, both professional and personal
- Share documents among members of your workgroup for collaborative projects

PWS also provides other information services and supports a variety of interfaces that you can use to develop other features for your Internet or intranet Web sites. Additional function include the following:

- Create high-performance client/server applications using the Microsoft Internet Server Application Programming Interface (ISAPI)
- Customize the WWW Service by creating ISAPI filter programs that listen to incoming or outgoing requests and automatically perform actions, such as enhanced logging
- Run Common Gateway Interface (CGI) applications or scripts
- Transmit or receive files using the FTP service
- Publish archives of information, spanning multiple computers, using the gopher service

Peer Web Services includes the following components:

- Internet services: WWW and FTP
- Internet Service Manager, the tool for administering the Internet services
- Internet Database Connector, the component for sending queries to databases
- Key Manager, the tool for installing Secure Sockets Layer (SSL) keys

WORKING WITH MICROSOFT PEER WEB SERVICES

An Intranet-Based Example

Peer Web Services was specifically developed to provide for interoperability across operating system platforms. Because Peer Web Services integrates Windows 2000 security and networking, you can add the software to an existing computer and use existing user accounts that are running either Windows 2000 Professional or Server.

In a small business, individuals can use Peer Web Services to publish information within the business, using an intranet as the publishing mechanism. In a larger business with multiple departments or workgroups, workgroup-specific information can be made available by using Peer Web Services on desktop computers rather than by putting workgroup documents on a central Web server. The workgroup's Web server can host personal Web-style pages, customize workgroup applications, serve as an interface to the workgroup's Structured Query Language (SQL) database, or use Remote Access Service (RAS) to provide dial-up access. Figure 11.1 shows how Peer Web Services can be used in conjunction with other applications, spanning functional boundaries within an organization.

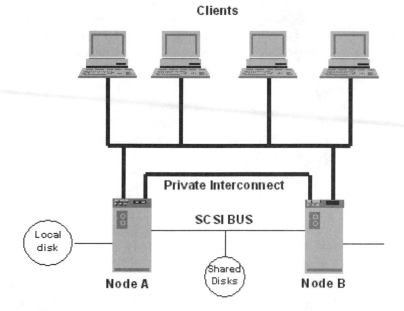

FIGURE 11.1   An example of a corporate intranet.

Installing Peer Web Services

Once you have installed Windows 2000 Professional, you can install Peer Web Services from the Windows 2000 Professional CD. If you already have the necessary Internet or intranet connection, you can accept all of the default settings during setup and then add your Hypertext Markup Language (HTML) content files to the Wwwroot folder. Your files will be immediately available to users. The default setup configurations are suitable for many publishing scenarios without any further modifications.

Installation Requirements

Microsoft Peer Web Services requires the following:

- A computer with at least the minimum configuration to support Windows 2000 Professional, ideally one that is listed on the Windows 2000 Hardware Compatibility List (HCL).
- Windows NT 4.0 or Windows 2000 Professional
- Transmission Control Protocol/Internet Protocol (TCP/IP) (included with Windows 2000 Professional and Windows 2000). Use the Network application in Control Panel to install and configure the TCP/IP protocol and related components.
- A CD-ROM drive for the installation compact disc.
- At least 110MB of disk space for your information content. It is recommended that all drives used with Microsoft Peer Web Services be formatted with the Windows 2000 File System (NTFS).

To publish on an Intranet, you will need the following:

- A network adapter card and local area network (LAN) connection.
- The Windows Internet Name Service (WINS) server or the Domain Name System (DNS) server installed on a computer in your intranet. This step is optional, but it does allow users to use "friendly names" instead of IP addresses.

WINDOWS 2000 CONFIGURATION AND SECURITY CHECKLISTS

Before PWS can work correctly, the Windows 2000 Professional on which it is installed needs to have network connections based on the TCP/IP protocol. Configuring security on Windows 2000 -based workstations and servers is also needed. Provided next is a series of checklists for handling the tasks that lead to getting Peer Web Services up and running.

Windows 2000 Configuration Checklist

Use the Network application in Control Panel for all configuration tasks mentioned in this section.

Obtain an Internet connection. Although Peer Web Services is designed mainly for publishing on a corporate intranet, you can set up a low-volume Web site on the Internet for testing or light use. To publish on the Internet, you must have a connection to the Internet from an ISP. To find an ISP, look in the telephone book under Computers—Networking or in your local newspaper's business or technology section.

If you want to set up a full-capacity Web site on a dedicated server, you should use Microsoft Internet Information Server running on Windows 2000 Server.

Install Windows 2000 Professional. Install Windows 2000 Professional and Microsoft Peer Web Services. If you have already installed Windows 2000 Professional or Server, you can install Peer Web Services using the Add/Remove Programs icon in the Control Panel. Using the options within this applet, select the Add/Remove Windows Components, then use the Windows Components wizard; install the Internet Information Services components, in addition to the components available through the Windows Components listing in the Windows Components wizard. Figure 11.2 shows the Windows Components available for installation on Windows 2000 Professional through the use of the Windows Components wizard.

Configure the TCP/IP protocol. Install the Windows 2000 TCP/IP protocol and connectivity utilities. If you are making your computer accessible from the Internet, your ISP must provide your server's IP address, subnet mask, and the default gateway's IP address. (The default gateway is the ISP computer through which your computer will route all Internet traffic.)

If the FTP service provided with Windows 2000 has been installed, remove it. Also remove any other previously installed Internet services.

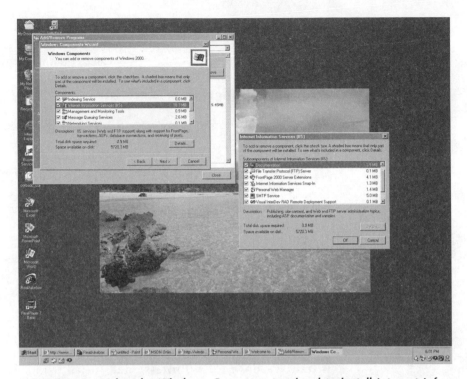

FIGURE 11.2 Using the Windows Components wizard to install Internet Information Services support.

Configure name resolution. You need a name resolution system to map IP addresses to computer names or domain names. On the Internet, Web sites usually use the Domain Name System. Once you have registered a domain name for your site, users can type your site's domain name in a browser to contact your site. ISPs can register domain names for you. On an intranet, you can use either DNS or the Windows Internet Name Service (WINS). Your network must have DNS or WINS servers to match IP addresses to host names, and client computers must know the IP address of the DNS or WINS server to contact.

An alternative to DNS is to use a HOSTS file. On intranets an alternative to WINS Servers is to use an LMHOSTS file. Use the Network application in Control Panel to make the appropriate Advanced TCP/IP

Configuration setting for this computer's name resolution. For more information on installing and configuring WINS or DNS, see the Windows 2000 online Help.

Windows 2000 Security Checklist

Using the following steps will ensure that your computers, workstations, and servers running Windows 2000 are secure from unauthorized access to the Web pages and their contents. Presented next are a series of guidelines to keep in mind when defining security for user accounts, NTFS file security, and the ongoing running of other network services.

User Accounts

Review the IUSR_*computername* account's rights.

Choose difficult passwords.

Manage strict account policies.

Limit the membership of the Administrators group.

NTFS File Security

Use Access Control Lists (ACLs), available with NTFS.

Enable auditing to track file access.

Running Other Network Services

Run only the services that you need.

Unbind unnecessary services from your Internet adapter cards.

Disabling Previous Web Services Prior to Installation

If your computer has another version of File Transfer Protocol (FTP), gopher, or World Wide Web (WWW) services installed, disable these services before you install the Microsoft Peer Web Services. Refer to the documentation for each service to see how to disable it.

Disabling FTP Guest Account Access

During the set-up process, a screen will appear, asking you whether you want to disable access by the Guest account to your FTP server. It is recommended that you select Yes to protect the contents of your system. If you choose the No option and enable guest access to your server, all existing files and any new files will be available to the Guest account through FTP. You will need to disable access to each file or folder individually to prevent unauthorized access. Disabling FTP access for the Guest account will not affect the IUSR_*computername* account that is created during set-up.

PUBLISHING A WEB PAGE USING PEER WEB SERVICES

Once PWS is up and running, you're ready to begin getting pages published in your intranet and Internet sites. If your files are in HTML format, just add them to the appropriate home directory. For example, to make files available to a Web browser using the WWW service, place the files in the Wwwroot folder.

Default.htm and the Peer Web Services Home Page

By default, Peer Web Services uses a file named Default.htm as the home page for the various samples, tools, and demonstrations that come with the product. If the Wwwroot folder of your Web server already contains a file named Default.htm when you install Peer Web Services, your file will not be overwritten with that file. As a result, you will not have immediate access to the sample home page and the links it provides when you run Peer Web Services.

In this case, to view our version of Default.htm and the links it provides, type the following Uniform Resource Locator (URL) in the Internet Explorer **Address** box.

```
http://computername/samples/default.htm
```

This command loads the file Default.htm from the Wwwroot\Samples folder.

You can also rename or move your version of Default.htm to a different folder and then copy the file Default.htm from the Samples folder. This approach will make our version of Default.htm your Web server's home page.

Testing Peer Web Services Installation

Using the steps here, you can test the installation of PWS on a workstation or server. To test a PWS connected to the Internet, do the following:

1. Ensure that your Web server has HTML files in the Wwwroot folder.
2. Start Internet Explorer on a computer that has an active connection to the Internet. This computer can be the computer you are testing, although using a different computer is recommended.
3. Type the URL for the home directory of your new Web server. The URL will be "http://" followed by the name of your Web server, followed by the path of the file you want to view. (Note the forward slash marks.) For example, if your server is registered in DNS as "www.company.com" and you want to view the file Homepage.htm in the root of the home directory, in the Address box you would type **http://www.company.com/homepage.htm**, then press the Enter key. The home page should appear on the screen.

To test PWS connected to your Intranet, do the following:

1. Ensure that your computer has an active network connection and that the WINS server service (or other name resolution method) is functioning.
2. Start Internet Explorer.
3. Type in the Uniform Resource Locator (URL) for the home directory of your new server. The URL will be "http://" followed by the Windows Networking name of your server, followed by the path of the file you want to view. (Note the forward slash marks.) For example, if your Web server is registered with the WINS server as "Admin1" and you want to view the file Homepage.htm in the root of the home directory, in the Address box you would type **http://admin1/homepage.htm**, then press the Enter key. The home page should appear on the screen.

SECURING YOUR SITE

Security is important even for a personal Web site that is accessed only by members of your workgroup. When you connect your computer to an intranet and make documents available to network users, your computer is accessible to individuals who might accidentally or deliberately damage files.

The Windows 2000 and Windows NT operating systems were designed to help you secure your system against intruders. Peer Web Services builds on the Windows 2000 security model and provides additional monitoring and security features. This section will help you effectively use Windows 2000 security and Peer Web Services security at your site.

How Peer Web Services Security Works

Peer Web Services is built on the Windows 2000 security model. Windows 2000 security helps you protect your computer and its resources by requiring assigned user accounts and passwords. You can control access to computer resources by limiting the user rights of these accounts. You can use the Windows 2000 New Technology File System (NTFS) to assign permissions to folders and files on your computer. You can control access to folders and files by preventing users from copying files to or from a folder or by preventing users from executing files in certain folders.

In addition to the Windows 2000 security features, you can set Read-only or Execute-only virtual directories by using Internet Service Manager. Peer Web Services supports the Secure Sockets Layer (SSL) protocol, which securely encrypts data transmissions between clients and servers.

When a computer running Peer Web Services receives a browser request for information, it determines whether the request is valid.

Controlling Anonymous Access via FTP Logins

On many Web servers, almost all WWW, FTP, and gopher access is anonymous; that is, the client request does not contain a username and password. This occurs in the following cases:

- An FTP client that logs on with the username "anonymous"
- A Web browser request that does not contain a username and password in the HTTP header (this is the default on new Web connections with most browsers)

Even though the user is not logged on with an individual username and password, you can still control and monitor anonymous access. Each Internet service maintains a Windows 2000 username and password that is used to process anonymous requests. When an anonymous request is received, the service "impersonates" the user configured as the "anonymous logon" user. The request succeeds if the anonymous logon user has permission to access the requested resource, as determined by the resource's Access Control List (ACL). If the anonymous logon user does not have permission, the request fails. You can configure the WWW service to respond to a failed anonymous request by requiring the user to provide a valid Windows 2000 username and password, a process called authentication.

CONFIGURING THE ANONYMOUS USER ACCOUNT

You can view and monitor the anonymous logon user account on the Service property sheets of Internet Service Manager (for the WWW, FTP, and gopher services). Each service running on the same workstation or server can use the same or different anonymous logon user accounts. Including the anonymous logon user account in file or folder ACLs enables you to precisely control the resources available to anonymous clients.

The anonymous logon user account must be a valid Windows 2000 user account on the computer providing the Web services, and the password must match the password for this user in that computer's user database. User accounts and passwords are configured in the Windows 2000 User Manager by setting User Rights in the Policies menu. The anonymous logon user account must have the Log on Locally user right.

The IUSR_*computername* account is automatically created (with a randomly generated password) on your computer during Peer Web Services set-up. For example, if the computer name is marketing1, then the anonymous access account name is IUSR_marketing1.

By default, all Web client requests use this account. In other words, Web clients are logged on to the computer by using the IUSR_*computername* account. The IUSR_*computername* account is permitted only to log on locally on the computer providing the Web services.

Note: The IUSR_*computername* account is also added to the group Guests. If you have changed the settings for the Guests group, those changes also apply to the IUSR_*computername* account. You should review the settings for the Guests group to ensure that they are appropriate for the IUSR_*computername* account.

For the WWW and FTP services, you can allow or prevent anonymous access (all gopher requests are anonymous). For each of the Web services (WWW, FTP, and gopher), you can change the user account used for anonymous requests and change the password for that account.

To allow anonymous access via FTP, follow these steps:

1. In Internet Service Manager, double-click the WWW service or the FTP service to display its property sheets, then click the Service tab.

2. For the WWW service, select the Allow Anonymous check box. For the FTP service, select the Allow Anonymous Connections check box.
3. Click OK.

To change the account or password used for anonymous access, follow these steps:

1. In Internet Service Manager, double-click the service to display its property sheets, then click the Service tab.
2. In the Anonymous Logon username box, type the new username.
3. The default user account is IUSR_*computername*, where *computername* is the name of your computer. This account is created automatically when you set up Peer Web Services.
4. In the Password box, type the new password. A randomly generated password is automatically created for the IUSR_*computername* account. Note: If you change the password for this account, you must also specify the new password for the account in User Manager.
5. Click OK.

CONTROLLING ACCESS BY USER OR GROUP

You can control access to your Web site by using the Windows 2000 User Manager to specify what certain users or groups of users are allowed to do on your computer. You can further control access by requiring Web client requests to provide a username and password that Peer Web Services confirms before completing the request.

Setting up User Accounts

Windows 2000 security helps you protect your computer and its resources by requiring assigned user accounts. Every operation on a computer running Windows 2000 identifies who is doing the operation. For example, the username and password that you use to log on to Windows 2000 identifies who you are and defines what you are authorized to do on that computer.

What a user is authorized to do on a computer is configured in User Manager by setting user rights in the Policies menu. User rights authorize a user to perform certain actions on the system, including the Log on Locally right, which is required for users to use Internet services if Basic authentication is being used.

If you are using Windows 2000 Challenge/Response Authentication, then the Access this computer from network right is required for users to use Internet services. By default, everyone has this right.

To increase security, follow these guidelines:

Do not give the IUSR_*computername* account, the Guests group, or the Everyone group any right other than the Log on Locally or the Access the computer from this network right.

Make sure that all user accounts on the system, especially those with administrative rights, have difficult-to-guess passwords. In particular, select a good administrator password (a long, mixed-case, alphanumeric password is best) and set the appropriate account policies. Passwords can be set by using User Manager or by typing at the system logon prompt.

Make sure that you specify how quickly account passwords expire (which forces users to regularly change passwords), and set other policies such as how many bad logon attempts will be tolerated before locking a user out. Use these policies to prevent exhaustive or random password attacks, especially on accounts with administrative access. You can set these policies by using User Manager.

Limit the membership of the Administrator group to trusted individuals.

If you use the predefined Windows 2000 user accounts INTERACTIVE and NETWORK for access control, make sure files in your Web site are accessible to these user accounts. In order for a file to be accessed by anonymous client requests or client requests using Basic authentication, the requested file must be accessible by the INTERACTIVE user. In order for a file to be accessible by a client request that uses Windows 2000 Challenge/Response authentication protocol, the file must be accessible by the NETWORK user.

Requiring a User Name and Password for Access

Most Web sites restrict access to authenticated clients; that is, Web clients that supply a valid Windows 2000 username and password. When you use authentication, no access is permitted unless a valid username and password are supplied. Password authentication is useful if you want only authorized individuals to access your Web site or specific portions controlled by NTFS. You can have both anonymous logon access and authenticated access enabled at the same time. The WWW service provides two forms of authentication: basic and Windows 2000 Challenge/Response (sometimes referred to as "NTLM").

Basic authentication does not encrypt transmissions between the client and server. Because Basic authentication sends the client's Windows 2000 username and password in essentially unencrypted form over the networks, intruders could easily learn usernames and passwords.

Windows 2000 Challenge/Response authentication, currently supported only by Microsoft Internet Explorer version 2.0 or later, protects the password, providing secure logon over the network. In Windows 2000 Challenge/Response authentication, the user account obtained from the client is that with which the user is logged on to the client computer. Because this account, including its Windows 2000 domain, must be a valid account on the computer running Peer Web Services, Windows 2000 Challenge/Response authentication is very useful in an intranet environment, where the client and server computers are in the same, or trusted, domains. Because of the increased security, Microsoft recommends using the Windows 2000 Challenge/Response method of password authentication whenever possible.

You have both basic and Windows 2000 Challenge/Response authentication enabled by default. If the browser supports Windows 2000 Challenge/Response, it uses that authentication method. Otherwise, it uses basic authentication. Windows 2000 Challenge/Response authentication is currently supported only by Internet Explorer 2.0 or later.

You can require client authentication for all FTP service requests or only for anonymous requests that fail. The FTP service supports only basic authentication; therefore, your site is more secure if you allow anonymous connections. Your site is most secure if you allow only anonymous FTP connections.

To enable authentication for the WWW Service, follow these steps:

1. In Internet Service Manager, double-click the WWW service to display its property sheets, then click the Service tab.
2. Select Basic (Clear Text), Windows 2000 Challenge/Response, or both.
3. Click OK.

To enable authentication for the FTP service, follow these steps:

1. In Internet Service Manager, double-click the FTP service to display its property sheets, then click the Service tab.
2. To enable authentication for failed anonymous connections, clear (delete) the Allow only anonymous connections check box.
3. To require all client requests to be authenticated, clear the Allow Anonymous Connections check box.

UNDERSTANDING HOW ANONYMOUS LOGONS AND CLIENT AUTHENTICATION INTERACT

You can enable both anonymous connections and client authentication for the WWW service and for the FTP service. This section explains how a PWS Web server responds to these access methods when both are enabled. Note that if client authentication is disallowed and anonymous connections are allowed, a client request that contains a username and password is processed as an anonymous connection, and the server ignores the username and password.

Authentication in WWW Service When the WWW service receives a client request that contains credentials (a username and password), the "anonymous logon" user account is not used in processing the request. Instead, the username and password received by the client are used by the service. If the service is not granted permission to access the requested resource while using the specified username and password, the request fails, and an error notification is returned to the client.

When an anonymous request fails because the "anonymous logon" user account does not have permission to access the desired resource, the response to the client indicates which authentication schemes the WWW service supports. If the response indicates to the client that the service is configured to support HTTP basic authentication, most Web browsers will display a username and password dialog box and reissue the anonymous request as a request with credentials, including the username and password entered by the user.

If a Web browser supports Windows 2000 Challenge/Response authentication protocol, and the WWW service is configured to support this protocol, an anonymous WWW request that fails due to inadequate permissions will result in automatic use of the Windows 2000 Challenge/Response authentication protocol. The browser will then send a username and encrypted password from the client to the service. The client request is reprocessed, using the client's user information.

If the WWW service is configured to support both basic and Windows 2000 Challenge/Response authentication, the Web server returns both authentication methods in a header to the Web browser. The Web browser then chooses which authentication method to use. Because the Windows 2000 Challenge/Response protocol is listed first in the header, a browser that supports the Windows 2000 Challenge/Response protocol will use it. A browser that does not support the Windows 2000 Challenge/Response protocol will use basic authentication. Currently, Windows 2000 Challenge/Response authentication is supported only by Internet Explorer 2.0 or later.

FTP Service

When the FTP service receives a client request that contains credentials (a username and password), the "anonymous logon" user account is not used in processing the request. Instead, the username and password received by the client are used by the service. If the service is not granted permission to access the requested resource while using the specified username and password, the request fails, and an error notification is returned to the client.

When an anonymous request fails because the "anonymous logon" user account does not have permission to access the desired resource, the server responds with an error message. Most Web browsers will display a username and password dialog box, and they will reissue the anonymous request as a request with credentials, including the username and password entered by the user. Because the FTP service (and WWW Basic authentication) sends usernames and passwords unencrypted over the network, intruders could use protocol analyzers to read the usernames and passwords.

CREATING CUSTOMIZED AUTHENTICATION SCHEMES

If you need a WWW request authentication scheme not supported by the service directly, obtain a copy of the Win32 Software Development Kit (SDK), and read the ISAPI Filters specification on how to develop user-written ISAPI Filter dynamic-link libraries (DLLs) that handle request authentication. The Win32 SDK is available through the Microsoft Developer Network. For more information, visit the Microsoft home page (http://www.microsoft.com).

Setting Folder and File Permissions

Every access to a resource, such as a file, an HTML page, or an Internet Server API (ISAPI) application, is done by the services on behalf of a Windows 2000 user. The service uses that user's username and password in the attempt to read or execute the resource for the client. You can control access to files and folders in two ways: by setting access permissions in the Windows 2000 New Technology File System (NTFS) or by setting access permissions in the Internet Service Manager. It's important to note that File Allocation Table (FAT) file system partitions do not support access control. However, a FAT partition may be converted to NTFS by using the convert utility.

Setting NTFS Permissions

It is highly recommended that you place your data files on an NTFS partition. NTFS provides security and access control for your data files. You can limit access to portions of your file system for specific users and services by using NTFS. In particular, it is a good idea to apply Access Control Lists (ACLs) to your data files for any Internet publishing service.

ACLs grant or deny access to the associated file or folder by specific Windows 2000 user accounts, or groups of users. When an Internet service attempts to read or execute a file on behalf of a client request, the user account offered by the service must have permission, as determined by the ACL associated with the file, to read or execute the file, as appropriate. If the user account does not have permission to access the file, the request fails, and a response is returned, informing the client that access has been denied.

Using the Windows 2000 Explorer you can configure file and folder ACLs. The NTFS file system gives you very fine control on files by specifying users and groups that are permitted access and what type of access they may have for specific files and directories. For example, some users may have Read-only access, while others may have Read, Change, and Write access. You should ensure that the IUSR_*computername* or authenticated accounts are granted or denied appropriate access to specific resources.

You should note that the group "Everyone" contains all users and groups, including the IUSR_*computername* account and the Guests group. By default the group everyone has full control of all files created on an NTFS drive. If there are conflicts between your NTFS settings and Microsoft Peer Web Services settings, the strictest settings will be used.

You should review the security settings for all folders in your Web site and adjust them appropriately. Generally you should use the settings in Table 11.1.

To secure your files on an NTFS drive, follow these steps:

1. Put your files on your NTFS drive and add them to your Web site by using the Directories property sheet in Internet Service Manager.
2. In Windows 2000 Explorer, right-click the folder (directory) you want to secure (select your site root to secure the entire site), and choose Properties.
3. In the Properties dialog box, choose the Security tab.
4. In the Security dialog box, choose Permissions.
5. In the Directory Permissions dialog box, click Add to add users and groups.
6. In the Add Users and Groups dialog box, add the users that should have access.
7. Click OK.
8. In the Directory Permissions dialog box, select the users and groups that should have permissions.
9. From the Type of Access list box, choose the permission level you want for the selected user or group.
10. Click OK.

TABLE 11.1   Recommended Security Settings

Directory Type	Suggested NTFS Access
content	Read access
programs	Read and Execute access
databases	Read and Write access

DEFINING WWW DIRECTORY ACCESS

When creating a Web publishing directory (folder) in Internet Service Manager, you can set access permissions for the defined home directory or virtual directory, and for all of the folders in it. These permissions are those provided by the WWW service and are in addition to any provided by the NTFS file system. The permissions are as follows:

Read. Read permission enables Web clients to read or download files stored in a home directory or a virtual directory. If a client sends a request for a file that is in a directory without Read permission, the Web server returns an error. Generally, you should give directories containing information to publish (HTML files, for example) Read permission. You should disable Read permission for directories containing Common Gateway Interface (CGI) applications and Internet Server Application Program Interface (ISAPI) DLLs to prevent clients from downloading the application files.

As an administrator you have most likely encountered CGI in the context of the intranet and Internet sites you've been working on. Just as a quick refresher, CGI stands for Common Gateway Interface, and is a specification for transferring information between a Web server and a CGI program. A CPI program is any program designed to accept and return data that conforms to the CGI specification. CGI programs, or scripts, can act on a database, perform functions on the server, and even generate pages on a website. CGI scripts are very common today and are used for the most common functions found on websites. Scripts are typically written in Perl or C++, Java or Visual Basic.

Execute. Execute permission enables a Web client to run programs and scripts stored in a home directory or a virtual directory. If a client sends a request to run a program or a script in a folder that does not have Execute permission, the Web server returns an error. For security purposes, do not give content folders Execute permission.

A client request can invoke a CGI application or an Internet Server Application Program Interface (ISAPI) application in one of two ways:

The file name of the CGI executable or the ISAPI DLL can be specified in the request (URL). An example URL would be:

http://inetsrvr.microsoft.com/scripts/httpodbc.dll/scripts/pubs.idc?lname=Smith

For this request to be valid, the file Httpodbc.dll must be stored somewhere in the Web "publishing tree" (the directory structure that contains your content files; in this example, in the Scripts folder), and the folder it is stored in must have the Execute permission selected. This way the administrator can permit applications (CGI or ISAPI) to be run from a small number of carefully monitored directories.

The other way to configure CGI and ISAPI applications is to use the Web File Extension Mapping feature, which allows your executables and DLLs to be stored somewhere other than the Web publishing tree. An example URL would be:

http://inetsrvr.microsoft.com/scripts/pubs.idc?lname=Smith

To set access permissions for a directory, follow these steps:

1. In Internet Service Manager, double-click the WWW service to display its property sheets, then click the Directories tab.
2. Select the folder for which you want to set permissions.
3. Click Edit Properties.
4. To allow Web clients to read and download the contents of a folder, select the Read check box.
5. To allow Web clients to run programs and scripts in a folder, select the Execute check box.
6. Click OK, then click OK again.

STEPS FOR MANAGING OTHER NETWORK SERVICES

As you get PWS up and running on your workstations and servers, be sure to keep in mind these following points about other services running on your systems:

Run Only the Services That You Need

The fewer services you are running on your system, the less likely a mistake will be made in administration that could be exploited. Use the Services application in the Control Panel to disable any services not absolutely necessary on your Internet server.

Unbind Unnecessary Services from Your Internet Adapter Cards

Use the Bindings feature in the Network application in the Control Panel to unbind any unnecessary services from any network adapter cards connected to the Internet. For example, you might use the Server service to copy new images and documents from computers in your internal network, but you might not want remote users to have direct access to the Server service from the Internet.

If you need to use the Server service on your private network, disable the Server service binding to any network adapter cards connected to the Internet. You can use the Windows 2000 Server service over the Internet; however, you should fully understand the security implications and comply with Windows 2000 Server Licensing requirements issues.

When you are using the Windows 2000 Server service you are using Microsoft networking (the server message block [SMB] protocol rather than the HTTP protocol), and all Windows 2000 Server Licensing requirements still apply. HTTP connections do not apply to Windows 2000 Server licensing requirements.

Check Permissions on Network Shares

If you *are* running the Server service on your Internet adapter cards, be sure to double-check the permissions set on the shares you have created on the system. You should also double-check the permissions set on the files contained in the shares' folders to ensure that you have set them correctly.

Do Not Enable Directory Browsing

Unless it is part of your strategy, you should not enable directory browsing on the Directories property sheet. Directory browsing potentially exposes the entire Web publishing file structure; if it is not configured correctly, you run the risk of exposing program files or other files to unauthorized access. If a default page (Default.htm) is not present and directory browsing is enabled, the WWW service will return a Web page containing a listing of files in the specified directory. It is always advisable to have a Default.htm page in any directory that you do not want to be browsed.

SECURING DATA TRANSMISSIONS WITH SECURE SOCKETS LAYER (SSL)

Previous sections of this chapter have dealt with securing your computer from unauthorized access. This section discusses protocols that use cryptography to secure data transmissions to and from your computer, workstation, or servcerf. Peer Web Services offers a protocol for providing data security layered between its service protocols (HTTP) and TCP/IP. This security protocol, called Secure Sockets Layer (SSL), provides data encryption, server authentication, and message integrity for a TCP/IP connection.

SSL is a protocol submitted to the W3C working group on security for consideration as a standard security approach for Web browsers and servers on the Internet. SSL provides a security "handshake" that is used to initiate the TCP/IP connection. This handshake results in the client and server agreeing on the level of security that they will use and fulfills any authentication requirements for the connection. Thereafter, SSL's only role is to encrypt and decrypt the byte stream of the application protocol being used (for example, HTTP). An SSL-enabled server can send and receive private communication across the Internet to SSL-enabled clients (browsers), such as Microsoft Internet Explorer version 2.0 or later.

SSL-encrypted transmissions are slower than unencrypted transmissions. To avoid reducing performance for your entire site, consider using SSL only for virtual folders that deal with highly sensitive information such as a form submission containing credit card information.

Enabling SSL security on a Web server requires the following steps:

1. Generate a key pair file and a request file.
2. Request a certificate from a certification authority.
3. Install the certificate on your server.
4. Activate SSL security on a WWW service folder.

Generating a Key Pair

As part of the process of enabling Secure Sockets Layer (SSL) security on your Web server, you need to generate a key pair and then acquire an SSL certificate. The new Key Manager application (installed with the product and located in the Internet Server program group) simplifies this procedure.

To generate a key pair, follow these steps:

1. In the Microsoft Peer Web Services submenu, click Key Manager, or click the Key Manager icon on the Internet Service Manager toolbar.
2. From the Key menu, click Create New Key.
3. In the Create New Key and Certificate Request dialog box, fill in the requested information, as follows:

Key Name
Assign a name to the key you are creating.

Password
Specify a password to encrypt the private key.

Bits
The size of each key you create is preset to 512 bits.

Organization
Preferably International Organization for Standardization (ISO)-registered, top-level organization or company name.

Organizational Unit
Your department within your company, such as Marketing.

Common Name
The domain name of the server, for example, www. *mycompany* .com.

Country
Two-letter ISO Country designation, for example, US, FR, AU, UK, and so on.

State/Province
For example, Washington, Alberta, California, and so on.

Locality
The city where your company is located, such as Redmond or Toronto.

Request File
Type the name of the request file that will be created.
After filling out the form, click OK.

4. When prompted, retype the password you typed in the form, and click OK.
5. An icon appears as the key is being created. When the key has been created, a screen appears giving you information about new keys and how to obtain a certificate.
6. After reading the New Key Information screen, click OK.
7. To save the new key, from the Servers menu choose Commit Changes Now.
8. When asked if you want to commit all changes now, click OK.

Your key will appear in the Key Manager window under the name of the computer for which you created the key. By default, a key is generated on your local computer.

Generating a Key Pair on Another Computer

You can set up a key pair on another computer and install the certificate there. From the Servers menu, click Connect to Server, and follow the previous procedure under "Generating a Key Pair." Once you have generated a key pair, you must get a certificate and then install that certificate with the key pair.

Acquiring a Certificate

The key generated by Key Manager is not valid for use on the Internet until you obtain a valid certificate for it from a Certificate Authority, such as VeriSign. Send the certificate request file to the Certificate Authority to obtain a valid certificate. Until you do so, the key will exist on its host computer, but it cannot be used. For instructions on acquiring a VeriSign certificate refer to VeriSign's Web site at http://www.verisign.com/microsoft/.

Installing a Certificate with a Key Pair

After you complete your certificate request, you will receive a signed certificate from the Certificate Authority (consult your Certificate Authority for complete details). The key manager program will create a file similar to the following example:

```
----BEGIN CERTIFICATE----
JIEBSDSCEXoCHQEwLQMJSoZILvoNVQECSQAwcSETMRkOA
MUTBhMuVrMmIoAnBdNVBAoTF1JTQSBEYXRhIFNlY3VyaXR
5LCBJbmMuMRwwGgYDVQQLExNQZXJzb25hIENlcnRpZmljY
XRlMSQwIgYDVQQDExtPcGVuIE1hcmtldCBUZXN0IFNlcn
ZlciAxMTAwHhcNOTUwNzE5MjAyNzMwWhcNOTYwNTE0MjAy
OTEwWjBzMQswCQYDVQQGEwJVUzEgMB4GA1UEChMXUlN
BIERhdGEgU2VjdXJpdHksIEluYy4xHDAaBgNVBAsTE1Blc
nNvbmEgQ2VydGlmaWNhdGUxJDAiBgNVBAMTG09wZW4gTWF
ya2V0IFRlc3QgU2VydmVyIDExMDBcMA0GCSqGSIb3DQE
BAQUAA0sAMEgCQQDU/7lrgR6vkVNX40BAq1poGdSmGkD1i
N3sEPfSTGxNJXY58XH3JoZ4nrF7mIfvpghNi1taYimvhbB
PNqYe4yLPAgMBAAEwDQYJKoZIhvcNAQECBQADQQBqyCpws
9EaAjKKAefuNP+z+8NY8khckgyHN2LLpfhv+iP8m+bF66H
NDUlFz8ZrVOu3WQapgLPV90kIskNKXX3a
----END CERTIFICATE----
```

Installing a certificate requires following these steps:

1. In the Internet Server program group, click Key Manager.
2. In the Key Manager window, select the key pair that matches your signed certificate. If you had backed up the key pair file, you have to load it first. For instructions, see "Loading a Key Pair File" earlier in this chapter.
3. From the Key menu, choose Install Key Certificate.
4. Select the Certificate file from the list (Certif.txt, for example), and click Open.
5. When prompted, type the password that you used in creating the key pair. The key and certificate are combined and stored in the registry of the server.
6. From the Servers menu, choose Commit Changes Now.
7. When asked if you want to commit all changes now, click OK.

You can back up a key and certificate combination by following the procedure under "Backing Up Keys" earlier in this chapter.

Configuring a Directory to Require SSL

Once you have applied the certificate, you must enable the SSL feature from Internet Service Manager. SSL can be required on any virtual folder available in your Web site and is configured on the Directories property sheet.

To require SSL, follow these steps:

1. In Internet Service Manager, double-click the WWW service to display its property sheets, then click the Directories tab.
2. Select the folder that requires SSL security, then click Edit Properties.
3. Select the Require secure SSL channel option, and then click OK.

Backing Up Keys

With Key Manager you download key information from the registry into a file on your hard disk and then copy this file or move it to a floppy disk or tape for safekeeping. You can back up a private key pair file or a key with an installed certificate.

To back up a key or a private key pair file, follow these steps:

1. From the Key menu in Key Manager, choose Export Key and then Backup File.
2. After reading the warning about downloading sensitive information to your hard disk, click OK.
3. Type the key name in the File Name box, and click Save.

The file is given a .req file-name extension and is saved to your hard disk drive. You can then copy it or move it to a floppy disk or magnetic tape.

Loading Backed-Up Keys

You can load backed-up keys or private key pair files into Key Manager with the Import command.

To load a backed-up key, follow these steps:

1. From the Key menu in Key Manager, choose Import Key and then Backup File.
2. Select the file name from the list, and click Open.

Loading a Key Created with Keygen.exe and Setkey.exe

If you have generated a key pair from the command line with the Keygen.exe command and installed a certificate with Setkey.exe, you can load them into Key Manager with the Import command.

To load a key, follow these steps:

1. From the Key menu in Key Manager, choose Import Key and then KeySet.
2. In the Private Key Pair File box, type the file name for the key pair or click Browse and select the file.
3. In the Certificate File box, type the file name for the certificate or click Browse and select the file.
4. Click OK.
5. Type the password for the private key in the Private Key Password box, and click OK.

Suggestions for SSL Configuration and Operation

It is recommended that you use separate content directories for secure and public content (for example, C:\InetPub\Wwwroot\Secure-Content and C:\InetPub\Wwwroot\Public-Content).

Save your key file in a safe place in case you need it in the future. It is a good idea to store your key file on a floppy disk and remove it from the local system after completing all setup steps. Do not forget the password you assigned to the key file.

CHAPTER SUMMARY

Developed specifically for the needs of organizations to create intranet and low-traffic Internet sites, Peer Web Services, which includes Internet Information Services, is available for use on Windows 2000 Professional & Server. The process of installing Peer Web Services is very intuitive; therefore its coverage in this chapter is exactly what you need to know. Focusing on the security aspects of logins, authentication, and steps you can take to ensure the security of your site is the chapter's purpose. Using Internet Information Services on the Windows 2000 platform provides a great publishing tool for intranet development, and one that is actively used by many companies today in the development of their intranets and Virtual Private Networks.

12 | Future Direction of Windows 2000 Networking

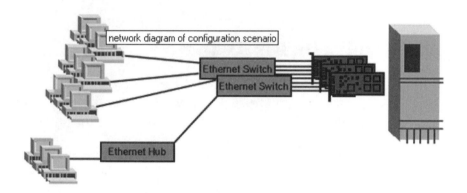

network diagram of configuration scenario

Ethernet Switch
Ethernet Switch
Ethernet Hub

Consider the state of networking even five years ago and its state today. You may be reading this after the year 2000 and see an entirely different world than was imagined in the first "years of the network," which many magazines proclaimed at the beginning of each year throughout the 1990s. It wasn't so much the proclamations but the technology advances and user need for quickly and efficiently sharing information that drove the widespread migration of networking into the Internet and extranets, and the rapid growth of virtual private networks. These developments made networking become so widespread that even the language of this technology has entered the mainstream of how people communicate.

How does Windows 2000 continually make a contribution to this growth? What are the market dynamics affecting the integration of TCP/IP and other protocols into operating systems? How do these fundamental design decisions surrounding a network and its connectivity affect operating system development, and in turn, how you use it? The answers to these questions and more are included in this chapter.

If you're an administrator you'll find this a useful chapter for getting others up to speed with the current state of networking technology and how it applies to Windows 2000- and NT-based networking. This chapter offers a glimpse into the future of how Windows 2000 will evolve to reflect the ongoing needs of users. If you're an intermediate to advanced user of Windows 2000 Professional or networks in general, you'll find this chapter useful as a tutorial on the latest changes in networking, thanks to both market and technology dynamics.

WHAT MAKES UP A NETWORK?

A local area network is used most of the time to share print and file resources, with Microsoft, Novell, Sun Microsystems, and other companies offering operating systems competing on the relative performance of handling these tasks. Local area networks are also extensively used for sharing applications through a client/server configuration. File and print services and application software are available either on a dedicated or a peer-to-peer network. A dedicated network is most often found in a client/server network, where one or more servers are defined as the servers that perform no other task but share their resources. Depending on the size of the network, potentially several types of servers can be included in the network configuration. A server that stores applications, data files, and other reference data is called a file server. A server that hosts the print resources for a network, sharing its printers with other computers in a network, is called a print server. A server that shares a large database is called a database server. Servers are designed to fulfill multiple roles, and they have features that make them fault tolerant and capable of handling multiple tasks simultaneously in a network environment.

A peer-to-peer network is one where the role of each system on the network fulfills the role of a server. Any computer can act as a server on the network. This approach to organizing a network originated with serial-based protocols and has continued into the Novell and Microsoft networking operating systems from the 1980s. The strengths of this approach include ease of integration, security (as protocols that support peer-to-peer are in many cases not routable), and availability of resources throughout the network being transparent between users. The weaknesses of this approach include the speed of the network, the lack of routability, and the limited nature of this approach in terms of the performance demands that can be made on a workstation that is also being used, as a server. Once the demands for applications and file and print services expand beyond the scope of workstations, servers are typically integrated into a network.

In the majority of networks in use today, there is a combination of dedicated and peer-to-peer networks. As more and more users are added to the network, new applications rely on dedicated network architecture. An example of an application that relies on this approach is Oracle's Manufacturing Modules, which run on a centralized server, where client workstations use the applications on the server. Data sets are sometimes housed on the server to enable other members of a design or engineering team to use them. Client/server networks are predominantly used for sharing larger data sets, where a centralized server is considered essential for the functioning of the network.

UNDERSTANDING HOW NETWORKS ARE ORGANIZED

Networks are really a group of systems gathered together by cables or wireless communications such as infrared or RF (radio frequency) communications. Each of the systems on a network is called a node, and it communicates with other systems using the physical components that make up the network itself. These components are in addition to the systems or nodes themselves, and they create the infrastructure needed for the network. Components in this category include bridges, routers, repeaters, network interface cards, and the other hardware components that make up a network. Taken together, these elements are considered the physical network. The following sections provide an overview of the key components in a physical network.

Introducing the Network Interface Card (NIC)

Network interface cards are essential components in every network; you'll find that every workstation attached to a network will have a network interface card, or NIC, installed. The NIC is the adapter card that is physically very similar in appearance to a video adapter, disk controller, or other adapter cards in your workstation, although obviously much more advanced in its function. Providing the essential link between the network and your workstation, the NIC is the intermediary that combines the software device drivers that make it recognizable by an operating system with the protocols being transmitted by a network. Increasingly, the protocol being transmitted is TCP/IP, so there has been a correspondingly high level of growth in the sales of NIC cards that are TCP/IP-compatible. As you would suspect, the NIC card acts as the information intermediary or translator, receiving, processing, and passing data to the network and the operating system. Its primary function is to work with data packets, or collections of data encapsulated for transmission. It's also important to note that the widespread adoption of the Remote Access Services (RAS) has made it possible for thousands of workstation and laptop users to gain access to the Internet, virtual private networks including intranets, and Internet company networks all through the use of modems. This doesn't alleviate the high growth of NICs, by the way, as they have much higher performance than their modem-based counterparts. Figure 12.1 shows an example of a NIC.

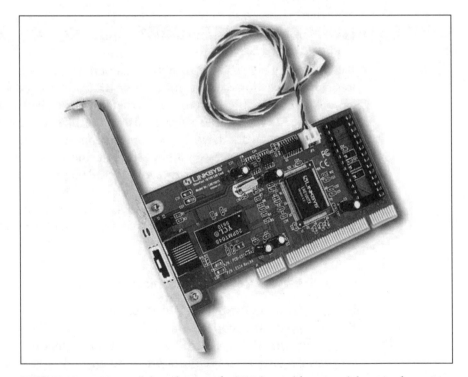

FIGURE 12.1 Network interface cards (NICs) provide essential network connectivity.

Each NIC has a specific and very unique network address assigned to it during production. This unique address makes it possible to enable network packets to be directed specifically to the NIC, much the way you would direct a letter to a specific business or home based on its street address or even fax number. The NIC has circuitry that monitors the packets coming from the network, and if the packet is meant for the workstation, the NIC processes the packet and transmits it up the OSI model (more on this later) to the operating system. If the packet is meant for another node, the NIC simply passes it on. The NIC doesn't perform this function by itself, however. The layers of the OSI mode, embedded in the operating system, serve as the foundation of a more global operating system, enabling the NIC to perform its tasks seamlessly between the network and the operating system.

Being at the location in the network connection where the network communicates with the workstation, the NIC plays a central role in how efficient the network performance is on a workstation and the performance of the network overall. Because data can be transmitted over the network faster than the NIC can process it, packets must first be buffered until the NIC can process them. A NIC that has slow performance can eventually slow an entire network down, as the packets are queued up, waiting to be read by the slower NIC. Conversely, the latest generation of NICs can dynamically flex between 10 and 100 Mbits/second, making it possible to accommodate both dominant speed ranges of networks available today.

Like other types of cards and adapters, NICs require an Interrupt Request Line (IRQ) to enable communication with the CPU. Except on workstations that enable IRQ sharing, this IRQ assignment has to be unique to the NIC. In many respects, IRQ assignments for NICs are synonymous with those for other card-based peripherals in that both need a specific identity for the workstation to use them reliably. By default, many NICs assign themselves IRQ2, while others default to IRQ 10. Chances are, if a NIC is being used in a Novell NetWare and Windows 2000 environment, either the IRQ is set to 10 or you as the system administrator have chosen to set the value to 10. All NICs during their setup routines do go out and sense the other IRQ values already in use, then recommend one that is not in conflict with any other IRQs on your workstation. Some older NICs require that you set jumpers or DIP switches on the NIC to set the IRQ, but others enable the operating system to automatically configure the IRQ setting. This latter type of NIC is often said to be software-configurable. Most 10/100 NICs have software configuration options associated with them.

NIC cards also rely on an I/O base address, which is also called the I/O port address, or base address. The NIC's base address defines the memory address, and this memory addresses the server as a port through which the NIC and operating system communicate. Network adapters use a base address in the range of 0x200 to 0x300, although a NIC can use a different base address.

If you've already installed a NIC in Windows 2000 Professional, you can easily view and change the IRQ and base address settings for the NICs in your system. Windows 2000 Professional is unlike many other network operating systems in that it can accommodate two NIC cards running simultaneously, providing compatibility with two distinct network operating systems. Using the Properties page of the Network application, you can see and edit the IRQ values associated with the NIC cards installed in your workstation. Chapter 7, "Configuring TCP/IP in Windows 2000 Professional," has specific hands-on tutorials on how to change the IRQ values for the NICs installed in your workstation.

EXPLORING LAN MEDIA

A local area network requires a mechanism for getting connections established between workstations and making sure packets get passed to their destinations. The term *media* actually refers to the type of connection between nodes on a network. Typically system administrators and support staff refer to cable as synonymous with media, with the exception of networks created using infrared and radio-based communications. There are several types of very common media in use today, with the most common being twisted pair, coaxial cable (also called coax), fiber-optic cable, and wireless.

Physical media such as twisted pair, coax, and fiber-optic cable are often referred to as *bounded media*. Wireless connections such as infrared and radio frequency are referred to as *unbounded media*.

INTRODUCING THE OSI MODEL

Understanding the Role of Client/Server Technology

The number of companies that continue to have stand-alone personal computers completing isolated tasks surprise you or, at the other end of the spectrum, that are focused on their mainframe systems. These two groups of users share a common thread: They both need to have easier access to rapidly changing data that often is difficult to deliver to the desktop.

TCP/IP Architecture Explained

The architecture of a computer system simply refers to how it is designed to work in conjunction with other parts of the system efficiently. Windows 2000 Server has been designed to provide a strong foundation so that protocols already supported in each of these operating systems can be in turn forwarded to the next generation of users. Windows 2000 Server specifically supports NetBEUI, Novell IPX/SPX, and the full suite of the TCP/IP command set. A strong differentiator that you can take advantage of when using NT Server is the support for the Dynamic Host Configuration Protocol, or DHCP for short. What is DHCP? An innovative approach to assigning IP addresses, the DHCP protocol works very much like a library that checks out books. Instead of issuing books, the NT Server actually dispenses addresses. The DHCP protocol is very popular with many electronic commerce vendors that serve multiple industries. Many Internet Service Providers (ISPs) also rely on this protocol for handling the task of getting IP addresses allocated to each customer. Specifics on how to configure DHCP for use in your organization are included later in this chapter.

TCP/IP is actually a suite of protocols that perform specific network communications tasks. You, as the implementer of electronic commerce solutions in your organization, have the flexibility of selecting from the suite of commands available in TCP/IP to accomplish networking goals. Network administrators specializing in the TCP/IP protocol can select the applicable commands from the TCP/IP protocol suite to meet specific needs. Throughout this chapter you'll see how to do the hands-on activities of creating a TCP/IP network.

Microsoft-based networks can involve multiple protocol suites, so developing a plan that clarifies the relations between the various protocols used in a specific situation and what needs they are addressing is paramount. The conventional approach to comparing network protocols is to use the OSI reference model as the backdrop or conceptual framework, which is an excellent approach to handling the layering of network protocols. The OSI Reference Model is an excellent place to begin learning how networks are created. The OSI model continues to survive despite the Internet's consolidation ofthe functionality of its layers; the Internet is a great tool for explaining how networks interact and function.

How TCP/IP Works

TCP/IP originally started out as a network protocol to support research between universities and government agencies, one that would enable accurate and secure transmission of data between locations. Integral to the development of the TCP/IP protocol was the evolution of *packet-switching* technology.

Research was begun in 1969 with the primary design goal to provide a common protocol for data transmission, ensuring commonality between vendors providing products during the initial procurement process.

The second design goal for TCP/IP is the creation of a networking protocol that ensures interoperability between hardware and software components. This second objective has continually generated competition in all arenas of the TCP/IP marketplace because interoperability levels the playing field between competitors. Having connectivity as a baseline forces the hardware vendors to compete on performance and ever more closely aligning themselves with the needs of customers integrating TCP/IP into their businesses, schools, universities, and governments.

Third, the need was for a network that has could handle a multitude of transactions and information requests transparently between thousands of users concurrently. The network protocol to orchestrate requests for what was first research information and is today used for handling transactions was originally developed to support a robust series of data transactions used in Department of Defense requests for information between network clients.

The fourth design objective for TCP/IP was to develop a protocol that had an ease of reconfiguration, and specifically a protocol that was able to assign addresses to each system on a network. Because the initial customers were with the Department of Defense and had a growing number of systems needing configuration, the design objective to dynamically assign IP addresses to each system was considered essential. With the need for security within Department of Defense installations at the time, it was essential to have a network protocol that could easily be configured and deconfigured easily.

EXPLORING THE OSI MODEL

What is the OSI Model? How do the layers contribute to more accurate information being shared between computers? How does the OSI Model make it possible to plan entire network topologies? The answers to these questions are provided in this section, along with insights into how the OSI Model and its layers are constantly changing to reflect the needs of electronic commerce. This section of the chapter is based on the OSI Model and its technology; it also covers the implications of using this model in the context of a business model and your specific electronic commerce initiatives.

Characteristics of Layered Protocols

Just like bricks in a building, the layers of the OSI Model build on top of the other, with each layer relying on the previous one to provide additional intelligence to the data as it travels to the application layer. What are the implications of layered protocols for electronic commerce? Plenty. Let's take a look at several of them right now.

Electronic commerce is streamlining the flow of data up and down the layers of the OSI Model. This is particularly seen in the development of Internet applications that span the entire range of OSI layers. The implications for companies building Web sites is that vendors selected for software development, service providers, and even the hardware selected need to have upward compatibility and connectivity with the latest additions to the TCP/IP command set.

Driving the adoption of TCP/IP in the electronic marketplace is the security inherent in its structure and the development of Internet standards using the OSI Model. By the very nature of the OSI Model, the various levels are protected from one another, with the lower levels passing data to the upper levels, with each successive layer adding more value and information to the message.

Electronic commerce specifically and the Internet in general are flattening the OSI Model by making increased functionality available more efficiently than ever before. The HTTP protocol is now the standard for sharing data and handling transactions globally over the Internet. Due to the growth of electronic commerce, the OSI Model is being reinvented in real-world terms with each succeeding product generation that addresses electronic commerce.

The key issue of a layered protocol such as the OSI Model is that each layer adds value to the incoming and outgoing messages received over a network. Think of each of the layers in a networking model as being a component that adds intelligence and value to networking messages.

THE INTERNET MODEL EXPLAINED

Applying the Internet to the OSI Model develops a different picture than the one shown in the classical seven-layer model discussed throughout this chapter. What does the Internet do to the OSI Model? How does this affect your decisions on electronic commerce over the Internet? In simple terms, the Internet is shrinking the OSI Model into a frame of reference where the combination of HTTP, SET, and SST technologies has actually flattened the OSI Model. The impact of the Internet on the OSI Model has been to actually traverse it, making networking available to more customers than ever before.

What then are the implications as you set up your Web site? What does this mean for you? Simply put, the OSI Model gives you a great cross-reference for explaining your plans and accomplishments with regard to electronic commerce. The OSI Model is also useful in pointing out how networking has progressed over time. In the mid-1980s the OSI Model showed a more disjointed approach to networking, while today the OSI Model is maturing at a rapid rate. This translates into a customer base with little or no barriers to purchasing electronically over time.

EXPLORING THE OSI MODEL LAYERS

Apart from being able to profile the rapid growth of the Internet by illustrating how the various levels of the OSI model are consolidating, what does the OSI Model provide for the digitally focused entrepreneur? While the OSI protocols have been surpassed by a growing interest in the TCP/IP command set, the OSI Model continues to be a great tool for comparing protocol suites. It is also very useful for explaining system-level strategies to potential investors in your electronic commerce plans. Being able to have a solid comparison tool for explaining your connectivity plans to others is the biggest payoff for understanding the OSI Model. You'll also be able to understand the concepts others in the industry describe as they define how they have structured their Web sites and entire electronic commerce offerings.

Notice that each of the layers has a predefined task and actually adds value by interpreting the messages being sent between systems on a network. Incoming messages over a network filter upward through the OSI Model and are then interpreted and acted on by the computer user at the presentation level. Once a command, query, request, or action is completed at the top level of the OSI Model, the message then flows back down the OSI Model, with increased intelligence added with every layer traversed. As you can see, this layering approach makes it easier to traverse a given network topology and even troubleshoot networking issues.

Exploring the Physical Layer of the OSI Model

The physical layer is the lowest layer, and it is focused on two key tasks: sending and receiving bits over the network. Network interface cards (NICs) are predominantly found at this level of the OSI Model, as are peripheral devices associated with electronic commerce. Levels above this one in the OSI model are responsible for collecting bits into a single message. As you would suspect, this layer is very elementary in its representation of data, to the point of seeing data in the pattern of state transitions in the bits being received over the network.

A wide variety of media is used for communicating bits to the physical layer. These include electric cable, fiber optics, light waves, radio, even microwave transmissions. The medium used can vary, yet the logic behind the physical layer does not change. The upper layers are completely independent from the particular process used to deliver bits through the network medium.

An important distinction is that the OSI physical layer does not, strictly speaking, describe the media attached to it. The physical layer describes the bit patterns to be used, but it does not define the medium; it describes how data is encoded into media signals and the characteristics of the media attachment interface. In actual practice, many physical layer standards cover characteristics of the OSI physical layer as well as characteristics of the medium.

The DataLink Layer

The function of the data link layer is to provide system-to-system communication on a single local network. Many times you'll also hear the term node-to-node communication pertaining to the data link layer. A node is actually a computer system. This layer of the OSI model focuses on enabling communication between nodes on a network. To provide this functionality, the data link layer needs to provide an address mechanism that enables messages to be delivered to the correct nodes, and it must also translate messages from the upper layers into bits that the physical layer can transit.

What happens when the data link layer receives a message to transmit? It first formats the message into a *data frame.*

Individual sections of a data frame are called fields. The fields that make up a data frame vary by network protocol, but the ones listed here have a high level of commonality between all protocols.

Start Indicator. This field contains a specific bit pattern that indicates the start of a data frame.

Source Address. The address of the sending node is also included so that replies to messages can be addressed properly.

Destination Address. Each node is identified by an address. The data link layer of the sending node adds the destination address to the frame. The data link layer of the receiver looks at the destination address to identify messages that it should receive.

Control. In many cases, additional control information must be included. This is the task of the control bit, and it provides specific data depending on the network protocol being used.

Data. This field contains all the data that was forwarded to the data link layer from the protocol layers located in the upper segments of the OSI model.

Error Control. This field contains information that enables the receiving node to determine whether an error occurred during transmission. A common approach to ensuring accuracy is to use cyclic redundancy checksum or checking (CRC), which is a calculated value that summarizes all of the data in the frame. The sending node calculates a checksum and stores it in the frame. The receiving system recalculates the checksum, and if the receiver's calculated CRC matches the CRC value in the frame, it can be assumed that the frame submitted over the network was without error.

In the case of TCP/IP-based networks, the transmission of a data frame is straightforward. Each of the systems on the network checks the destination address at the beginning of the frame. If the destination address in the frame matches the node's address, the data link layer at the node receives the frame and forwards it up the protocol stack.

How the Network Layer Adds Value

In the world of electronic commerce, there is rarely a situation where your Web site will be seen only on a single network. Most probably the Web sites and revenue-generating applications you are creating will be accessed from several different networks. This is where the network layer becomes a valuable asset in your plans for electronic commerce. In the case of a series of distributed networks, called an internetwork, the network layer provides network addressing for each network with which it communicates. This does not apply to the Internet itself; rather in terms of speaking of the network layer and internetworking, the discussion revolves around sending and receiving a packet that has source and destination network addresses.

The network layer completes to get a data packet to the correct location on the network is called routing. You may have heard the term router; this is the device used for getting a data frame from one location to another on a network. There are two types of nodes on internetworks. These nodes are the targets for data frames originating from the originating nodes on a network.

End nodes use the network layer to add network address information to packets, but they do not perform routing. End nodes are sometimes called end systems or hosts. In the case of TCP/IP implementations the latter name is most typically used.

Routers use a special mechanism that checks the data frame that then sends them to the target destination. Routers are actually stand-alone systems in their own right, and they are called intermediate systems in the OSI Model; in the world of TCP/IP they are also called gateways.

The network layer relies on the data link layer to organize incoming bits into an assemblage that is usable. Because routers are network layer devices, they are used to forward packets between physically different networks. A router can join an Ethernet or Token Ring network. Routers also are often used to connect systems to local area networks, whether they are Ethernet, Token Ring, or ATM based.

Transport Layer

Almost every protocol defines a size for the packets being sent from one location to another on a network. Ethernet, for example, sets the size of the data field at 1500 bytes. Limiting the size of the packets is essential for the following two reasons:

Smaller frames tend to improve network efficiency when many devices share the network. This is comparable to partitioning messages into smaller segments so that they can be successfully sent over a network.

Smaller frames lead to less translation during the process of being sent from the network carrier to the destination system.

The transport layer is primarily focused on taking messages and in effect creating smaller packets for ease of transmission and the accuracy of the received messages. At the originating workstation each message is broken into a series of fragments, with the transport layer at the destination system responsible for re-creating the messages that have been packetized.

Session Layer

Controlling the dialog between two systems and making sure nodes on a network agree to exchange data through the network are the responsibilities of the session layer. This layer of the OSI Model is one of the most critical, with a series of communications or, in the case of the OSI Model, dialogs being completed. A dialog is the term used by LAN engineers to define the communication between two systems on a network. The Session Layer's mission is to enable the dialog between multiple systems. These dialogs can take place in three dialog modes, specifically listed here:

Simplex. This is a dialog between systems where one node transmits exclusively, while another receives exclusively.

Half duplex. Only one node may send at a given time, and the modes take turns transmitting.

Full duplex. Nodes may transmit and receive simultaneously. Full-duplex communication requires some form of flow control to ensure that neither device sends faster than the other device can receive.

Regardless of the type of communication being completed, each type of dialog consists of the following three phases. These phases or steps in a dialog within the session layer are defined here:

Creating a connection. This is the introductory phase where one node initiates and holds communication with another system. During this phase the two systems negotiate for rules of communication, including the types of communication rules or parameters to be used and the parameters for communication.

Data transfer. This is the second step in the process where the actual communication between each of the systems on the network occurs.

Connection release. When the communication between network nodes has been completed, the connection between systems is completed, and each system completes a series of steps to stop communication.

Looking at these three steps from the standpoint of commercially based transactions flowing through an OSI Model, you can see the clear need for efficiency on the connection establishment and connection release.

Presentation Layer

The key value-add of the presentation layer of the OSI Model is to gather the frames of data and create a cohesive grouping or set of data for use at the application layer. As will be explained in the next section, the application layer is the level of the OSI Model at which most electronic commerce users interact with their systems.

Let's explore the presentation layer in more depth. A good example is the network traffic for a Web site composed of several types of systems. In the case of UNIX-based workstations running Sun's Solaris operating system, some systems running Windows 2000 and NT, and a mainframe running a mainframe operating system, the presentation layer completes the role of gateway in this specific network topology. The presentation layer is then actually responsible for acting as the gateway in distributed and, often in the world of electronic commerce, heterogeneous environments.

Many companies involved with electronic commerce are using translation tools on various systems in their Web complexes (or groups of systems composing their Web site) to create a consistency of data formats to ensure the efficiency of transactions. The need for streamlining transactions in electronic commerce is becoming commonplace as the profitability for electronic initiatives is often measured in the number of inventory turns accomplished during a given period of time. To digress briefly, the role of the application layer in the OSI Model is really as a catalyst for changing the approach companies use to attack the issue of inventory turns in an electronic commerce model. Most often, companies use the External Data Representation (XDR) format, which is used with the Network File System (NFS).

Due to the increasing focus on inventory turns in electronic commerce business models and the forethought that goes into data transmission between systems before actual Web sites are even created, device drivers for the presentation layer are rarely created today. The effects of inventory efficiencies, data transportability, and the Internet itself with its own formats are all factors that provide the functionality of the presentation layer.

Application Layer

The application layer is the most prevalently used layer of the OSI Model, as applications' interfaces reside at this topmost position of the OSI Model. From an electronic commerce perspective, this is the layer of the OSI Model where most of programming of electronic tools is located. Specifically, tools for completing auctions on the Internet, setting up and maintaining electronic storefronts, and creating interactive purchasing sites that store the preferences and tastes of shoppers as they browse an online mall are all created at and reside in this layer of the OSI Model.

Let's take a look at the more common applications found at the application layer of the OSI Model, their implications on the future of Windows 2000, and their role in electronic commerce.

Electronic mail transport. Inherent in the application layer are protocols that enable electronic mail between networked systems.

Remote file access. There continues to be a dominant trend in operating systems design of enabling both applications within operating systems and applications themselves to support network connectivity. Inherent in seamless interoperability is the need for electronic commerce applications and sites to provide users with access to any location on a Web site, private Internet, or intranet they choose. Remote file access that is platform independent is critical for the success of an electronic commerce application or offering.

Remote job execution. Long the stronghold of the UNIX power user as a reason not to migrate to Windows 2000 and NT, remote job execution is supported in NT and other network-based operating systems from within the application layer. In the context of electronic commerce, remote job execution is not used very often, as many applications and their resulting transactions are meant to be transparent to the user.

Directories. Central to the ongoing debate over directories is the role Active Directory in Windows 2000 will have relative to the Novell Directory Services (NDS) loyal customer base. An integral part of the application layer, directories are used by many companies creating electronic commerce Web sites using both Novell for file and print services and Windows 2000 servers for their adeptness at applications server capabilities. The point of this ongoing debate is the dilemma facing the Novell NetWare-based companies and the opportunity to continue using Novell Directory Services (NDS) or migrate to Windows 2000's Active Directory. This decision is anything less than black-and-white; rather, many companies are comparing each of the directory architectures to see which is better suited for their specific needs. Which is the best for electronic commerce? There is no right answer; rather you need to ask yourself what is the best architecture for serving your customers.

WHAT'S NEW IN MICROSOFT NETWORK PROTOCOLS?

In moving into an entirely new market area with Windows 2000 and NT, Microsoft focused on differentiation through close alignment with customers' needs. Delivering an operating system robust enough to support multiple subsystems for multiple application types, have TCP/IP networking built in, and meet the design objectives Microsoft had for NT was a significant task. In studying the potential customers for NT, Microsoft realized that the entire arena of networking gave it a significant opportunity to differentiate its upcoming operating system from Novell NetWare, UNIX, and the Macintosh platform. Microsoft decided that as a basis for building ever-stronger differentiation into Windows 2000 and NT, a solid foundation of networking functionality was crucial. Throughout this part of the chapter, you'll get a chance to explore in detail the aspects of Windows 2000's networking and TCP/IP connectivity features.

What are the differentiators Microsoft has relied on to position NT relative to UNIX? How about NetWare? In the case of Novell, one in three NetWare users is migrating to NT due mainly to the robust, native support for TCP/IP found in Windows 2000 Professional and Server. Simply put, the differentiator of having native TCP/IP built directly into the operating system was strong enough relative to NetWare to make migrating to NT very attractive to NetWare users. The strength of differentiation on TCP/IP continues in the Novell arena versus IntranetWare, with a showdown coming over Active Directory in NT versus the Novell Directory Service (NDS) structure so pervasively communicated by Novell as its differentiator in the next release of NetWare, code-named Moab. The dominance of an operating system in the future is very much determined by its success at differentiating itself relative to its competition.

Microsoft realized with BackOffice that it having a strong positioning statement based on differentiation of features versus UNIX was essential if NT was to survive and thrive. Setting out to create an operating system that could stand on its own merits worldwide focused the entire design process on creating solid differentiators. Here are the primary differentiation points for Windows 2000 primarily versus UNIX and secondarily versus NetWare.

The modularity of the Windows 2000 kernel architecture relative to UNIX. This modularity of the NT kernel has very much distanced Microsoft over time, as it has made this an aspect of its operating system.

Strong differentiation in the designation of directory architectures. Prior to Windows 2000 5.0 there existed a strong difference between Novell and NT. With the advent of Active Directory in NT 5.0 Server, many of the features included in Novell NetWare are now part of the Windows 2000 Server 5.0 Active Directory.

Integration of the Dynamic Host Configuration Protocol (DHCP).

Windows Internet Naming Service in conjunction with the Domain Naming Service.

Support for several types of applications through the use of subsystems. These include Windows 16-bit applications, Windows 32-bit applications that include multithreading capability support for OS/2 character-based applications, and support for POSIX applications.

Integration of the IEEE 1394 standard directly into the Windows 2000 Kernel architecture.

EXPLORING CHANGES IN THE DHCP PROTOCOL

Being able to configure your network to check IP addresses in and out of a central library just as you'd check a book in and out of a library is really what the Dynamic Host Configuration Protocol (DHCP) is all about. What are the implications for electronic commerce? What can you do to make your Web strategies more efficient electronic commerce strategies? In short, how can you use the capabilities in DHCP to your competitive advantage? These questions and more are addressed here. Let's get started with an overview of the technology behind DHCP and then look at how you can use this approach to IP address management to your strategic advantage in serving your customers.

DHCP in Depth

One of the reasons why DHCP was created in the first place was to stream-line managing IP addresses over a network. As its name suggests, DHCP is focused on how to dynamically assign identities or IP addresses directly to systems, as they become members of a network or domain. This is partic-ularly relevant to companies building intranets, which have populations of users who are literally in motion many days of the month. The DHCP pro-tocol is really a flexible protocol that suits the needs of a large, rapidly growing, and mobile population of users. Companies that have standard-ized on this approach to handling IP addresses in the context of their elec-tronic commerce strategies have focused on creating mini "stores" or "outlets" by first enabling them with IP addresses, thereby creating the necessary links to centralized servers using the TCP/IP protocol. As a com-pany providing electronic commerce tools to widely distributed members of the sales force, Data General employs the DHCP protocol to enable its sales force to gain access to the latest corporate pricing and product infor-mation. Electronic commerce for DG is the delivery of timely product and pricing data to its sales force, ensuring more accurate orders and more ef-ficient communication to the end customer. DHCP is then a strong con-tributor to DG's communication process with the customer. The efficient distribution of information to the sales person is so transparent that the sales person doesn't even need to worry about configuring the right IP pa-rameters; all is handled by the DHCP Server component at headquarters.

DHCP relies on a client/server architecture in which the DHCP server component provides the actual IP addresses used throughout a network. Client systems in a DHCP-based network use a variety of mechanisms to gain access to IP addresses, with the BOOTP protocol being the most pervasive.

THE FUTURE OF WINDOWS 2000 NETWORKING

Instead of actively promoting operating systems well into the future, Mi-crosoft today has a singular focus on the goal of getting Windows 2000 Professional and Server integrated into as many environments as possible. Many organizations are also singularly focused on how they can improve the security, efficiency, stability, manageability, and responsiveness of their sites relative to those of their competitors. Microsoft sees this dynamic throughout the industries it serves, hearing this from existing users through their sales force, through Internet news groups, and even from letters from their customers asking for specific features. Intel, a long-time strategic partner with Microsoft, is likewise enamored with the idea of being closer to present and potential customers than anyone else. Both companies have cultures that reward aggressiveness and in-depth knowledge of customers' needs. With Intel and Microsoft focusing on the future, the key attributes of a future operating system can be ascertained from the Intel roadmaps by division and their public announcements of entering the network hardware and electronic commerce consulting businesses. Here's a list of the key attributes that Microsoft will be including in either a Windows 2000 Service Pack or the next-generation operating system following Windows 2000:

Strengthening of Kerberos security and integration of Microsoft wallet-like purchasing capabilities in client-based systems. Server-based OS counterparts will include electronic commerce components to relieve potential customers from having to look at secondary software providers for their tools on the Windows 2000 platform.

True 64-bit operating system performance. This will complement the Merced and McKinley processor introductions due from Intel in the 2000 timeframe.

Further improvement and simplification of the user interface.

Increased intelligence in the interaction of the user interface. This will allow the PC to handle even more tasks for the user.

A focus on natural user interfaces that enable the user to speak to the PC and the PC to read aloud.

Complete elimination of the differences in interfaces with various types of communication (such as mapping drives for servers and using hypertext on the Internet.

Continued focus on the Zero-Administration Windows (ZAW) end goal. The goal is to establish the administrative-free client. ZAW is a new technology in Windows 2000 for improved software installation in Windows-based environments.

Automatic load balancing across large clusters.

Self-repairing network topologies.

Regular development of the distributed facilities and integration to the Internet.

Seamless integration of TCP/IP on Windows 2000 with Sun Microsystems implementation of DNS, and integration with Novell NetWare.

CHAPTER SUMMARY

The future of Windows 2000 Professional and Windows 2000 Server is clearly in enabling electronic commerce through the many enhancements to the networking functionality included in this latest version. The OSI Model is used throughout this chapter as a reference point to how the Internet, and with it electronic commerce, is changing how systems will communicate in the future. The role of Internet-based protocols in the future direction of Windows 2000 is unmistakable. Continually changing the direction of networking connectivity is taking HTTP over TCP/IP into the new standard for operating system connectivity. S-HTTP and secured protocols are increasingly important and will also begin to be included as standard within the next generation of operating systems, with the intent of bringing down the barriers of completing electronically enabled transactions.

13 Learning to Use Windows 2000's Administrative Tools

Being able to troubleshoot the performance of a network, a workgroup, or even an individual system is made possible by the tools available in Performance Monitor. If you've used this tool in previous versions of Windows NT Workstation or Server, you'll find that the core functionality is the same. The graphical interface has become much more Web-like, using the Microsoft Management Console (MMC) interface to organize Performance Monitor relative to other applications. The structure of the MMC looks like a browser interface, making the navigation fairly intuitive. Figure 13.1 shows an example of the Microsoft Management Console interface with Performance Monitor running.

One of the major differences between Windows 2000 Professional and Windows NT 4.0 Workstation is that in Windows 2000, the Administrative Tools are now found in the Control Panel within their own group. In previous versions of NT the Administrative Tools were directly accessible from the Start menu. In Windows 2000 Professional, Administrative Tools are now located in the Control Panel, within the Administrative Tools icon. Figure 13.2 shows the Administrative Tools located in the Control Panel.

The tools included in the Administrative Tools include Component Services, Computer Management, Data Source (ODBS), Event Viewer, Internet Services Manager, Local Security Policy, Performance Monitor, Personal Web Manager, Server Extensions, Services, and Telnet Server Administration. These 11 applications were first introduced during the Windows 2000 betas, and they are available from the Windows 2000 Professional CD during installation. Each of these applications is briefly illustrated here and shown in Figure 13.3.

Component Services. This application in represented by an icon with a hammer and a chart in the background. One of the primary tasks that this application completes is the configuring and managing of COM+ applications. Figure 13.4 shows the COM+ Applications that are being profiled using this application.

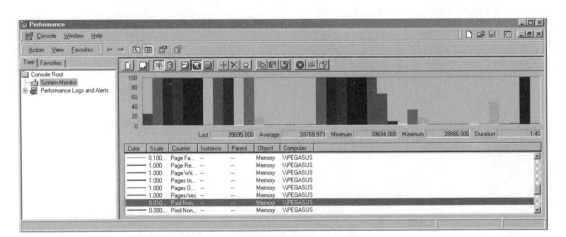

FIGURE 13.1 Using Performance Monitor in Windows 2000.

FIGURE 13.2 Windows 2000 Professional includes the Administrative Tools within the Control Panel.

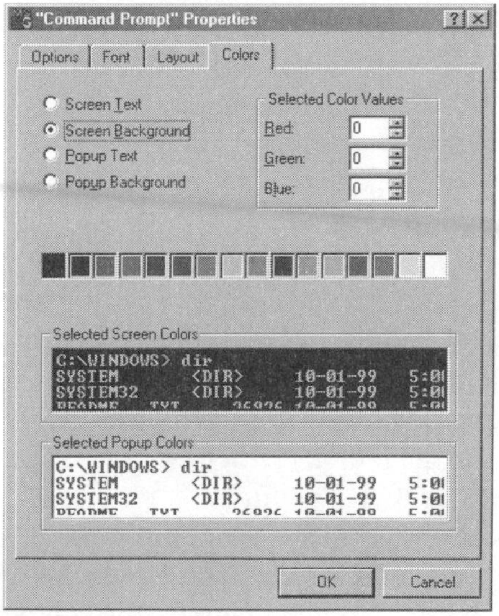

FIGURE 13.3 Introducing the Administrative Tools in Windows 2000 Professional.

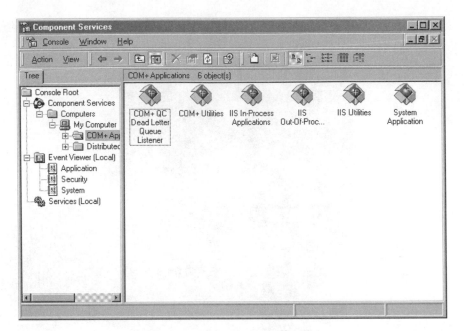

FIGURE 13.4 Tracking COM+ application performance using Component Services.

Computer Management. Represented by a PC with two boxes on-screen, this application manages disks and provides access to other tools to manage local and remote computers. The extensive use of navigational tools in this application, coupled with its access to system monitoring and system performance metrics, makes this application one of the most useful in the set.

Using the MMC interface, it's possible to check the status of all major performance metrics. You can also use the Computer Management application for inquiring as to the file system that is being used by disk volume on a workstation. The MMC is the basis for all future systems management functions within Windows 2000 Professional and Server, so be sure to get familiar with how to navigate its interface and use the commands included within it.

This is particularly useful if as an administrator you inherit a series of workstations from another department and want to see how the file system by disk drive has been configured. The Explorer-like approach to managing system information is invaluable in that many of the system attributes and characteristics are available from the single Computer Management interface. Like all interfaces on these applications, Computer Management uses the MMC interface for viewing and working with system performance tools and analytical applications for checking system performance. Figure 13.5 shows the Computer Management application.

Data Sources (ODBC). Represented by an icon showing a PC, server, and spreadsheet, the Data Sources (ODBC) application is used for defining ODBC drivers for databases on workstations relative to servers, defining ODBC drivers for tracking in Performance Monitor, and options for defining DNS values that enable communication between databases (see Figure 13.6).

Event Viewer. Represented as a notebook with a series of grammatical symbols on it, the Event Viewer is one of the most valuable applications for tracking system activity. The Event Viewer includes Application, Security, and System logs. Windows 2000 sees each discrete event as the operating system boots up and runs, recording events of interest into one of the three categories of log files. The analysis of log files is one of the most valuable analytical tools for checking the performance of Windows 2000. The Security Log is also invaluable for checking if there have been security breaches to the workstations and

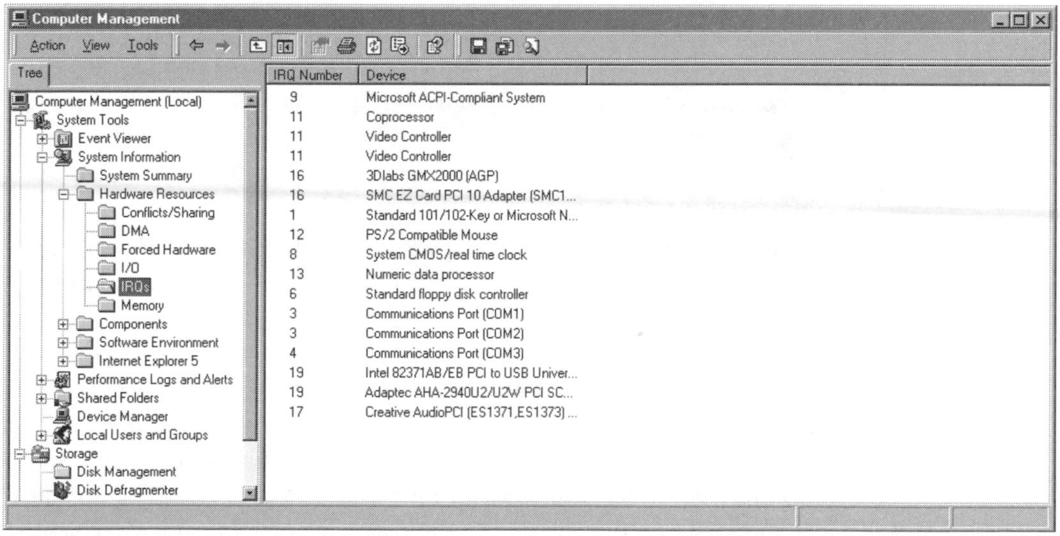

FIGURE 13.5   Using the Computer Management interface to find tools and utilities for analyzing system performance.

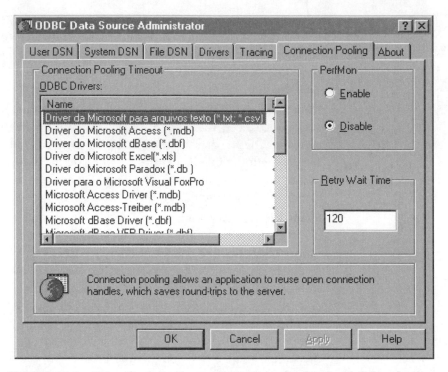

FIGURE 13.6 Using the Data Source (ODBC) application for defining database links.

servers for which you are responsible. There is also the opportunity to save log files in text, comma-delimited, or event log format. If you plan to use these files in Microsoft Excel, be sure to export them in comma-delimited format for ease of importing. Figure 13.7 shows the Event Viewer used for checking the Application Log of a Windows 2000 workstation.

Internet Services Manager. Represented by a globe with a server in the foreground, the Internet Service Manager handles the task of managing the Internet Information Server, the Web server for Internet and intranet Web sites. The interface for this application is intuitive and easy to navigate. Within the Internet Service Manager interface, the default FTP Site, Default Web Site, and Default SMTP Virtual Server are all shown along the left side of the page. Checking on the ASP scripts for your Web site, for example, is possible using this application as the subdirectory structure of your site is shown on-screen. Using the

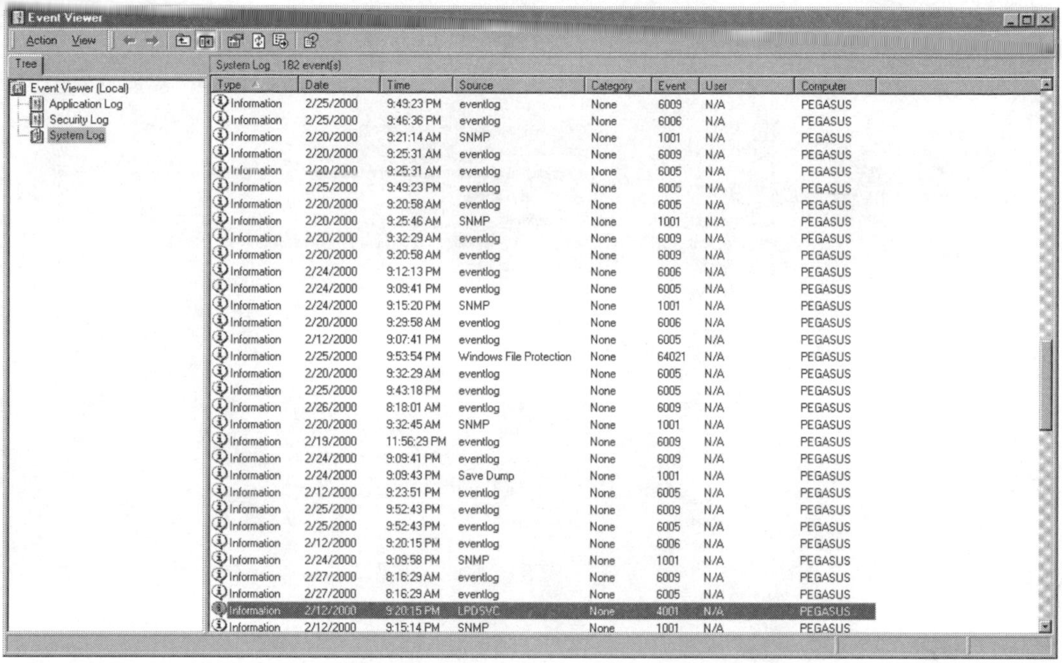

FIGURE 13.7   Using the Event Viewer for checking how applications are running.

Explorer-like interface for navigating the FTP site, Web site, or SMTP Virtual Server makes it possible to quickly edit files regardless of their location in the hierarchy. Figure 13.8 shows an example of the Internet Service Manager application.

Local Security Policy. Represented by a server with a padlock on it, the purpose of this application is to view and modify local security policy, such as user rights and audit policies. There are comprehensive series of tools available for managing Account Policies, Local Policies for security, Public Key Policies, and IP Security Polices for the local workstation. You can also monitor both successful and unsuccessful event completions by login. Figure 13.9 shows the contents of the Local Security Policy application.

Performance Monitor. Starting with the first versions of Windows NT, Performance Monitor has consistently been one of the most used applications for tracking system performance and analyzing the behavior of workstations and servers as modifications are made to them.

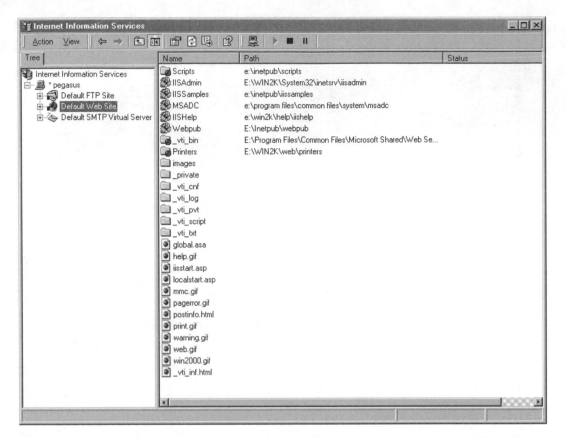

FIGURE 13.8 Using the Internet Services Manager application for managing FTP and Web sites.

There is a great deal of depth in the Performance Monitor as the object:counter relationships are used to quantify overall system performance by functional area. Interpreting the results that Performance Monitor provides in terms of metrics is one of the most useful troubleshooting tools available in Windows 2000. Figure 13.10 shows the Performance Monitor in use.

Personal Web Manager. Represented by a hand holding a piece of paper with an illuminated globe, the purpose of this application is to assist in the Web page publishing process. There are selections for Main, Tour, and Advanced aspects of handling Web page publishing. Figure 13.11 shows an example of the Personal Web Manager application.

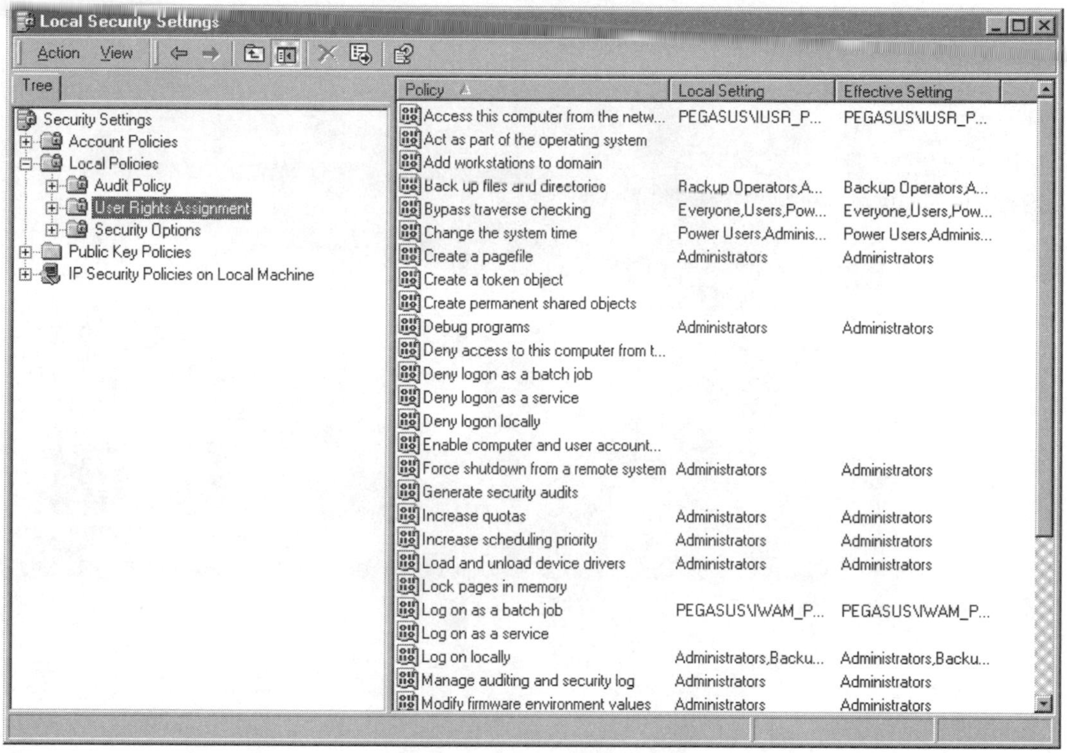

FIGURE 13.9 The Local Security Policy dialog box provides a series of tools for monitoring system events and their implications for the security of the system.

Server Extensions Administrator. Like Component Services, Server Extensions Administrator is used for handling the FrontPage extensions for your Web site. Integral to Microsoft's Site Server and overall Web strategy is the development of Web sites, which rely on FrontPage Extensions. Figure 13.12 shows an example of the Server Extensions Administrator application.

Services. This application icon shows two gears that interlock with each other. The Services application is used to define which operating systems start and stop, and which classes of users can have access to system resources for full control of all events on the Windows 2000 system, the ability to modify only, read and execute only, just read, or just write to the networked volume. There is also extensive support for

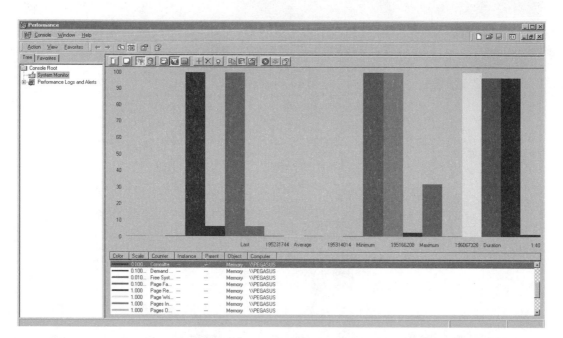

FIGURE 13.10 Performance Monitor is one of the best tools for analyzing the performance of Windows 2000 workstations and servers.

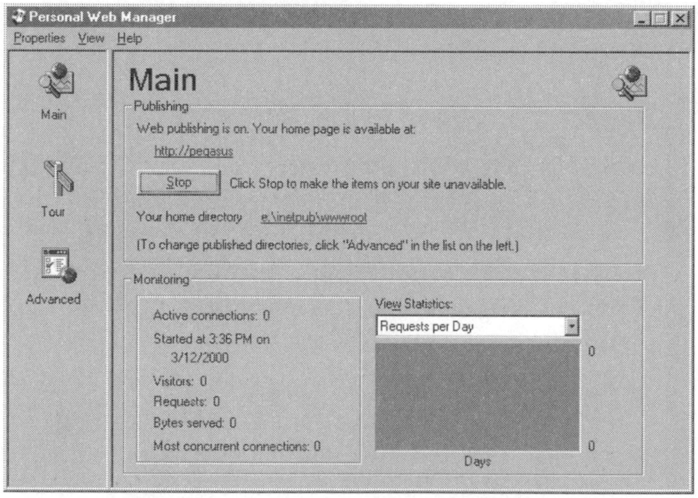

FIGURE 13.11 Using the Personal Web Manager to publish pages on your Web site.

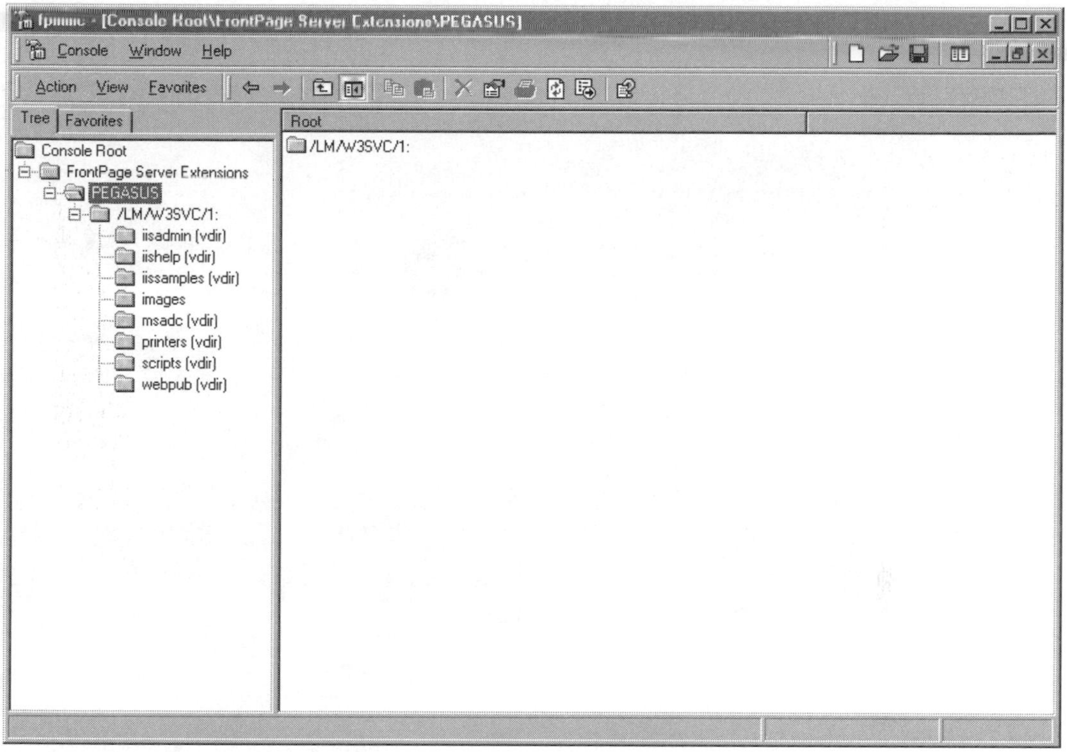

FIGURE 13.12 Managing FrontPage extensions using the Server Extensions Administrator.

customizing each login type on a Windows 2000 system, giving or re-stricting access to 13 advanced permissions that can be applied to each class of user. The biggest contribution of the Services application is that it lists the name, description, status, startup type, and logon, which triggers the service to start. Like Performance Monitor, this is a very comprehensive application. Figure 13.13 shows an example of the Ser-vices being tracked through the MMC interface.

Telnet Server Administration. A small PC in front of a server repre-sents Telnet Server Administration. The Telnet Server Administration application defines the communications parameters for handling tel-net connections between workstations and servers on your network. The graphically based nature of the interface also makes it possible to view multiple telnet sessions at the same time occurring on multiple points over the network. The interface itself resembles the beginning of

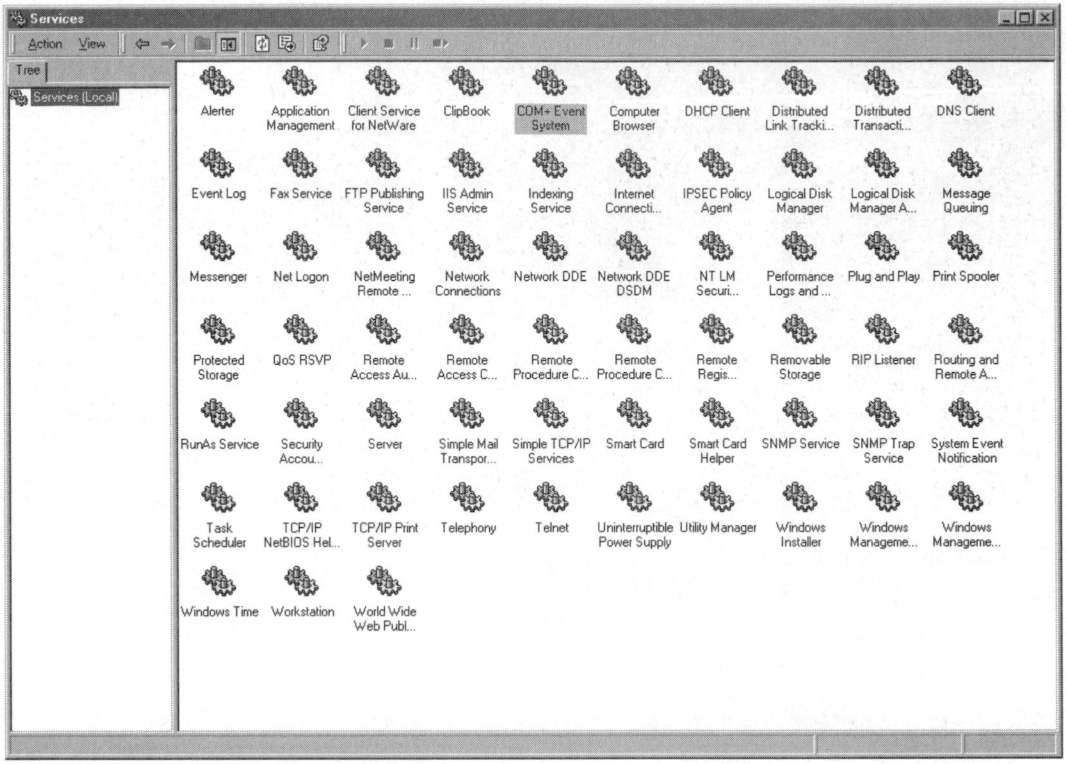

FIGURE 13.13 Services provides tools for defining how services are used in Windows 2000.

a telnet session in progress, resembling disk-formatting utilities from previous versions of Windows95/98 and Windows NT. The interface is easily navigated and quick to implement. In the case of organizations using telnet connections, the need for handling the transfer of large files quickly andthe need for coordinating access with the ftp command are also apparent. Many system administrators build a shell script for handling routine transfers of files via ftp, and they occasionally use telnet sessions for enabling text-based applications on remote systems. The convenience of starting a shell script via telnet session is one of the best uses of this application, especially for the system administrator responsible for a wide array of workstations and servers. Figure 13.14 shows an example of the Telnet Server Administration screen.

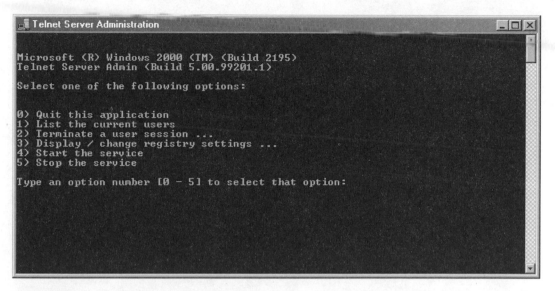

```
Telnet Server Administration                                    _ □ ×

Microsoft (R) Windows 2000 (TM) (Build 2195)
Telnet Server Admin (Build 5.00.99201.1)

Select one of the following options:


0) Quit this application
1) List the current users
2) Terminate a user session ...
3) Display / change registry settings ...
4) Start the service
5) Stop the service

Type an option number [0 - 5] to select that option:
```

FIGURE 13.14   Using the Telnet Server Administration application for handling remote application tasks.

DAY-IN-THE-LIFE: COMPASSPOINTE TRAINING

Teaching the fundamentals of networking and operating systems to potential MCSEs and CNEs is the mission of CompassPointe Training. In addition to courses on TCP/IP, the differences between UNIX and Windows NT, troubleshooting network connections, and load balancing servers are all taught. What makes CompassPointe Consulting different is that the majority of the teaching is done directly over the Internet, making the role of servers critical in the delivery of connections and content. In effect, CompassPointe has built a Virtual Private Network, which makes it possible to serve its students globally via any browser at any time. The role of Performance Monitor is critical in ensuring continued video and audio feeds to students who are using streaming media at client companies, which include Chrysler, Canon Computer, and several universities.

With the primary intent of using Performance Monitor to predict when a server could potentially slow the transmission of materials or stop functioning, the system administrators at CompassPointe Training have developed a series of profiles in Performance Monitor. These profiles predict when a server is getting too much traffic, is having trouble with disk space, or is about to run out of memory for typical workloads. The profiles are made up of object:counter relationships that define a performance variable and the value or event for the variable.

The system administrators also extensively use the Add to Chart option for accessing servers throughout the network using their DNS equivalent names. Using the Computer: entry across the Add to Chart dialog box, which is shown in Figure 13.15, administrators can access servers located throughout the network and profile their performance as well.

In addition, the IS department at CompassPointe is using the Log feature of Performance Monitor. The Log interface for Performance Monitor is accessible from the View menu or the icon located in the Performance Monitor toolbar, and it is represented as a spiral-bounded notebook. Using the export options, the team at CompassPointe also catalogs the reports in Microsoft Excel format and uses them for comparing performance of the Web sites during the high-use timeframes.

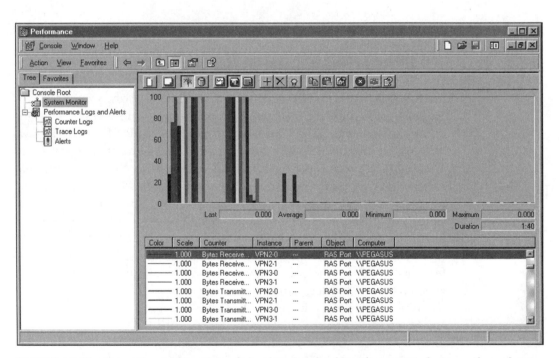

FIGURE 13.15 Using the Add to Chart dialog box for checking the status of servers throughout a network.

WHAT IS PERFORMANCE MONITOR?

The most widely used tool for managing a Windows 2000 Professional and Server's performance is Performance Monitor. What is truly significant about Performance Monitor are the analytical insights it can provide as to systems' performance. Performance Monitor enables you to track a variety of items and display information on their relative performance in different ways. Performance Monitor comes with Windows 2000 Professional and Server, and it is designed to provide feedback on the relative level of performance on systems. Performance Monitor is capable of monitoring performance of systems throughout a network as well, using the DNS naming conventions as defined within Windows 2000 and previous generations of Windows NT Workstation and Server. Many companies that have standardized on Windows 2000 find Performance Monitor to be a very useful, capable tool for troubleshooting performance issues and alleviating system bottlenecks.

Performance Monitor is made up of a series of object:counter relationships that are used to track the health of the system you're working with. Each object is, in effect, a characteristic of the system, and the counters are the classes of variable value by which a system's performance is measured. The following is a list of the objects, or performance attributes, that are measured in Performance Monitor.

You'll notice that as additional software is loaded onto your workstations and servers, new objects not defined here will appear in the Performance Monitor. That's because applications like SQL Server, RAS, and other network-based applications insert and install their own objects within Performance Monitor.

The Performance Monitor application is actually a series of four different views on systems' performance. These four views are as follows:

Chart. The Chart view is the default view in Performance Monitor. Chart view enables objects to be graphically displayed. This view enables you to view the monitored items over a short period of time (as short as every second) by choosing the options from within the dialog boxes available in this view. This view is best for actively watching the performance of a Windows 2000 Professional system, and it is illustrated in Figure 13.16.

Alert. The alert view enables you to monitor the system in the background while working with other applications. You can define a threshold for a counter, and if it is reached, an alert will be triggered.

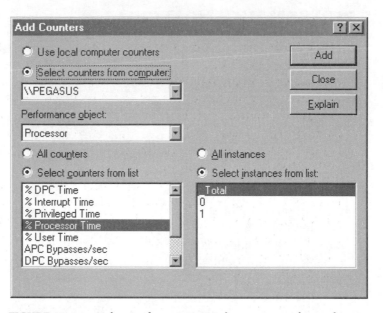

FIGURE 13.16   Using Performance Monitor to gauge the performance of systems on a network.

The form of the alert can be specified to switch to the alert view, log the event, or send an alert message directly to the person whose system is experiencing the alert condition.

Counter and Trace Logs. This log view enables Performance Monitor to record the selected counters into a file, known as a log file, while the user runs applications on the Windows 2000 system. The log file can later be examined to find potential and existing bottlenecks.

Report View. Using the View Report icon in the main Performance Monitor window, the objects and counters defined in the Performance Monitor chart view are provided in tabular view.

Using Performance Monitor

Learning how to use Performance Monitor definitely helps you to track and analyze performance data. Performance Monitor has a variety of settings that can be used to help you understand how the Windows 2000 operating system is running. Performance Monitor's windows can also be arranged to better suit the needs of the user.

Launching Performance Monitor

Using Performance Monitor does not require any special administrative passwords or privileges, and it is accessible to any of the user classes that are defined when a new login is defined. The Performance Monitor program is accessible through the Control Panel of Windows 2000. In previous versions of Windows NT, Performance Monitor was found in the Administrative Tools section, located off the Start button. You can also launch Performance Monitor by using the file PERFMON.EXE in the WINNT/SYSTEM32 subdirectory.

If you're going to use disk monitoring in conjunction with Performance Monitor, be sure to also use the DISKPERF command. Without the DISKPERF command, the Physical and Logical Disk counters do not display any values. To use the DISKPERF command, you have to log in with administrative rights. To activate the counter using the DISKPERF command, follow these steps:

1. Log in to the Windows 2000 workstation as a member of the Administrators group.
2. Select Start, Programs, Command Prompt to get to the command prompt.
3. At the command prompt, type DISKPERF to see whether the disk performance counters have been turned on. If they are not turned on, type DISKPERF-Y to turn the counters on. Also, using the E switch with -Y allows performance monitoring of striped disk sets. You do not use striped disk sets, so you do not need to use the E parameter.
4. Shut down and restart the Windows 2000 workstation. The disk counters are now available to be tracked in Performance Monitor.
5. You can turn off the Performance Monitor by using the command DISKPERF -N to turn off all counters when monitoring is complete.

Quick tip: For more information about the DISKPERF *command-line switches, type DISKPERF /? at the command prompt.*

You can change certain display options to better organize the Performance Monitor's various graphical interfaces. You can perform actions such as hiding the menu and title bar, hiding the toolbar, and hiding the status bar. Hiding these items enables you to have a larger viewing area for your monitored data.

To hide or show the menu and title bar, toolbar, or status bar of Performance Monitor, follow these steps:

1. If it is not already started, start Performance Monitor.
2. Choose Options from the Performance Monitor menu. In the menu, items that are shown in the Performance Monitor window have a checkmark beside the menu item.
3. To hide an item that has a checkmark beside it, select that item. You can select the following menu items: Menu and Title, Toolbar, and Status Bar. You can also hide the menu and title bar by pressing Ctrl+M. You can hide the toolbar by pressing Ctrl+T and the status bar by pressing Ctrl+S. You can also use the Ctrl+H for highlighting selected counters to make them more visible when many objects and events are being tracked.
4. To show the items after they are hidden, perform steps 2 and 3 so the item has the checkmark beside it. If the title bar and menu are hidden, you can show those items by double-clicking on a nontext area of the window. If the title bar and menu are shown, you can hide them by double-clicking in a nontext area in the Performance Monitor window.

Switching Between Views

Performance Monitor has a total of three different views: Chart, Log, and Report. Each of these views is represented with icons on the toolbar. Notice the similarity of the icons' locations on the toolbar. This has been migrated from Windows NT 4.0 Workstation and Server.

When views are changed, menu items are automatically changed to handle the options available in each view. Some of these menu items are specific to a particular view. Other menu items will change part of the item to reflect the view.

Managing Workspace and Settings Files

You'll find when using Performance Monitor that the same series of objects and associated counters is used for tracking overall system performance. These object:counter relationships are used for troubleshooting common system-level and network-wide performance issues affecting overall productivity and throughput in a given company. For example a company may be doing quite a bit of network traffic with servers located

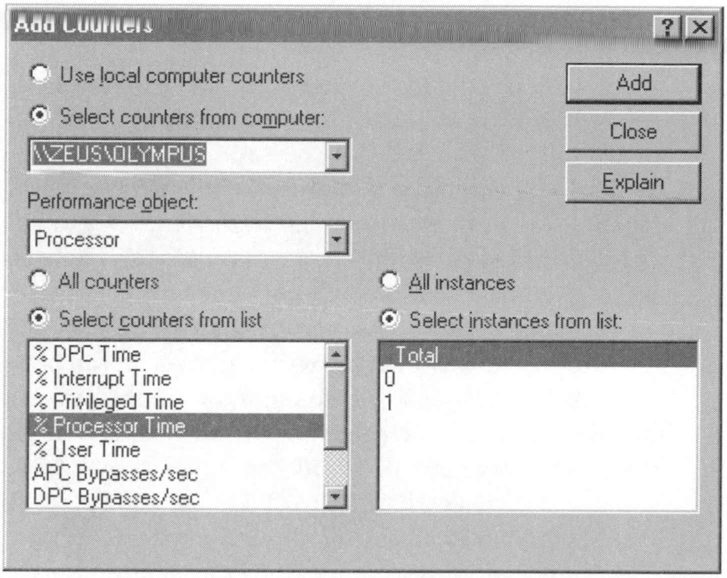

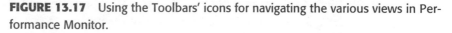

FIGURE 13.17 Using the Toolbars' icons for navigating the various views in Performance Monitor.

throughout a campus (as is the case with Microsoft), and the need for capturing performance "profiles" in the case of settings files is strong.

To accommodate this need, the Performance Monitor has settings files that enable you to save the counters for the view. A similar type of file called a workspace file saves all the counters for all the views. Using a workspace file prevents the user from always adding counters to track each Performance Monitor session. Instead, the settings or workspace file is opened using the File, Open command in Performance Monitor. Follow these steps:

1. If Performance Monitor is not already running, open the Control Panel, then select Administrative Tools. You'll find Performance Monitor in the Administrative Tools group.
2. Choose the view window for which you want to save the settings. Add any counters to the view that you want to monitor.
3. From the Console menu, select Save Settings to save the settings file or Save Workspace to save the settings in the MMC format compatible with this interface. The Save As dialog box will appear.

4. In the File Name box of the Save As dialog box, type the file name you want for the settings or workspace file.

5. Choose the Save button. The file name appears in the title bar of the Performance Monitor window.

Each of the settings or workspace files is saved with an extension that depends on the type of view that was saved. For example, chart view settings files are saved with the extension.MMC. These extensions assist in defining the type of settings or workspace file being saved. Table 13.1 provides a listing of file extensions and counters saved.

After a settings or workspace file is saved, it can be loaded into Performance using the File, Open command from the pull-down menus. Additionally the file can be copied to disks or across the network for use on other Windows 2000 or Windows NT machines that use Performance Monitor. This makes it easier to obtain the same performance data on different Windows 2000 and Windows NT systems.

Settings and workspace files can be even more convenient by specifying them on the command line of a shortcut or a menu item. This makes it easier to start Performance Monitor with the proper settings already loaded. To load a settings file when Performance Monitor has already been started, type the following command line in the command prompt window:

```
C:\<Windows NT directory>\system32\perfmon.exe
c: <settings file path>
```

Here the <Windows NT directory> is the location of the Windows 2000 directory and <settings file path> is the directory location of the settings file.

TABLE 13.1 File Extensions and Counters

Extension	View Counters Saved
MMC	Microsoft Management Console (default)
PMC	Chart View counters
PMR	Report View counters
PMA	Alert View counters
PML	Log View counters
PMW	All views counters (workspace)

For example, to load PROCESS.PMA located in the WINNT directory, you would type c:\winnt\system32\perfmon.exe c:\winnt\process.pma in the command line box of the Run dialog box, the shortcut properties dialog box, or the menu item properties dialog box.

Using and Monitoring Counters

As you can see from the series of counters included in Windows 2000, the Performance Monitor can actively track the performance of many aspects of a workstation. Performance Monitor categorizes them into objects. These object types relate to actual devices, sections of memory, or processes. Objects contain items known as counters. These counters are the specific items to be measured using Performance Monitor. For example, under the Processor object, a counter called % Processor Time is used to monitor the percentage amount of total processor time that is being used by the system.

Object types also can have several instances. Instances do not already appear as objects per se, but object types such as the Processor object have an instance for each processor in a workstation or server. Instances represent an individual object out of multiple objects of the same type. Other object types, for example, the memory object type, do not have any instances.

To begin monitoring a Windows 2000 Professional-based workstation, you need to add counters to monitor system performance. You'll see that adding Counter is identical no matter which of the three views in Performance Monitor are being used. To add Counters in Performance Monitor, do the following:

1. Launch Performance Monitor from the command line or within Control Panel in Windows 2000.
2. Select the view you want to display by clicking one of the view buttons on the toolbar.
3. From the Performance Monitor menu, select the "+" sign to have the Add Counters dialog box appear. One of the major differences between Windows NT and Windows 2000 is that the latter adds a counter to each view when selected; in previous versions the object:counter relationships would need to be redefined. Figure 13.18 shows an example of the Add Counters dialog box.
4. The Computer box displays the computer that Performance Monitor is tracking. The Object item in the dialog box is a drop-down

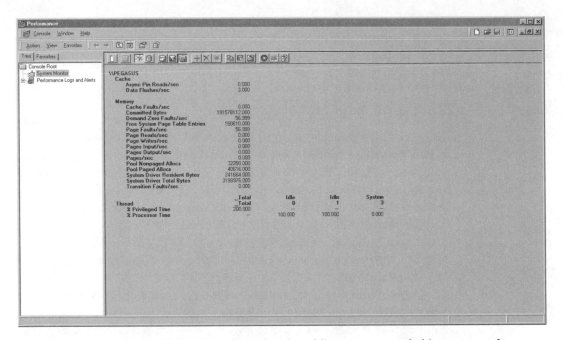

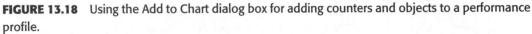

FIGURE 13.18 Using the Add to Chart dialog box for adding counters and objects to a performance profile.

list displaying the entire series of object types currently available on the system. Select the object type for which you want to see counters.

5. In the Counter list, select a counter you want to track. If you are unsure of what a counter does and you would like more information on the specific object:counter relationship, select the counter and click once on the Explain button. The Explain button defines the purpose of the counter. The Explain dialog box is also detached from the main dialog box as well, which makes it possible to move and copy the definition for use by other members of your team in learning how object:counter relationships work.

6. If multiple objects exist in the Instance list, select the instance of the object you want to monitor.

7. Click Done after you have added all of the counters you want to monitor. This will close the Add to window.

8. Be sure to save these settings as an MMC file so the counters are accessible across all views in Performance Monitor.

Using Performance Monitor with Other Workstations on Your Network

One of the best benefits of Performance Monitor is its ability to monitor object:counter relationships on systems located throughout the network. Any other Windows 2000 and Windows NT workstations connected to the same network can be monitored using Performance Monitor. Multiple systems, each with their own set of counters, can be monitored in the same view. This makes Performance Monitor a useful tool for network administrators who need to obtain performance data from multiple systems on the network, in addition to seeing how the load on workstations and servers across the network influences the overall network performance.

Here are the steps you'll need to use for monitoring the performance of a system across a network using the Performance Monitor:

1. Launch Performance Monitor if it is not already running.
2. Select the view you want to use in Performance Monitor.
3. Select the Add To command from the Toolbar across the top of the main Performance screen.
4. Monitor screen. This displays the Add to dialog box. You can also click once on the "+" button within each specific view to gain access to the Add to Chart dialog box or the comparable Add to dialog box, depending on the view you're in.
5. If you know the name and Universal Naming Connection or UNC path of the computer to monitor, type the UNC path and computer name in the select counters on computer: field. If you want to browse the network, click the button with the three ellipses.
6. Select the counters you want to monitor for that specific workstation. Also, select the options for those counters. Click Add to add those counters to the current view.
7. You can select additional workstations on the network and choose counters for those systems to be monitored.
8. When you are finished selecting counters, click once on Done to close the Add to Dialog box.

TIP

Quick tip: Be sure to keep the profiles being used throughout the network as streamlined as possible, to ensure that the network traffic for other tasks enterprise-wide is not impacted. Having too many networked workstations trying to send performance information has a negative effect on the Windows 2000 and Windows NT computers running Performance Monitor. Don't try to monitor too many network computers or counters at once. Also, consider modifying the update intervals for the different views being used or use manual updates for counter information. This helps reduce the amount of data being transmitted across the network and to the machine running Performance Monitor.

EXPLORING EVENT VIEWER

Windows 2000 interprets commands as individual events, recording them in the log types as defined earlier in this chapter. What's significant about this is that these events—seemingly unrelated and even trivial when a workstation or server is running well—can often provide clues to a solution when a problem arises. The time series nature of the information provides a glimpse of how systems are performing over time. Event Viewer is the operating system's way of telling you about events. It functions like a report card and status report by storing lists of eventsin log files that you can review, archive, or transfer to a database or spreadsheet for analysis.

Windows 2000 recognizes three broad categories of events: system events, security events, and application events. Events of each type are recorded in log files.

System events are generated by Windows 2000 itself and by installed components, such as services and devices. They are recorded in a file called a system log. Windows 2000 classifies system events according to their severity as errors, warnings, or information events as follows.

Errors are system events that represent possible loss of data or functionality. Examples of errors include events related to network contention or a malfunctioning network card, and loss of functionality caused by a device or service that doesn't load at startup.

Warnings are system events that represent less significant or less immediate problems than errors. Examples of warning events include a nearly full disk, a timeout by the network redirector, and data errors on a backup tape.

Information events are all other system events that Windows 2000 logs. Examples of information events include someone using a printer connected to your computer or the successful loading of a database program.

Security events are generated by Windows 2000 when an activity you choose to audit succeeds (a success audit) or fails (a failure audit). Security events are recorded in a file called a security log. These include file-related events, such as attempts to access files or change permissions (NTFS volumes only), and other security-related events, such as logon/logoff events and changes to security policies. By default, Windows 2000 auditing is turned off, so you will likely see no events in the security log. To enable event auditing, open Local Security Policy from Administrative Tools in Control Panel. In the left pane, open Local Policies and select Audit Policy. In the right pane, right-click each event type you want to audit, and then select Success, Failure, or both Success and Failure in the following dialog box.

Application events are generated by applications and are recorded in a file called the application log. The application developer determines which events to monitor and how those events will be recorded in the application log. Windows 2000 Backup, for example, records an application event whenever you erase a tape or run a backup.

The importance of the Windows 2000 log depends on your situation. If you work in a security-sensitive environment or one in which users freely access resources on each other's workstations you'll find the event logs useful in helping you keep track of who, what, when, and where the system has been used. If you don't care about details of usage on workstations and servers, then the security log will probably be of little interest to you. The system log can still be helpful in diagnosing performance problems and hardware errors, and the application log can give you insight into how certain applications are working. Only applications designed to record their "events or thoughts" in the application log will appear there, but those that are so designed provide an obvious benefit—to you, your technical support person, and even the developer working on applications—for identifying and resolving problems that may arise.

If your workstation or servers are set up to share files or a printer with other users, checking the system log for print jobs and the security logon/logoff access will give you a feel for how and when your computer's resources are being used. Although the information might simply make you feel more in control of your system, you might also find patterns that help you determine better ways to manage it.

Viewing a Log File

You can easily see what a log looks like even if you never before thought of monitoring your system.

1. To view a log, open Event Viewer.
2. Open the Start menu and choose Settings, Control Panel.

3. In Control Panel, open Administrative Tools.
4. In Administrative Tools, open Event Viewer.
5. Select a log to view. The system log is shown in Figure 13.19.

Using the Task Manager as a Performance Analysis Tool

The long-standing usefulness of the Task Manager as a tool for "checking under the hood" on what is actually happening with processes and multiple processor usage (in the case of multiprocessor workstations and servers), CPU time taken by task, memory usage, and priorities of tasks has been proven. The shortcut access to the Task Manager from the Taskbar also makes this a useful tool in its convenience and succinctness of data presentation.

Right-clicking on the Taskbar presents a pop-up menu that includes Task Manager as the second entry from the bottom. Selecting Task Manager... provides the graphical interface shown in Figure 13.20. Notice that along the top of the Task Manager window there are tabs for Applications, Processes, and Performance.

What you'll notice as you go through the tabs for Task Manager is that the glimpse you get of system performance is short time. For an immedi-

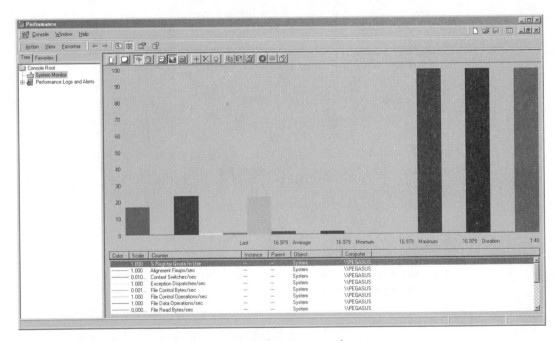

FIGURE 13.19 Using the Event Viewer to analyze an event log.

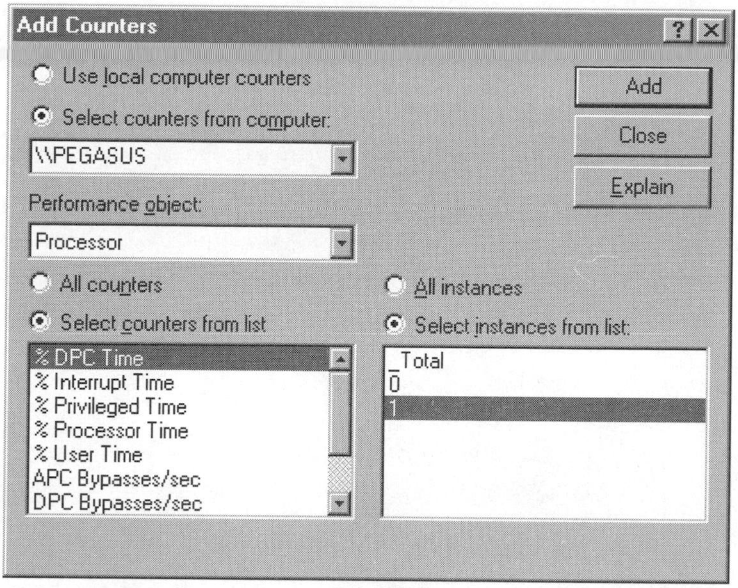

FIGURE 13.20   The Windows NT Task Manager interface provides a quick glimpse of how workstations are functioning.

ate problem, this is a great tool for troubleshooting. For longer-range and sporadic problems, which take time series data to solve, the Performance Monitor is by far the better tool to use.

The first time you use Task Manager, the Applications tab is selected and shows the applications that are running. The benefit of this first view in Task Manager is that if you have an application behaving sporadically you can quickly see if it is indeed running. The quick view capability of Task Manager gives you the chance to troubleshoot the immediate issues that could be impeding the performance of a Windows 2000 Professional or Server-based system.

Exploring the Performance Tab

The closest tool in the Task Manager that approaches the functionality of Performance Monitor, the Performance tab provides four graphical indicators of relative system performance for the point in time you are viewing them: Charting CPU Usage, Usage History for the last few minutes, Memory Usage, and Memory Usage History over the last five minutes. Figure 13.21 shows the Performance tab in Task Manager. Notice the feedback

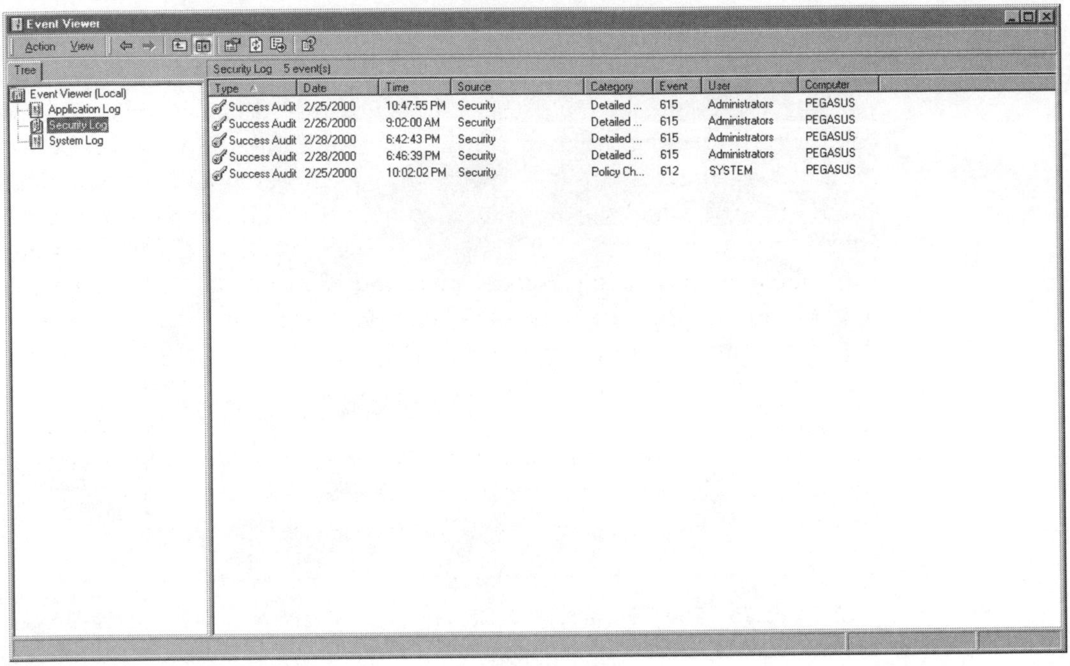

FIGURE 13.21 The Performance tab of the Task Manager provides a quick glimpse of system performance.

provided in both graphical and textual format, with the key variables across the bottom of the screen.

Each of the measures is relatively self-explanatory, with CPU Usage showing the amount of current processor usage, expressed as a percentage of total capacity. The CPU Usage History shows the percentage of CPU capacity used over time. The MEM Usage images show the amount of virtual memory that has been "committed" or used, and the Memory Usage History images chart virtual memory usage over time.

In the textual area of the dialog box, the Commit Charge (K) area is one of the most valuable in that it defines the maximum committed virtual memory on the workstation or server. This metric provides a glimpse of current, maximum allowable, and maximum committed virtual memory.

The Physical Memory tab is also useful as a metric for determining if the total available memory on your workstation is being used by applications currently running. This is very similar to checking the total available capacity of a hard disk after installing applications—it's a measure of "head room" for future expansion and use of applications.

Understanding the Processes Tab

If you're a system administrator who has worked on UNIX systems before, you'll recognize the format of this command, as it replicates the processes command on many of these systems. It's a graphical interpretation, complete with sorting capabilities, of the old UNIX command for querying about to the health of a given system. Just as in UNIX, a process is an executable program (such as PowerPoint or Explorer), a service (a function controlled by Services in Control Panel, such as Event Log or Messenger), or a subsystem (such as one from Windows 3.X applications). You can see that this tab provides a glimpse of which applications are taking what percentage of processor and virtual memory usage.

Figure 13.22 shows the Processes tab selected with processes sorted in CPU Time order. You can also sort processes ascending or descending by clicking the appropriate column heading the Processes tab. By default, the Process tab displays for each process the following variables:

Image name. The application or process name.

PID. The process ID, a number that uniquely identifies a process while it runs.

FIGURE 13.22   Here: Using the Processes Tab to see which applications are using allocations of memory, CPU time, and process ID assignment.

CPU. The percentage of elapsed time that the process used the processor (CPU) to execute instructions.

CPU Time. The elapsed time (in hours, minutes, and seconds) that the process has been running.

MEM Usage. The number of kilobytes of virtual memory used by the process.

Other process-related columns are also available for display. To make your selections, choose Select Columns from the View menu while the Processes tab is displayed.

On a final note about the Task Manager, you can also minimize it to the size of a small icon on the Taskbar along the right corner. This icon is a representation of CPU Usage only. The Task Manager is minimized by clicking on the horizontal bar in the upper right corner of the dialog box. You can reinitialize the full size of the Task Manager by double-clicking on the small icon in the Task Bar.

CHAPTER SUMMARY

The intent of this chapter is to provide an overview of the performance monitoring tools available in Windows 2000. The Performance Monitor for tracking system performance is one of the most effective tools available in Windows 2000 as the object:counter relationships are useful for troubleshooting system-level issues. As a system administrator, trying to uncover sporadic problems is one of the more challenging aspects of that role. Taking the historical perspective on objects and their associated counters is invaluable in spotting performance-limiting aspects of overall system variable interactions. The Task Manager is also a convenient tool you can use to teach many of your internal customers how they can quickly spot and troubleshoot the performance of their workstations on their own. The Task Manager's metrics, though only a snapshot, are extremely useful for quickly resolving performance issues.

14 The Future of Windows 2000 and Interoperability

Creating intellectual capital, and empowering others to do so via their tools are the most valuable contributions software companies have made to the industries they participates in and sells to. Never before has the leverage of human creativity and talent been translated into tools that assist others in leveraging their own core competencies. Microsoft may be selling software on the surface, but at a deeper level, it is selling the ability to be more efficient, more adept at handling an increasingly competitive and complex world of business. This is its value proposition, and the future continues to raise challenges about how the equation of delivering value to clients changes and includes new variables, the latest being Linux and the concept of network and Internet appliances.

The intent of this chapter is to delve into the issues surrounding Windows 2000 and, in the process, gauge its impact on the arena of network operating systems when it is released.

EXPLORING MICROSOFT'S PRODUCT DIFFERENTIATION

Microsoft's differentiation of the Windows NT operating systems into four products makes predicting the future of the entire Windows NT situation slightly more difficult. Of Microsoft's recent changes, the following six variations of Windows NT (now Windows 2000) exist:

- Windows 2000 Professional Workstation
- Windows 2000 Server
- Windows 2000 Advanced Server Enterprise Edition

- Windows 2000 DataCenter Server
- BackOffice Small Business Server
- BackOffice Server, Enterprise Edition

FOCUSING ON WINDOWS 2000 PROFESSIONAL

Beginning in development as part of the OS/2 operating system Microsoft was building for IBM, Windows NT 3.1 was the first full operating system to reflect the partnership between IBM and Microsoft. Discussions with IBM began to change the nature of the operating system, and it's been widely reported that OS/2 was to be an outsourced product from Microsoft in the IBM strategy.

The structure of Windows NT (now Windows 2000) closely parallels a server-based product design, specifically down to the modularity of the actual structure of the kernel. Windows 2000 does not have the level of applications and server-level performance that Windows 2000 Server does, as Microsoft realized that there would be a gradual migration away from PCs at the low end of the market and UNIX workstations in the midrange and high end. One of the troubling aspects of Windows 2000 is the limitation of only 10 workstation computers being able to be connected to a system running Windows 2000 Workstation at any time. Microsoft's efforts to differentiate Windows 2000 Server with Workstation have made this 10-license issue a major discussion point with each new version. Possibly Microsoft will relent and give clients the flexibility to create lower-end servers under licensing options in the future, allowing for more than 10 connections to a Windows 2000 Workstation.

DIRECTION OF THE MARKET

The ongoing education and growing sophistication of network operating systems customers are driving the innovation and future direction of Windows NT, Linux, NetWare, and every other key operating system available today. The future of network operating systems is dictated by the needs of the customers each company is striving to serve and win over. Looking into the future from the perspective of the customer brings the functionality instantaneously into focus as this: bringing solutions to problems network-based organizations face. Competition when focused on a comparable set of customer needs drives innovation and price/performance—just as sprinters in a race, each by virtue of his or her passion for performance, drives the others to higher performance. It's a good market dynamic that Linux is actually on the computing landscape; the same holds true for NetWare. The world would be pretty boring if there were no variety of network operating systems, peripherals, systems, and network solutions. The alignment of each network operating systems' strengths relative to the needs of the customers is what competition in this arena is all about. The intent of this section is to define the market dynamics and user needs that are indicative of the future direction of network operating systems. Let's begin with an overview of the path to Windows 2000 on which Microsoft has been embarked since its first release of NT in 1994.

THE PATH TO WINDOWS 2000

Why will customers upgrade to Windows 2000? A recent survey completed by Microsoft's Research Department in conjunction with International Data Corporation (IDC) asked this question. Figure 14.1 displays the reasons why companies are likely to upgrade. The number one reason given is the new features and capabilities of Windows 2000, and the number two reason is the desire to stay up-to-date with current technology. These reasons emphasize why it is important that you and your staff are trained and ready to meet the upcoming customer demand for Windows 2000.

While many respondents identified other or nonspecific reasons for upgrading, those reasons mentioned specifically, in order of frequency, include Directory Services, security, better/more tools, better integration, and fault recovery.

Solution providers developing solutions centered on planning for and deploying Windows 2000 will perhaps be the first to benefit from new business opportunities. The new Active Directory services may require a rethinking of current domain models, and your customers will need your expertise in evaluating the options. They will need solutions that address staged upgrades and migrations, allowing their current infrastructure to operate and coexist with the new operating system.

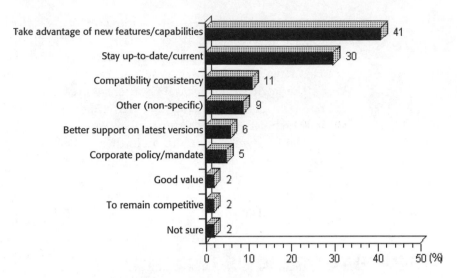

FIGURE 14.1 Reasons for upgrading Windows 2000, courtesy of International Data Corporation.

Which Applications Are Being Planned?

Research shows that customers now evaluating network operating systems for future deployment want to create line-of-business applications that are available to all types of systems, whether they are PCs or terminals. Often these applications must also be available via the Web. As we expand the reach of these applications, businesses become more and more reliant on them. Thus, they must be built on a scalable platform—one that gives end users fast response times but also allows the IT staff to exploit the latest in hardware and software technologies. Windows 2000 Professional is the best platform on which to run desktop applications. Windows 2000 Server is the best platform on which to build and run server-based, mission-critical business applications, with features such as large memory support, cluster services, terminal services, and integrated application and directory services.

INTERNET AND INTRANET MARKET REQUIREMENTS REFLECTED IN WINDOWS 2000

Today more than ever, organizations need to implement solutions that will enable users to easily share and access information. Windows 2000 makes information publishing and dissemination via the Internet and intranets manageable with its Internet Information Services (IIS) version 5.0, Net-Show services, and Internet printing. The intent of this section is to briefly review each of these areas and discuss the research completed by IDC in conjunction with Microsoft.

Internet and Intranet Technology Trends

According to the IDC report findings, 70 percent of organizations planning to upgrade to Windows 2000 are also planning to implement or adopt a Web site. Among current environments, customers using Windows NT have higher Intranet usage than those using NetWare. In addition, customers using Windows NT are using more types of technologies overall, most noticeably in terms of virtual meetings and VPNs.

Figure 14.2 presents the percentage of sites employing a wide variety of server-related technologies. Overall, these organizations are implementing an average of four technologies. Among the list, intranets (80 percent) are the most common, followed by backup tape libraries (64 percent) and CD-ROM jukeboxes (51 percent). Almost half (44 percent) are hosting a Web site for electronic commerce or customer support, and slightly more than

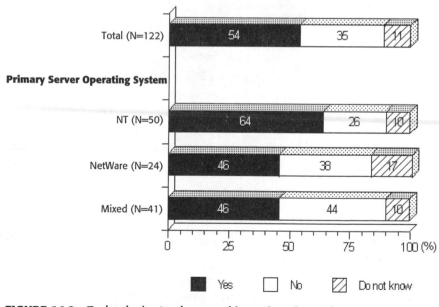

FIGURE 14.2 Technologies Implemented in conjunction with Windows 2000.

one-third (37 percent) have extranets in order to conduct business with partners and/or vendors. Many of these organizations have plans to adopt new technologies, with the most interest in Virtual Private Networks (VPN) (35 percent).

Windows 2000 Internet-based Product Direction

One of the true barometers of change in the area of network operating systems is the competition to add Internet/intranet/extranet functionality as part of the operating system. Linux, through the ongoing and proactive support of a global village of developers, continually adds functionality through a community-like contribution of effort. In response to the legitimization of the Internet as a distribution channel by the likes of Dell, Gateway, Amazon.com, and others, the need to provide a higher level of functionality than ever before has been Microsoft's challenge during the last three years, leading to the Windows 2000 launch in February 2000. Table 14.1 explores the key features of Windows 2000 in response to the explosive growth of the Internet.

TABLE 14.1 Key Features of Windows 2000

Internet Information Services (IIS)	IIS 5.0 technology extends the file-sharing capabilities of Windows 2000 Server to the Web and allows you to easily create transparent access to file servers on Windows 2000-, Novell NetWare-, and UNIX-based servers.
Internet Printing Protocol (IPP)	IPP is the latest Internet standard that lets users print directly to a URL, view printer status using a browser, and install drivers from a URL.
NetShow Services	NetShow Services is a scalable streaming media solution handling up to 1400 simultaneous streams from a single-CPU server. It supports connectionless multicast, giving customers the ability to deliver a single stream to hundreds or thousands of people simultaneously; thereby reducing the network bandwidth utilization normally used in Unicast streaming.
Internet Explorer 5.0	Microsoft Internet Explorer version 5.0 is the next version of Microsoft's Web browser software. Integrated into Windows 2000, some of its features are intelligent caching, enhanced security, and faster performance.

(Continues)

TABLE 14.1 (*Continued*)

DirectX 7.0	Microsoft DirectX® API is the multimedia driver architecture for the Windows platforms. DirectX on Windows95 and Windows98 enjoys pervasive support across all major display, audio, and input devices and compatibility with thousands of applications. Because Windows 2000 Professional requires its own set of drivers, a key focus is to encourage IHVs to write compatible drivers.
Open GL 1.2	Windows 2000 Professional will support the new OpenGL 1.2 specification. OpenGL is the standard 3-D graphics-programming interface for professional CAD and scientific visualization applications.

CONNECTIVITY WITH OTHER NETWORK OPERATING SYSTEMS

If you've been a system administrator or worked within an environment where multiple operating systems are being used, tools included in an operating system to enable one to speak with another can be invaluable. TCP/IP's ability to traverse operating systems and provide fundamental file-level compatibility is the foundation of many client/server networks throughout organizations. With the advent of specific server-based applications on UNIX and NetWare that are compatible with Windows 2000, the need for cross-compatibility tools in the latter operating system has become more pronounced. Let's briefly explore each of the tools included in Windows 2000 that ensure connectivity with other networking operating systems. For example, the Services for UNIX provide cross-compatibility between Windows 2000 and UNIX-based workstations and servers. The need to connect with legacy systems is a challenge many system administrators face. The options listed in Table 14.2 were specifically developed to alleviate the dilemmas of working with legacy applications and data.

TABLE 14.2   Windows 2000 Connectivity Tools

Active Directory Service Interface (ADSI)	ADSI abstracts the capabilities of directory services from different network providers to present a single set of directory service interfaces for managing network resources.
Domain Name System (DNS)	DNS is the distributed namespace used on the Internet to resolve computer and service names to TCP/IP addresses. Dynamic DNS reduces network administration costs by reducing the need to manually edit and replicate the DNS database each time a change occurs in a DNS client's configuration.
The Directory Service Migration Tool	This tool allows for offline shaping, modeling, and exporting of the NDS tree from the NetWare environment to the Active Directory environment.
Services for UNIX	This is an add-on product for Windows 2000 Server and Windows 2000 Professional that provides core network interoperability with existing UNIX environments and synchronization of user passwords between Windows 2000-based systems and UNIX. Users can map to UNIX resources using Microsoft Explorer just as they would to connect to any Windows-based resources—no cryptic UNIX commands are required.

NETWORK OPERATING SYSTEM CONNECTIVITY TOOLS IN WINDOWS 2000

Working through the issues of supporting dual platforms and, as a result, dual operating systems in any organization can be a challenge. The need to provide seamless interoperability between multiple systems was first bridged by TCP/IP's becoming the dominant connectivity solution, followed by the third-party tools available from companies specializing in connectivity tools to legacy systems. In focusing on aligning Windows 2000 to the needs of clients, Microsoft has over time introduced tools into Windows NT 4.0 through Service Packs and as stand-alone components available over its Web site.

EXPLORING THE FUTURE OF NETWORKING AND COMMUNICATIONS

Efficient, secure, and cost-effective networking and communications are vital to the smooth operation of any business. Managing networking and communications is the cornerstone of any company's information infrastructure. As the number of remote users that an IT staff must support increases monthly, laptops and notebook computers figure strongly into organizations' hardware infrastructures. Companies need guidance to plan for and support the technological challenges of remote computing.

Windows 2000 technologies can be used to enhance the networking environment, providing greater bandwidth control, secure remote network access, and native support for a new generation of communications solutions. Solution providers with expertise in Windows 2000 Server and Windows 2000 Professional will find enormous opportunities for new revenue by bringing networking, communications, and routing solutions to customers.

Enabling Secure, Cost-effective Mobile and Remote Computing

With the majority of systems in many companies being portable computers, the need for more efficient networking is essential for providing the security and reliability of Windows 2000 on portables. Table 14.3 shows the features included in Windows 2000 in response to the needs of portable or laptop.

TABLE 14.3 Windows 2000 Features for Mobile Computing

Offline file and folder access	Users can have full, automatic access (cached) to network files, folders, entire mapped network drives, and UNC paths when fully disconnected from the network (similar to Microsoft Outlook® folders).
Offline content synchronization	Synchronization Manager manages file synchronization. Synchronization works for offline files and folders, Web pages, and applications exploiting Synchronization Manager extensions.

(Continues)

TABLE 14.3   *(Continued)*

Encrypting Files System	The Encrypting File System (EFS) is a feature of the Windows NT File System 5.0 (NTFS). It provides a level of security above standard NTFS permissions by using public/private key encryption.
IP Security protocol (IPSEC)	IPSEC is an Internet Engineering Task Force (IETF) standard for encrypting TCP/IP traffic across a network using public-key encryption methods. Windows 2000 tightly integrates IPSEC with system policy management to enforce encryption between systems, transparent to the end user. IPSEC can be used for both private and virtual private network (VPN) communications.
Smart Card infrastructure	Smart cards are a key component of the public-key infrastructure that Microsoft is integrating into the Windows platform because smart cards enhance software-only solutions, such as client authentication, single sign-on, secure storage, and system administration.
Layer 2 Tunneling Protocol (L2TP)	L2TP is an IETF draft specification for encapsulating and transmitting non-IP traffic through TCP/IP networks. Because L2TP utilizes IPsec for encrypted transmission, L2TP requires a public-key infrastructure for VPN use.
Point-to-Point Tunneling Protocol (PPTP)	PPTP is a multivendor-defined protocol that has been widely adopted for use in creating virtual private networking solutions. PPTP offers tunneling services to support non-TCP/IP based protocols.
Extensible Authentication Protocol (EAP)	This is an extensible addition to the PPP authentication methods for dial-up and VPN connections. It allows ISVs to supply new authentication modules for client and server for token cards, smart cards, biometric hardware, and so on.

Gaining Control of Network Bandwidth and Services

Network bandwidth is at a premium, and the demand for bandwidth keeps growing. The IT staff wants to control network bandwidth usage to ensure that the truly critical services have high-quality bandwidth when they need it. In order to build the next generation of bandwidth-intensive applications it is imperative that functionality is integrated with the operating system and uses network bandwidth as efficiently as possible. Table 14.4 lists Windows 2000's networking features.

TABLE 14.4   Windows 2000 Network Services

Windows Quality of Service (QoS)	Windows QoS is a set of service requirements that the network must meet while transmitting a flow of data. QoS-based services and protocols provide a guaranteed, end-to-end, express delivery system for Internet Protocol (IP) traffic.
RSVP	RSVP enables applications (primarily multimedia) to obtain the necessary service quality from the network. It also enables network administrators to manage the effect of these applications on network resources. Interoperates with Diff-serv.
Directory-Enabled Networking (DEN)	DEN is the product of work championed by Microsoft and Cisco to define and implement networking features that are tightly integrated with directory service functionality.

Network Appliances and Terminals: Implications for Windows 2000

In looking at the role of Windows 2000 relative to the highly hyped network computer (NC), there is an interesting dichotomy for everyone who relies on networks for their jobs: Is this NC a new, innovative approach to rapidly driving TCO out of products? Or is the NC a solution in search of a problem? It seems like the Chinese symbol that represents both disaster and opportunity: It's all in your perspective of the issue at hand. Clearly, the NC has been heralded as having very high, fast-growing forecasts, which have failed to come to fruition.

Let's look at the key aspects of the NC, and discuss the response from the industry to these key trends.

Total cost of ownership. You could argue that the key aspects of the NC are its very low entry costs and limited number of components, which would feasibly lead to a lower cost of ownership. The industry has found Java to be intriguing and interesting, finding it used throughout the Internet and some of the world's most traveled sites. Yet, the industry has yet to fully adopt the argument of lowered TCO with network computers. Instead, there will continue to a recognition that TCO is a very small part of the total cost of running an enterprise, and while an NC is inexpensive, the training and support costs of yet another platform are not. The industry seems to be cautious about the idea of another platform with its own operating system.

Low cost to purchase. The NC was originally priced below $600, then fell below $300 before becoming less and less featured. The NC's low cost is one of its strongest features. With the operating system built in on the motherboard, there is a very low cost to manufacture.

No operating system royalties. One of the biggest benefits of the NC is that it doesn't require an operating system royalty, which would affect its price and resulting attractiveness to organizations. While the royalties have been minimal, so has the willingness of organizations to take on an embedded operating system for accessing the Internet. It's the same dynamic with Linux as IS organizations question the payoff of yet another operating system when they have enough to do with existing operating systems already.

TCP/IP compatible in a small form factor. The continued drop in prices on the corporate workstation, server, and desktop markets is going to aggressively continue through 2002 and beyond. What was originally the foundation of the NC focus of providing fully featured Pentium II class and better systems has been replaced with a very low-cost platform without a disk drive and expensive (then) graphics. The rapid drop in price bands for Intel-based desktops, workstations, and servers has hastened the competition between NCs and traditional Intel-based systems.

KEY TECHNOLOGY TRENDS AND PREDICTIONS

In looking into the future of the Internet and the role of Windows 2000 within it, the following predictions are made. What's apparent is that the Internet is changing the nature of how companies and organizations will interact with each other, and the nature of mergers in the coming years. In integrating Windows 2000 into your organization, it's a good idea to check the trends shown here to make sure your organizations' strategy can accommodate the magnitude of the changes that will come. These predictions are the result of discussions with Microsoft, International Data Corporation, Gateway, and other leading companies in the industry.

Product Implications for Windows 2000 The trends will have implications for Windows 2000:

PCs below $1,000 will run Windows 2000. With Microsoft's Windows NT-Everywhere strategy, there will be PCs that cost $1K or less with Windows 2000 running on them during 2000. The ability to manage these systems through Exchange Server, Terminal Server, and Systems Management Server will continue to make obsolete the concept of network computers.

Home networking will grow quickly. The need to share multiple systems throughout a household will become increasingly important, as more and more companies purchase portable computers for their employees to use during the evening and on weekends. The role of Windows 2000 networking will continue to stress smaller, TCP/IP-compliant networks to ensure data reliability and security.

Personalization/customization will be the "ante" for successful commercial sites. IDC predicts that users who have personalized a site visit with frequency two to four times that of users who have not. IDC also predicts that over 60 percent of the leading Web commerce sites already have some form of personalization in place.

Personal merchandising redefines Web sites. There will be an increasing need on the part of organizations to provide a highly customized shopping experience for their customers, as evidenced by the approach amazon.com took to customizing the front page to reflect the books and CDs you have purchased in the past.

Internet stock correction will occur. Those organizations that have, as an explicit strategy, leveraged the market valuations of Internet stocks will find the results for the future mixed. As the focus moves away from being "live" on the Internet, companies that take the next step and provide personalization and stability will reap the rewards of continued market valuations.

Reliance on Site Server functionality will grow. Handling the traffic on Web sites, then customizing the overall appearance of the HTML pages generated, is possible with Site Server running on Windows 2000 Server, Enterprise Edition today. The need for this level of customization at the operating system level is coming.

Networking connectivity will become effortless. With the majority of new Internet sites coming from small businesses during the coming years, the need to generate Web sites and accommodate networking needs within an organization is going to be paramount. The productizing of networking experience and service is coming, and it will be aggressively sold over the Internet to small organizations on a global level.

Focus on global reach will expand. The long-heralded internationalization of commerce and the erasing of geographic boundaries will be more pronounced in the coming three to five years than has ever before. Over 51 percent of global electronic commerce will be generated outside the United States within the coming three years. This shift makes localization of content, the ability to quickly localize Web sites, and the function of absorbing the needs of foreign customers, then integrating them into a global product offering, the difference between those companies who survive and those that don't. Companies will need to be aggressively positioned on the global level to continually outpace market growth and therefore grow over the next three years.

Transition of two-tier to direct channels, especially in education, will occur. Just as the computing industry was transformed from a dual-tier channel to a direct one within a matter of months instead of years, the "Dellizing" of industries will happen with increasing frequency and velocity. The education industry will be the most transformed area of all, due to the advent of the Internet, and the role of Windows 2000 in supporting this market dynamic is already apparent. Suddenly, educational institutions unaccustomed to seeing each class in the context of return on investment (ROI) will refocus and apply business concepts to their curricula, aligning knowledge conveyed and generated with the needs of the marketplace. This alignment is now underway, and it will become increasingly apparent in the coming years. Windows 2000 will be taught more than ever before as the ability to manage Web-based initiatives will become as common as learning the fundamentals of accounting in core business classes.

CHAPTER SUMMARY

Knowledge gives us the ability to seize opportunities on the horizon, both at a personal and organizational level; specifically, knowledge of Windows 2000 Professional and Server will be in high demand today and in the coming years. This chapter has focused on the alignment of customer needs with the future direction of Windows 2000, and although there are needs still not met within this operating system, the focus on the future provides a roadmap for planning in adopting Windows 2000 into the formative and emerging business areas your organization is pursuing.

Of particular relevance in the context of Windows 2000 is the continual role of the Internet in the business plans of organizations globally. The result of Windows 2000 will be many enhancement products not even considered by Microsoft yet. The market dynamics surrounding the high-growth markets in which Windows 2000 is participating will drive subindustries that will, in turn, drive the "Dellizing" of entire industries in a matter of months instead of the years other industry transformations have taken in the past. Education and its refocus on the bottom line is just one example of how the Internet is silently yet quickly bringing competition in an entirely new level of customer responsiveness. Industries will begin to consolidate in response to the quickness of change, aligning more accurately with user needs than before. The Internet's implications promise to sharpen competition, responsiveness, and the depth of applications delivered over the Web. The result? More value to customers at lower cost.

Windows 2000 is uniquely positioned as a network operating system adaptable to organizations looking to seize the future rather than wait for the future to happen. Looking at Windows 2000 as a foundation for the future of your organization's information strategy provides legacy support for previous applications and their data, and a fourth-generation operating system that is flexible enough for the demands of an increasingly borderless global economy.

Troubleshooting TCP/IP Configurations

When troubleshooting any problem, it is helpful to use a logical approach. Some questions to ask are these:

- What can be reliably done?
- Where do the TCP/IP connections stop working?
- What is the characteristic of the problem?
- Have the things that do not work ever worked on this computer/ network?
- If so, what has changed since they last worked?

Troubleshooting a problem from the bottom up is often a good way to isolate the problem quickly. The tools listed here are organized for this approach.

IPCONFIG

IPConfig is a command-line utility that prints out the TCP/IP-related configuration of a host. When used with the **/all** switch, it produces a detailed configuration report for all interfaces, including any configured serial ports (RAS). Output can be redirected to a file and pasted into other documents:

    C:\>ipconfig /all

Windows 2000 IP Configuration

    Host Name . . . . : HERCULES
    Primary DNS Suffix . . . . : test.emarkets.com

Node Type . . . . : Hybrid
IP Routing Enabled. . . . . : No
WINS Proxy Enabled. . . . . : No

Ethernet adapter Local Area Connection 2:
Connection-specific DNS Suffix . :
Description . . . . : Intel EtherExpress
Physical Address. . . . . : 00-20-AF-1D-2B-91
DHCP Enabled. . . . . : No
IP Address. . . . . : 10.57.8.190
Subnet Mask . . . . : 255.255.255.0
Default Gateway . . . . :
   DNS Servers . . . . : 10.57.9.254
   Primary WINS Server . . . . : 10.57.9.254
Ethernet adapter Local Area Connection:Connection-specific DNS Suffix . :
Description . . . . : AMD PCNET Family PCI Ethernet Adapter
Physical Address. . . . . : 00-80-5F-88-60-9A
DHCP Enabled. . . . . : No
IP Address. . . . . : 199.199.40.22
Subnet Mask . . . . : 255.255.255.0
Default Gateway . . . . : 199.199.40.1
DNS Servers . . . . : 199.199.40.254
Primary WINS Server . . . . : 199.199.40.254

PING

Ping is a tool that helps to verify IP-level reachability. The **ping** command can be used to send an ICMP echo request to a target name or IP address. First you should ping the IP address of the target host to see if it responds because this is the simplest test. If that succeeds, try pinging the name. **Ping** uses Windows Sockets-style name resolution to resolve the name to an address; therefore, if pinging by address succeeds but pinging by name fails, the problem lies in name resolution, not network connectivity.

Type **ping -?** to see what command-line options are available. **Ping** allows you to specify the size of packets to use, how many to send, whether to record the route used, what TTL value to use, and whether to set the **don't fragment** flag. See the PMTU discovery section of this document for details on using ping to manually determine the PMTU between two computers.

The following example illustrates how to send two pings, each 1450 bytes in size, to address 10.99.99.2:

```
C:\>ping -n 2 -l 1450 10.99.99.2

Pinging 10.99.99.2 with 1450 bytes of data:

Reply from 10.99.99.2: bytes=1450 time<10ms TTL=32
Reply from 10.99.99.2: bytes=1450 time<10ms TTL=32

Ping statistics for 10.99.99.2:
    Packets: Sent = 2, Received = 2, Lost = 0 (0% loss),
Approximate round trip times in milli-seconds:
    Minimum = 0ms, Maximum =  0ms, Average =  0ms
```

By default, **ping** waits one second for each response to be returned before timing out. If the remote system being pinged is across a high-delay link, such as a satellite link, responses could take longer to be returned. The -**w** (wait) switch can be used to specify a longer time-out. Computers using IPSec may require several seconds to set up a security association before they respond to a ping.

ARP

The **arp** command is useful for viewing the ARP cache. If two hosts on the same subnet cannot ping each other successfully, try running the **arp -a** command on each computer to see if the computers have the correct MAC addresses listed for each other. Use **IPConfig** to determine a host's media access control address. If another host with a duplicate IP address exists on the network, the ARP cache may have had the media access control address for the other computer placed in it. Use **arp -d** to delete an entry that may be incorrect. Add entries by using **arp -s**.

TRACERT

Tracert is a route-tracing utility. **Tracert** uses the IP TTL field and ICMP error messages to determine the route from one host to another through a network. Sample output from the **tracert** command is shown in the ICMP section of this document.

ROUTE

Route is used to view or modify the route table. **Route print** displays a list of current routes known by IP for the host. Sample output is shown in the IP section of this document. Note that in Windows 2000 the current active default gateway is shown at the end of the list of routes. **Route add** adds routes to the table. **Route delete** removes routes from the table.

Routes added to the table are not made persistent unless the -**p** switch is specified. Nonpersistent routes last only until the computer is rebooted. For two hosts to exchange IP datagrams, they must both have a route to each other, or they must use a default gateway that knows of a route. Normally, routers exchange information with each other by using a protocol such as Routing Information Protocol (RIP) or Open Shortest Path First (OSPF). Silent RIP is available for Windows 2000 Professional, and full routing protocols are supported by Windows 2000 Server in the Routing and Remote Access Service.

NETSTAT

Netstat displays protocol statistics and current TCP/IP connections. **Netstat -a** displays all connections, and **netstat -r** displays the route table and any active connections. The -**n** switch tells **netstat** not to convert addresses and port numbers to names, which speeds up execution. The -**e** switch displays Ethernet statistics and may be combined with the -**s** switch, which shows protocol statistics. Sample output is shown here:

```
C:\>netstat -e
Interface Statistics

                              Received              Sent

Bytes                       372959625         123567086
Unicast packets                134302            145204
Non-unicast packets             55937               886
Discards                            0                 0
Errors                              0                 0
Unknown protocols             1757381


C:\>netstat -an
```

Active Connections

```
Proto  Local Address         Foreign Address
   State
TCP    0.0.0.0:1723          0.0.0.0:0
   LISTENING
   TCP     0.0.0.0:3268         0.0.0.0:0
   LISTENING
TCP    10.99.99.1:389        10.99.99.1:1092
   ESTABLISHED
D:\>netstat -s
IP Statistics
  Packets Received                      = 3175996
  Received Header Errors                = 0
  Received Address Errors               = 38054
  Datagrams Forwarded                   = 0
  Unknown Protocols Received            = 0
  Received Packets Discarded            = 0
  Received Packets Delivered            = 3142564
  Output Requests                       = 3523906
  Routing Discards                      = 0
  Discarded Output Packets              = 0
  Output Packet No Route                = 0
  Reassembly Required                   = 0
  Reassembly Successful                 = 0
  Reassembly Failures                   = 0
  Datagrams Successfully Fragmented     = 0
  Datagrams Failing Fragmentation       = 0
  Fragments Created                     = 0
```

ICMP Statistics

	Received	Sent
Messages	462	33
Errors	0	0
Destination Unreachable	392	4
Time Exceeded	0	0
Parameter Problems	0	0
Source Quenchs	0	0
Redirects	0	0
Echos	1	22
Echo Replies	12	1
Timestamps	0	0
Timestamp Replies	0	0
Address Masks	0	0
Address Mask Replies	0	0

TCP Statistics

Active Opens	= 12164
Passive Opens	= 12
Failed Connection Attempts	= 79
Reset Connections	= 11923
Current Connections	= 1
Segments Received	= 2970519
Segments Sent	= 3505992
Segments Retransmitted	= 18

UDP Statistics

Datagrams Received	= 155620
No Ports	= 16578
Receive Errors	= 0
Datagrams Sent	= 17822

NBTSTAT

NBSTAT is a useful tool for troubleshooting NetBIOS name-resolution problems. **NBSTAT -n** displays the names that applications, such as the server and redirector, registered locally on the system. **NBSTAT -c** shows the NetBIOS name cache, which contains name-to-address mappings for other computers. **NBSTAT -R** purges the name cache and reloads it from the Lmhosts file. **NBSTAT –RR** (new in Windows 2000) reregisters all names with the name server. **NBSTAT -a** *name* performs a NetBIOS adapter status command against the computer that is specified by *name*. The adapter status command returns the local NetBIOS name table for that computer and the media access control address of the adapter card. **NBSTAT -s** lists the current NetBIOS sessions and their status, including statistics.

NSLOOKUP

Nslookup was added in Windows NT 4.0 and is a useful tool for troubleshooting DNS problems, such as host name resolution. When you start **nslookup**, it shows the host name and IP address of the DNS server that is configured for the local system and then displays a command prompt. If you type a question mark (?), **nslookup** shows the different commands that are available.

To look up the IP address of a host, using the DNS, type the host name and press Enter. **Nslookup** defaults to the DNS server that is configured for the computer that it is running on, but you can focus it on a different DNS server by typing **server** *name* (*name* is the host name of the server that you want to use for future lookups).

When you use **Nslookup**, you should be aware of the domain name devolution method. If you type in just a host name and press Enter, **nslookup** appends the domain suffix of the computer (such as cswatcp.microsoft .com) to the host name before it queries the DNS. If the name is not found, the domain suffix is devolved by one label (in this case, cswatcp is removed, and the suffix becomes microsoft.com). Then the query is repeated. Windows 2000 Professional workstations only devolve names to the second-level domain (microsoft.com in this example), so if this query fails, no further attempts are made to resolve the name. If a fully qualified domain name is typed in (as indicated by a trailing dot), the DNS server is queried only for that name and no devolution is performed. To look up a host name that is completely outside of your domain, you must type in a fully qualified name.

An especially useful troubleshooting feature is debug mode, which you can invoke by typing **set debug**, or for even greater detail, **set d2**. In debug mode, **nslookup** lists the steps being taken to complete its commands, as shown in this example:

```
C:\>nslookup
(null)    davemac3.cswatcp.microsoft.com
Address: 10.57.8.190

> set d2
> rain-city
(null)    davemac3.cswatcp.microsoft.com
Address: 10.57.8.190

_____

SendRequest(), len 49
    HEADER:
        opcode = QUERY, id = 2, rcode = NOERROR
        header flags: query, want recursion
        questions = 1,  answers = 0,  authority records
= 0,  additional = 0

    QUESTIONS:
        rain-city.cswatcp.microsoft.com, type = A, class
= IN

    _____

    _____

Got answer (108 bytes):
    HEADER:
        opcode = QUERY, id = 2, rcode = NOERROR
        header flags: response, auth. answer, want re-
cursion, recursion avail.
        questions = 1,  answers = 2,  authority records
= 0,  additional = 0

    QUESTIONS:
        rain-city.cswatcp.microsoft.com, type = A, class
= IN
    ANSWERS:
    -> rain-city.cswatcp.microsoft.com
        type = CNAME, class = IN, dlen = 31
        canonical name = seattle.cswatcp.microsoft.com
        ttl = 86400 (1 day)
```

```
    ->   seattle.cswatcp.microsoft.com
         type = A, class = IN, dlen = 4
         internet address = 10.1.2.3
         ttl = 86400 (1 day)


    (null)   seattle.cswatcp.microsoft.com
    Address: 10.1.2.3
    Aliases: rain-city.cswatcp.microsoft.com
```

In this example, **set d2** was issued to set **nslookup** to debug mode, then address lookup was used for the host name *rain-city*. The first two lines of output show the host name and IP address of the DNS server to which the lookup was sent. As the next paragraph shows, the domain suffix of the local machine (cswatcp.microsoft.com) was appended to the name *rain-city*, and **nslookup** submitted this question to the DNS server. The next paragraph indicates that **nslookup** received an answer from the DNS and that there were two answer records in response to one question.

MICROSOFT NETWORK MONITOR

Microsoft Network Monitor is a tool developed by Microsoft to make the task of troubleshooting complex network problems easier and more economical. It is packaged as part of the Microsoft Systems Management Server product, but it can be used as a stand-alone network monitor. In addition, Windows NT and Windows95 include Network Monitor Agent software, and Windows NT Server and Windows 2000 include a limited version of Network Monitor. Stations running Network Monitor can attach to stations running the agent software over the network or by using dial-up (RAS) to perform monitoring or tracing of remote network segments. This can be a very useful troubleshooting tool.

Network Monitor works by placing the NIC on the capturing host into promiscuous mode so that it passes every frame on the wire up to the tracing tool. (The limited version of Network Monitor that ships with Windows NT Server allows only traffic to and from the computer to be traced.) Capture filters can be defined so that only specific frames are saved for analysis. Filters can be defined based on source and destination NIC addresses, source and destination protocol addresses, and pattern matches. Once the frames have been captured, display filtering can be used to further narrow down a problem. Display filtering allows specific protocols to be selected as well.

Windows 2000–based computers use the Server Message Block (SMB) protocol for many functions, including file and print sharing. The smb.hlp file in the Netmon parsers directory is a good reference for interpreting this protocol.

Understanding TCP/IP Configuration Parameters

T he TCP/IP protocol suite implementation for Windows 2000 obtains all its configuration data from the Registry. This information is written to the Registry by the Setup program. Some of this information is also supplied by the Dynamic Host Configuration Protocol (DHCP) client service, if it is enabled. This appendix defines all of the Registry parameters used to configure the protocol driver, Tcpip.sys, which implements the standard TCP/IP network protocols.

The implementation of the protocol suite should perform properly and efficiently in most environments using only the configuration information gathered by Setup and DHCP. Optimal default values for all other configurable aspects of the protocols for most cases have been encoded into the drivers. Some customer installations may require changes to certain default values. To handle these cases, optional Registry parameters can be created to modify the default behavior of some parts of the protocol drivers.

Note: The Windows NT TCP/IP implementation is largely self-tuning. Adjusting Registry parameters may adversely affect system performance.

All of the TCP/IP parameters are Registry values located under the Registry key:

```
HKEY_LOCAL_MACHINE
    \SYSTEM
        \CurrentControlSet
            \Services:
                \Tcpip
                    \Parameters
```

Adapter-specific values are listed under subkeys for each adapter. Depending on whether the system or adapter is DHCP configured and/or static override values are specified, parameters may have both DHCP and statically configured values. If any of these parameters are changed using the Registry editor, a reboot of the system is generally required for the change to take effect. A reboot is usually not required if values are changed using the network connection interface.

PARAMETERS CONFIGURABLE USING THE REGISTRY EDITOR

The following parameters receive default values during the installation of the TCP/IP components. To modify any of these values, use the Registry Editor (Regedt32.exe). A few of the parameters are visible in the Registry by default, but most must be created to modify the default behavior of the TCP/IP protocol driver. Parameters configurable from the user interface are listed separately.

ArpAlwaysSourceRoute
```
Key: Tcpip\Parameters
Value Type: REG_DWORD—Boolean
Valid Range: 0, 1, or not present (False, True, or not
present)
Default: not present
```

Description: By default, the stack transmits ARP queries without source routing first and retries with source routing enabled if no reply is received. Setting this parameter to 0 causes all IP broadcasts to be sent without source routing. Setting this parameter to 1 forces TCP/IP to transmit all ARP queries with source routing enabled on Token Ring networks. (A change to the definition of the parameter was introduced in Windows NT 4.0 SP2.)

ArpCacheLife
```
Key: Tcpip\Parameters
Value Type: REG_DWORD—Number of seconds
Valid Range: 0–0xFFFFFFFF
```

Default: In absence of an *ArpCacheLife* parameter, the defaults for ARP cache time-outs are a two-minute time-out on unused entries and a ten-minute time-out on used entries.

Description: See *ArpCacheMinReferencedLife*

ArpCacheMinReferencedLife (note - not working properly in Beta-3)
```
Key: Tcpip\Parameters
Value Type: REG_DWORD—Number of seconds
Valid Range: 0-0xFFFFFFFF
Default: 600 seconds (10 minutes)
```

Description: *ArpCacheMinReferencedLife* controls the minimum time until a referenced ARP cache entry expires. This parameter can be used in combination with the *ArpCacheLife* parameter, as follows:

If *ArpCacheLife* is greater than or equal to *ArpCacheMinReferencedLife* referenced and unreferenced ARP cache entries expire in *ArpCacheLife* seconds.

If *ArpCacheLife* is less than *ArpCacheMinReferencedLife*, unreferenced entries expire in *ArpCacheLife* seconds, and referenced entries expire in *ArpCacheMinReferencedLife* seconds.

Entries in the ARP cache are referenced each time that an outbound packet is sent to the IP address in the entry.

ArpRetryCount
```
Key: Tcpip\Parameters
Value Type: REG_DWORD—Number
Valid Range: 1-3
Default: 3
```

Description: This parameter controls the number of times that the computer sends a gratuitous ARP for its own IP address (es) while initializing. Gratuitous ARPs are sent to ensure that the IP address is not already in use elsewhere on the network. The value controls the actual number of ARPs sent not the number of retries.

ArpTRSingleRoute
```
Key: Tcpip\Parameters
Value Type: REG_DWORD—Boolean
Valid Range: 0, 1 (False, True)
Default: 0 (False)
```

Description: Setting this parameter to 1 causes ARP broadcasts that are source-routed (Token Ring) to be sent as single-route broadcasts, instead of all-routes broadcasts.

ArpUseEtherSNAP
```
Key: Tcpip\Parameters
```

```
Value Type: REG_DWORD—Boolean
Valid Range: 0, 1 (False, True)
Default: 0 (False)
```

Description: Setting this parameter to 1 forces TCP/IP to transmit Ethernet packets using 802.3 SNAP encoding. By default, the stack transmits packets in DIX Ethernet format. It always receives both formats.

DatabasePath
```
Key: Tcpip\Parameters
Value Type: REG_EXPAND_SZ—Character string
Valid Range: A valid Windows NT file path
Default: %SystemRoot%\system32\drivers\etc
```

Description: This parameter specifies the path to the standard Internet database files (Hosts, Lmhosts, Network, Protocols, and Services). It is used by the Windows Sockets interface.

DefaultTOSValue
```
Key: Tcpip\Parameters
Value Type: REG_DWORD—Number
Valid Range: 0-255
Default: 0
```

Description: Specifies the default Type of Service (TOS) value set in the header of outgoing IP packets. See RFC 791 for a definition of the values. This may be overridden by a program using option IP_TOS (IPPROTO_IP level), provided that *DisableUserTosSetting* is not set, or by enabling QoS policy on the network.

DefaultTTL
```
Key: Tcpip\Parameters
Value Type: REG_DWORD—Number of seconds/hops
Valid Range: 1-0xff (1-255 decimal)
Default: 128
```

Description: Specifies the default time-to-live (TTL) value set in the header of outgoing IP packets. The TTL determines the maximum amount of time that an IP packet may live in the network without reaching its destination. It is effectively a limit on the number of routers that an IP packet may pass through before being discarded.

DisableDHCPMediaSense
```
Key:  Tcpip\Parameters
Value Type: REG_DWORD—Boolean
```

Valid Range: 0, 1 (False, True)
Default: 0 (False)

Description: This parameter can be used to control DHCP Media Sense behavior. If set to 1, the DHCP client will ignore Media Sense events from the interface. By default, Media Sense events trigger the DHCP client to take an action, such as attempting to obtain a lease (when a connect event occurs) or invalidating the interface and routes (when a disconnect event occurs).

DisableIPSourceRouting
```
Key:   Tcpip\Parameters
Value Type: REG_DWORD—Boolean
Valid Range: 0, 1 (False, True)
Default: 1 (True)
```

Description: IP source routing is a mechanism allowing the sender to determine the IP route that a datagram should take through the network, used primarily by tools such as tracert.exe. This parameter was added to Windows NT4.0 in Service Pack 5 per KB article Q217336. Windows NT 2000 disables IP source routing by default.

DisableMediaSenseEventLog
```
Key:   Tcpip\Parameters
Value Type: REG_DWORD—Boolean
Valid Range: 0, 1 (False, True)
Default: 0 (False)
```

Description: This parameter can be used to disable logging of DHCP Media Sense events. By default, Media Sense events (connection/disconnection from the network) are logged in the event log for troubleshooting purposes.

DisableTaskOffload
```
Key:   Tcpip\Parameters
Value Type: REG_DWORD—Boolean
Valid Range: 0, 1 (False, True)
Default: 0 (False)
```

Description: This parameter instructs the TCP/IP stack to disable offloading of tasks to the network card for troubleshooting and test purposes.

DisableUserTOSSetting
```
Key:   Tcpip\Parameters
```

```
Value Type: REG_DWORD—Boolean
Valid Range: 0, 1 (False, True)
Default: 1 (True)
```

Description: This parameter can be used to allow programs to manipulate the Type Of Service (TOS) bits in the header of outgoing IP packets. In Windows 2000, this defaults to True. In general, individual applications should not be allowed to manipulate TOS bits because this can defeat system policy mechanisms. See also *DefaultTOSValue*.

DontAddDefaultGateway

```
Key:  Tcpip\Parameters \Interfaces\<interface>
Value Type: REG_DWORD—Boolean
Valid Range: 0, 1 (False, True)
Default: 0
```

Description: When you install PPTP, a default route is installed for each LAN adapter. You can disable the default route on one of them by adding this value and setting it to 1. After doing so, you may need to configure static routes for hosts that are reached using a router other than the default gateway.

EnableAddrMaskReply

```
Key: Tcpip\Parameters
Value Type: REG_DWORD—Boolean
Valid Range: 0, 1 (False, True)
Default: 0 (False)
```

Description: This parameter controls whether the computer responds to an ICMP address mask request.

EnableBcastArpReply

```
Key: Tcpip\Parameters
Value Type: REG_DWORD—Boolean
Valid Range: 0, 1 (False, True)
Default: 1 (True)
```

Description: This parameter controls whether the computer responds to an ARP request when the source Ethernet address in the ARP is not unicast. Network Load Balancing Service (NLBS) will not work properly if this value is set to 0.

EnableDeadGWDetect

```
Key: Tcpip\Parameters
Value Type: REG_DWORD—Boolean
```

Valid Range: 0, 1 (False, True)
Default: 1 (True)

Description: When this parameter is 1, TCP is allowed to perform dead-gateway detection. With this feature enabled, TCP may ask IP to change to a backup gateway if a number of connections are experiencing difficulty. Backup gateways may be defined in the Advanced section of the TCP/IP configuration dialog in the Network Control Panel.

EnableFastRouteLookup
Key: Tcpip\Parameters
Value Type: REG_DWORD—Boolean
Valid Range: 0, 1 (False, True)
Default: 0 (False)

Description: Fast route lookup is enabled if this flag is set. This can make route lookups faster at the expense of nonpaged pool memory. This flag is used only if the computer is a Windows NT Server and falls into the Medium or Large class (contains at least 64 MB of memory). This parameter is created by the Routing and Remote Access Service.

EnableMulticastForwarding
Key: Tcpip\Parameters
Value Type: REG_DWORD—Boolean
Valid Range: 0, 1 (False, True)
Default: 0 (False)

Description: The routing service uses this parameter to control whether IP multicasts are forwarded. This parameter is created by the Routing and Remote Access Service.

EnablePMTUBHDetect
Key: Tcpip\Parameters
Value Type: REG_DWORD—Boolean
Valid Range: 0, 1 (False, True)
Default: 0 (False)

Description: Setting this parameter to 1 (True) causes TCP to try to detect *black hole* routers while doing Path MTU Discovery. A black hole router does not return ICMP Destination Unreachable messages when it needs to fragment an IP datagram with the Don't Fragment bit set. TCP depends on receiving these messages to perform Path MTU Discovery. With this feature enabled, TCP tries to send segments without the Don't Fragment bit set if several retransmissions of a segment go unacknowledged. If the segment is acknowledged as a result, the MSS is decreased and the Don't Fragment bit is set in future packets on the connection. Enabling black hole detection increases the maximum number of retransmissions that are performed for a given segment.

EnablePMTUDiscovery
```
Key: Tcpip\Parameters
Value Type: REG_DWORD—Boolean
Valid Range: 0, 1 (False, True)
Default: 1 (True)
```

Description: When this parameter is set to 1 (True) TCP attempts to discover the Maximum Transmission Unit (MTU or largest packet size) over the path to a remote host. By discovering the Path MTU and limiting TCP segments to this size, TCP can eliminate fragmentation at routers along the path that connect networks with different MTUs. Fragmentation adversely affects TCP throughput and network congestion. Setting this parameter to 0 causes an MTU of 576 bytes to be used for all connections that are not to hosts on the local subnet.

FFPControlFlags
```
Key: Tcpip\Parameters
Value Type: REG_DWORD—Boolean
Valid Range: 0, 1 (False, True)
Default: 1 (True)
```

Description: If this parameter is set to 1, Fast Forwarding Path (FFP) is enabled. If it is set to 0, TCP/IP instructs all FFP-capable adapters not to do any fast forwarding on this computer. Fast Forwarding Path-capable network adapters can receive routing information from the stack and forward subsequent packets in hardware without passing them up to the stack. FFP parameters are located in the TCP/IP Registry key, but they are actually placed there by the Routing and Remote Access Service (RRAS) service. See the RRAS documentation for more details.

FFPFastForwardingCacheSize
```
Key: Tcpip\Parameters
Value Type: REG_DWORD—Number of bytes
Valid Range: 0–0xFFFFFFFF
Default: 100,000 bytes
```

Description: This is the maximum amount of memory that a driver that supports fast forwarding (FFP) can allocate for its fast-forwarding cache if it uses system memory for its cache. If the device has its own memory for fast-forwarding cache, this value is ignored.

ForwardBroadcasts
Key: Tcpip\Parameters
Value Type: REG_DWORD—Boolean
Valid Range: 0, 1 (False, True)
Default: 0 (False)

Description: Forwarding of broadcasts is not supported. This parameter is ignored.

ForwardBufferMemory
Key: Tcpip\Parameters
Value Type: REG_DWORD—Number of bytes
Valid Range: <network MTU>- <some reasonable value
  smaller than 0xFFFFFFFF>
Default: 74240 (enough for fifty 1480-byte packets,
  rounded to a multiple of 256)

Description: This parameter determines how much memory IP allocates initially to store packet data in the router packet queue. When this buffer space is filled, the system attempts to allocate more memory. Packet queue data buffers are 256 bytes in length, so the value of this parameter should be a multiple of 256. Multiple buffers are chained together for larger packets. The IP header for a packet is stored separately. This parameter is ignored, and no buffers are allocated if the IP routing function is not enabled. MaxForwardBufferMemory controls the maximum amount of memory that can be allocated for this function.

GlobalMaxTcpWindowSize
Key: Tcpip\Parameters
Value Type: REG_DWORD—Number of bytes
Valid Range: 0-0x3FFFFFFF (1073741823 decimal; however,
  values greater than 64 KB can be achieved only when
  connecting to other systems that support RFC 1323 Window scaling, which is discussed in the TCP section of
  this document. Additionally, Window scaling must be
  enabled using the Tcp1323Opts Registry parameter.)
Default: This parameter does not exist by default.

Description: The *TcpWindowSize* parameter can be used to set the receive window on a per-interface basis. This parameter can be used to set a global limit for the TCP window size on a system-wide basis. This parameter is new in Windows 2000.

IPAutoconfigurationAddress
Key: Tcpip\Parameters\Interfaces\<interface>

```
Value Type: REG_SZ—String
Valid Range: A valid IP address
Default: None
```

Description: The DHCP client stores the IP address chosen by auto-configuration here. This value should not be altered.

IPAutoconfigurationEnabled
```
Key: Tcpip\Parameters, Tcpip\Parameters\Interfaces\<in-
terface>
Value Type: REG_DWORD—Boolean
Valid Range: 0, 1 (False, True)
Default: 1 (True)
```

Description: This parameter enables or disables IP autoconfiguration. This parameter can be set globally or per interface. If a per-interface value is present, it overrides the global value for that interface.

IPAutoconfigurationMask
```
Key: Tcpip\Parameters, Tcpip\Parameters\Interfaces\
  <interface>
Value Type: REG_SZ—String
Valid Range: A valid IP subnet mask
Default: 255.255.0.0
```

Description: This parameter controls the subnet mask assigned to the client by autoconfiguration. This parameter can be set globally or per interface. If a per-interface value is present, it overrides the global value for that interface.

IPAutoconfigurationSeed
```
Key: Tcpip\Parameters, Tcpip\Parameters\Interfaces\
  <interface>
Value Type: REG_DWORD—Number
Valid Range: 0-0xFFFF
Default: 0
```

Description: This parameter is used internally by the DHCP client and should not be modified.

IPAutoconfigurationSubnet
```
Key: Tcpip\Parameters, Tcpip\Parameters\Interfaces\<in-
terface>
Value Type: REG_SZ—String
Valid Range: A valid IP subnet
Default: 169.254.0.0
```

Description: This parameter controls the subnet address used by auto-configuration to pick an IP address for the client. This parameter can be set globally or per interface. If a per-interface value is present, it overrides the global value for that interface.

IGMPLevel
```
Key: Tcpip\Parameters
Value Type: REG_DWORD—Number
Valid Range: 0,1,2
Default: 2
```

Description: This parameter determines to what extent the system supports IP multicasting and participates in the Internet Group Management Protocol. At level 0, the system provides no multicast support. At level 1, the system can send IP multicast packets but cannot receive them. At level 2, the system can send IP multicast packets and fully participate in IGMP to receive multicast packets.

IPEnableRouter
```
Key: Tcpip\Parameters
Value Type: REG_DWORD—Boolean
Valid Range: 0, 1 (False, True)
Default: 0 (False)
```

Description: Setting this parameter to 1 (True) causes the system to route IP packets between the networks to which it is connected.

IPEnableRouterBackup
```
Key: Tcpip\Parameters
Value Type: REG_DWORD—Boolean
Valid Range: 0, 1 (False, True)
Default: 0 (False)
```

Description: Setup writes the previous value of *IPEnableRouter* to this key. It should not be adjusted manually.

KeepAliveInterval
```
Key: Tcpip\Parameters
Value Type: REG_DWORD—Time in milliseconds
Valid Range: 1–0xFFFFFFFF
Default: 1000 (one second)
```

Description: This parameter determines the interval between keep-alive retransmissions until a response is received. Once a response is received, the delay until the next keep-alive transmission is again controlled by the value of *KeepAliveTime*. The connection is aborted after the number of retransmissions specified by *TcpMaxDataRetransmissions* has gone unanswered.

KeepAliveTime
```
Key: Tcpip\Parameters
Value Type: REG_DWORD—Time in milliseconds
Valid Range: 1-0xFFFFFFFF
Default: 7,200,000 (two hours)
```

Description: The parameter controls how often TCP attempts to verify that an idle connection is still intact by sending a keep-alive packet. If the remote system is still reachable and functioning, it acknowledges the keep-alive transmission. Keep-alive packets are not sent by default. This feature may be enabled on a connection by an application.

MaxForwardBufferMemory
```
Key: Tcpip\Parameters
Value Type: REG_DWORD—Number of bytes
Valid Range: <network MTU>-0xFFFFFFFF
Default: 2097152 decimal (2 MB)
```

Description: This parameter limits the total amount of memory that IP can allocate to store packet data in the router packet queue. This value must be greater than or equal to the value of the *ForwardBufferMemory* parameter.

MaxForwardPending
```
Key:  Tcpip\Parameters\Interfaces\<interface>
Value Type: REG_DWORD—Number of packets
Valid Range: 1-0xFFFFFFFF
Default: 0x1388 (5000 decimal)
```

Description: This parameter limits the number of packets that the IP forwarding engine can submit for transmission to a specific network interface at any time. Additional packets are queued in IP until outstanding transmissions on the interface complete. Most network adapters transmit packets very quickly, so the default value is sufficient. A single RAS interface, however, may multiplex many slow serial lines. Configuring a larger value for this type of interface may improve its performance. The appropriate value depends on the number of outgoing lines and their load characteristics.

MaxFreeTcbs
```
Key: Tcpip\Parameters
```

```
Value Type: REG_DWORD—Number
Valid Range: 0-0xFFFFFFFF
Default: The following default values are used. (Note
   that Small is defined as a computer with less than19 MB
   of RAM, Medium is 19-63 MB of RAM, and Large is 64 MB
   or more of RAM. Although this code still exists, nearly
   all computers are Large now.)
```

For Windows NT Server:

```
Small system—500
Medium system—1000
Large system—2000
For Windows NT Workstation:
Small system—250
Medium system—500
Large system—1000
```

Description: This parameter controls the number of cached (preallocated) Transport Control Blocks (TCBs) that are available. A Transport Control Block is a data structure that is maintained for each TCP connection.

MaxFreeTWTcbs
```
Key: Tcpip\Parameters
Value Type: REG_DWORD-Number
Valid Range: 1-0xFFFFFFFF
Default: 1000
```

Description: This parameter controls the number of Transport Control Blocks (TCBs) in the TIME-WAIT state that is allowed on the TIME-WAIT state list. Once this number is exceeded, the oldest TCB will be scavenged from the list. In order to maintain connections in the TIME-WAIT state for at least 60 seconds, this value should be >= (60 * (the rate of graceful connection closures per second)) for the computer. The default value is adequate for most cases.

MaxHashTableSize
```
Key: Tcpip\Parameters
Value Type: REG_DWORD—Number (must be a power of 2)
Valid Range: 0x40-0x1000 (64-65536 decimal)
Default: 512
```

Description: This value should be set to a power of 2 (for example, 512, 1024, 2048, and so on.) If this value is not a power of 2, the system configures the hash table to the next power of 2 value (for example, a setting of

513 is rounded up to 1024.) This value controls how fast the system can find a TCP control block and should be increased if *MaxFreeTcbs* is increased from the default.

MaxNormLookupMemory
```
Key: Tcpip\Parameters
Value Type: REG_DWORD—Number
Valid Range: Any DWORD (0xFFFFFFFF means no limit on
memory)
Default: The following default values are used. (Small
  is defined as a computer with less than19 MB of RAM,
  Medium is 19-63 MB of RAM, and Large is 64 MB or more
  of RAM. Although this code still exists, nearly all
  computers are Large now.)
```

For Windows NT Server:
```
Small system—150,000 bytes, which accommodates 1000
  routes.
Medium system—1,500,000 bytes, which accommodates 10,000
  routes.
Large system—5,000,000 bytes, which accommodates 4,0000
  routes.
Windows NT Workstation:
150,000 bytes, which accommodates 1000 routes.
```

Description: This parameter controls the maximum amount of memory that the system allows for the route table data and the routes themselves. It is designed to prevent memory exhaustion on the computer caused by adding large numbers of routes.

MaxNumForwardPackets
```
Key: Tcpip\Parameters
Value Type: REG_DWORD—Number
Valid Range: 1-0xFFFFFFFF
Default: 0xFFFFFFFF
```

Description: This parameter limits the total number of IP packet headers that can be allocated for the router packet queue. This value must be greater than or equal to the value of the *NumForwardPackets* parameter.

MaxUserPort
```
Key: Tcpip\Parameters
Value Type: REG_DWORD—Maximum port number
Valid Range: 5000-65534 (decimal)
Default: 0x1388 (5000 decimal)
```

Description: This parameter controls the maximum port number used when an application requests any available user port from the system. Normally, short-lived ports are allocated in the range from 1024 through 5000. Setting this parameter to a value outside the valid range causes the nearest valid value to be used (5000 or 65534).

MTU
```
Key:   Tcpip\Parameters\Interfaces\<interface>
Value Type: REG_DWORD—Number
Valid Range: 68-<the MTU of the underlying network>
Default: 0xFFFFFFFF
```

Description: This parameter overrides the default Maximum Transmission Unit (MTU) for a network interface. The MTU is the maximum packet size, in bytes, that the transport can transmit over the underlying network. The size includes the transport header. An IP datagram can span multiple packets. Values larger than the default for the underlying network cause the transport to use the network default MTU. Values smaller than 68 cause the transport to use an MTU of 68.

Windows NT TCP/IP uses PMTU detection by default and queries the NIC driver to find out what local MTU is supported. Altering the MTU parameter is generally not necessary and may result in reduced performance.

NumForwardPackets
```
Key: Tcpip\Parameters
Value Type: REG_DWORD—Number
Valid Range: 1-<some reasonable value smaller than
0xFFFFFFFF>
Default: 0x32 (50 decimal)
```

Description: This parameter determines the number of IP packet headers that are allocated for the router packet queue. When all headers are in use, the system attempts to allocate more, up to the value configured for *MaxNumForwardPackets*. This value should be at least as large as the *ForwardBufferMemory* value divided by the maximum IP data size of the networks that are connected to the router. It should be no larger than the *ForwardBufferMemory* value divided by 256 because at least 256 bytes of forward buffer memory are used for each packet. The optimal number of forward packets for a given *ForwardBufferMemory* size depends on the type of traffic that is carried on the network and is somewhere between these two values. This parameter is ignored and no headers are allocated if routing is not enabled.

NumTcbTablePartitions
```
Key:   Tcpip\Parameters\
Value Type: REG_DWORD—Number of TCB table partitions
Valid Range: 1-0xFFFF
Default: 4
```

Description: This parameter controls the number of TCB table partitions. The TCB table can be partitioned to improve scalability on multiprocessor systems by reducing contention on the TCB table. This value should not be modified without careful performance study. A suggested maximum value is (Number of CPUs) * 2.

PerformRouterDiscovery
```
Key:   Tcpip\Parameters\Interfaces\<interface>
Value Type: REG_DWORD—BOOLEAN
Valid Range: 0, 1 (False, True)   (At RC1, will be 0, 1,
2)
Default: 1 (True) for Beta 3. Slated to change to in RC1
  to 2, DHCP-controlled but off by default.
```

Description: This parameter controls whether Windows NT attempts to perform router discovery per RFC 1256 on a per-interface basis. See also *SolicitationAddressBcast.*

PerformRouterDiscoveryBackup
```
Key:   Tcpip\Parameters\Interfaces\<interface>
Value Type: REG_DWORD—Boolean
Valid Range: 0, 1 (False, True)
Default: none
```

Description: This parameter is used internally to keep a backup copy of the *PerformRouterDiscovery* value. It should not be modified.

PPTPTcpMaxDataRetransmissions
```
Key: Tcpip\Parameters
Value Type: REG_DWORD—Number of times to retransmit a
  PPTP packet
Valid Range: 0-0xFF
Default: 5
```

Description: This parameter controls the number of times that a PPTP packet is retransmitted if it is not acknowledged. This parameter was added to allow retransmission of PPTP traffic to be configured separately from regular TCP traffic.

SackOpts
```
Key: Tcpip\Parameters
Value Type: REG_DWORD—Boolean
Valid Range: 0, 1 (False, True)
Default: 1 (True)
```

Description: This parameter controls whether Selective Acknowledgment (SACK, specified in RFC 2018) support is enabled.

SolicitationAddressBcast
```
Key:  Tcpip\Parameters\Interfaces\<interface>
Value Type: REG_DWORDBoolean
Valid Range: 0, 1 (False, True)
Default: 0 (False)
```

Description: This parameter can be used to configure Windows NT to send router discovery messages as broadcasts, instead of multicasts, as described in RFC 1256. By default, if router discovery is enabled, router discovery solicitations are sent to the all-routers multicast group (224.0.0.2). See also *PerformRouterDiscovery*.

SynAttackProtect
```
Key:  Tcpip\Parameters
Value Type: REG_DWORD—Boolean
Valid Range: 0, 1 (False, True)
Default: 0 (False)
```

Description: When enabled, this parameter causes TCP to adjust the retransmission of SYN-ACKS to cause connection responses to time out more quickly if it appears that there is a SYN-ATTACK in progress. This determination is based on the *TcpMaxPortsExhausted* parameter.

Tcp1323Opts
```
Key: Tcpip\Parameters
Value Type: REG_DWORD—Number (flags)
Valid Range: 0,1,2,3
0 (disable RFC 1323 options)
1 (window scale enabled only)
2 (timestamps enabled only)
3 (both options enabled)
Default: 0 (RFC 1323 options disabled)
```

Description: This parameter controls RFC 1323 Timestamps and window-scaling options. Timestamps and window scaling are enabled by default, but they can be manipulated with flag bits. Bit 0 controls window scaling, and bit 1 controls Timestamps.

TcpDelAckTicks
```
Key: Tcpip\Parameters\Interfaces\<interface>
Value Type: REG_DWORD—Number
Valid Range: 0-6
Default: 2 (200 milliseconds)
```

Description: This parameter specifies the number of 100-millisecond intervals to use for the delayed-ACK timer on a per-interface basis. By default, the delayed-ACK timer is 200 milliseconds. Setting this value to 0 disables delayed acknowledgments, which causes the computer to immediately ACK every packet it receives. Microsoft does not recommend changing this value from the default without careful study of the environment.

TcpInitialRTT
```
Key: Tcpip\Parameters\Interfaces\<interface>
Value Type: REG_DWORD—Number
Valid Range: 0-0xFFFF
Default: 3 seconds
```

Description: This parameter controls the initial time-out used for a TCP connection request and initial data retransmission on a per-interface basis. Use caution when tuning with this parameter because exponential backoff is used. Setting this value larger than 3 results in much longer time-outs to nonexistent addresses.

TcpMaxConnectResponseRetransmissions
```
Key: Tcpip\Parameters
Value Type: REG_DWORD—Number
Valid Range: 0-255
Default: 2
```

Description: This parameter controls the number of times that a SYN-ACK is retransmitted in response to a connection request if the SYN is not acknowledged. If this value is greater than or equal to 2, the stack employs SYN-ATTACK protection internally. If this value is less than 2, the stack does not read the Registry values at all for SYN-ATTACK protection. See also *SynAttackProtect*, *TCPMaxPortsExhausted*, *TCPMaxHalfOpen*, and *TCPMaxHalfOpenRetried*.

TcpMaxConnectRetransmissions

```
Key: Tcpip\Parameters
Value Type: REG_DWORD—Number
Valid Range: 0-255 (decimal)
Default: 3
```

Description: This parameter determines the number of times that TCP retransmits a connect request (SYN) before aborting the attempt. The retransmission time-out is doubled with each successive retransmission in a given connect attempt. The initial time-out is controlled by the *TcpInitialRtt* Registry value.

TcpMaxDataRetransmissions

```
Key: Tcpip\Parameters
Value Type: REG_DWORD—Number
Valid Range: 0-0xFFFFFFFF
Default: 5
```

Description: This parameter controls the number of times that TCP retransmits an individual data segment (not connection request segments) before aborting the connection. The retransmission time-out is doubled with each successive retransmission on a connection. It is reset when responses resume. The Retransmission Timeout (RTO) value is dynamically adjusted, using the historical measured round-trip time (Smoothed Round Trip Time, or SRTT) on each connection. The starting RTO on a new connection is controlled by the *TcpInitialRtt* Registry value.

TcpMaxDupAcks

```
Key: Tcpip\Parameters
Value Type: REG_DWORD—Number
Valid Range: 1-3
Default: 2
```

Description: This parameter determines the number of duplicate ACKs that must be received for the same sequence number of sent data before fast retransmit is triggered to resend the segment that has been dropped in transit.

TcpMaxHalfOpen

```
Key: Tcpip\Parameters
Value Type: REG_DWORD—Number
Valid Range: 100-0xFFFF
Default: 100 (server), 500 (workstation)
```

Description: This parameter controls the number of connections in the SYN-RCVD state allowed before SYN-ATTACK protection begins to operate. See the *SynAttackProtect* parameter for more details.

TcpMaxHalfOpenRetried
```
Key: Tcpip\Parameters
Value Type: REG_DWORD—Number
Valid Range: 80-0xFFFF
Default: 80 (server), 400 (workstation)
```

Description: This parameter controls the number of connections in the SYN-RCVD state for which there has been at least one retransmission of the SYN sent, before SYN-ATTACK attack protection begins to operate. See the *SynAttackProtect* parameter for more details.

TcpMaxPortsExhausted
```
Key: Tcpip\Parameters
Value Type: REG_DWORD—Number
Valid Range: 0-0xFFFF
Default: 5
```

Description: This parameter controls the point at which SYN-ATTACK protection starts to operate. SYN-ATTACK protection begins to operate when *TcpMaxPortsExhausted* connect requests have been refused by the system because the available backlog for connections is at 0.

TcpNumConnections
```
Key: Tcpip\Parameters
Value Type: REG_DWORD—Number
Valid Range: 0-0xFFFFFE
Default: 0xFFFFFE
```

Description: This parameter limits the maximum number of connections that TCP can have open simultaneously.

TcpTimedWaitDelay
```
Key: Tcpip\Parameters
Value Type: REG_DWORD—Time in seconds
Valid Range: 30-300 (decimal)
Default: 0xF0 (240 decimal)
```

Description: This parameter determines the length of time that a connection stays in the TIME_WAIT state when being closed. While a connection is in the TIME_WAIT state, the socket pair cannot be reused. This is also known as the 2MSL state because the value should be twice the maximum segment lifetime on the network. See RFC 793 for further details.

TcpUseRFC1122UrgentPointer
```
Key: Tcpip\Parameters
Value Type: REG_DWORD—Boolean
Valid Range: 0, 1 (False, True)
Default: 0 (False)
```

Description: This parameter determines whether TCP uses the RFC 1122 specification for urgent data or the mode used by BSD-derived systems. The two mechanisms interpret the urgent pointer in the TCP header and the length of the urgent data differently. They are not interoperable. Windows NT defaults to BSD mode.

TcpWindowSize
```
Key: Tcpip\Parameters, Tcpip\Parameters\Interface\
  <interface>
Value Type: REG_DWORD—Number of bytes
Valid Range: 0-0x3FFFFFFF (1073741823 decimal). Values
  greater than 64 KB can be achieved only when connect-
  ing to other systems that support RFC 1323 Window
  Scaling.
Default: The smaller of the following values:
0xFFFF
```

GlobalMaxTcpWindowSize (another Registry parameter)

The larger of four times the maximum TCP data size on the network 16384 rounded up to an even multiple of the network TCP data size.

The default can start at 17520 for Ethernet, but it may shrink slightly when the connection is established to another computer that supports extended TCP head options, such as SACK and TIMESTAMPS, because these options increase the TCP header beyond the usual 20 bytes, leaving slightly less room for data.

Description: This parameter determines the maximum TCP receive window size offered. The receive window specifies the number of bytes that a sender can transmit without receiving an acknowledgment. In general, larger receive windows improve performance over high-delay, high-bandwidth networks. For greatest efficiency, the receive window should be an even multiple of the TCP Maximum Segment Size (MSS). This parameter is both a per-interface parameter and a global parameter, depending on where the Registry key is located. If there is a value for a specific interface, that value overrides the system-wide value. See also *GobalMaxTcpWindowSize*.

TrFunctionalMcastAddress
```
Key: Tcpip\Parameters
Value Type: REG_DWORD—Boolean
Valid Range: 0, 1 (False, True)
Default: 1 (True)
```

Description: This parameter determines whether IP multicasts are sent using the Token Ring Multicast address described in RFC 1469 or using the subnet broadcast address. The default value of 1 configures the computer to use the RFC1469 Token Ring Multicast address for IP multicasts. Setting the value to 0 configures the computer to use the subnet broadcast address for IP multicasts.

TypeOfInterface
```
Key: Tcpip\Parameters\Interfaces\<interface>
Value Type: REG_DWORD
Valid Range: 0, 1, 2, 3
Default: 0 (allow multicast and unicast)
```

Description: This parameter determines whether the interface gets routes plumbed for unicast, multicast, or both traffic types, and whether those traffic types can be forwarded. If it is set to 0, both unicast and multicast traffic are allowed. If it is set to 1, unicast traffic is disabled. If it is set to 2, multicast traffic is disabled. If it set to 3, both unicast and multicast traffic are disabled. Because this parameter affects forwarding and routes, it may still be possible for a local application to send multicasts out over an interface, if there are no other interfaces in the computer that are enabled for multicast, and a default route exists.

UseZeroBroadcast
```
Key:  Tcpip\Parameters\Interfaces\<interface>
Value Type: REG_DWORD—Boolean
Valid Range: 0, 1 (False, True)
Default: 0 (False)
```

Description: If this parameter is set to 1 (True), IP will use 0s broadcast (0.0.0.0) instead of 1s broadcast (255.255.255.255). Most systems use 1s broadcasts, but some systems derived from BSD implementations use 0s broadcasts. Systems that use different broadcasts do not interoperate well on the same network.

PARAMETERS CONFIGURABLE FROM THE USER INTERFACE

The following parameters are created and modified automatically by the NCPA as a result of user-supplied information. There should be no need to configure them directly in the Registry.

DefaultGateway

```
Key:  Tcpip\Parameters\Interfaces\<interface>
Value Type: REG_MULTI_SZ—List of dotted decimal IP ad-
dresses
Valid Range: Any set of valid IP addresses
Default: None
```

Description: This parameter specifies the list of gateways to be used to route packets that are not destined for a subnet to which the computer is directly connected, and for which a more specific route does not exist. This parameter, if it has a valid value, overrides the *DhcpDefaultGateway* parameter. There is only one active default gateway for the computer at any time, so adding multiple addresses is done only for redundancy.

Domain

```
Key: Tcpip\Parameters\Interfaces\<interface>
Value Type: REG_SZ—Character string
Valid Range: Any valid DNS domain name
Default: None
```

Description: This parameter specifies the DNS domain name of the interface. In Windows 2000, this and *NameServer* are per-interface parameters, rather than system-wide parameters. This parameter overrides the *DhcpDomain* parameter (filled in by the DHCP client), if it exists.

EnableDhcp

```
Key:  Tcpip\Parameters\Interfaces\<interface>
Value Type: REG_DWORD—Boolean
Valid Range: 0, 1 (False, True)
Default: 0 (False)
```

Description: If this parameter is set to 1 (True), the DHCP client service attempts to use DHCP to configure the first IP interface on this adapter.

EnableSecurityFilters

```
Key:  Tcpip\Parameters
Value Type: REG_DWORD—Boolean
```

```
Valid Range: 0, 1 (False, True)
Default: 0 (False)
```

Description: If this parameter is set to 1 (True), IP security filters are enabled. See *TcpAllowedPorts*, *UdpAllowedPorts*, and *RawIPAllowedPorts*. To configure these values, on the Start menu, point to Settings, click Network and Dial-up Connections, right-click Local Area Connection, and then click Properties. Select Internet Protocol (TCP/IP), click Properties, and then click Advanced. Click the Options tab, select TCP/IP filtering, and click Properties.

Hostname
```
Key: Tcpip\Parameters
Value Type: REG_SZ—Character string
Valid Range: Any valid DNS hostname
Default: The computer name of the system
```

Description: This parameter specifies the DNS host name of the system, which is returned by the hostname command.

IPAddress
```
Key: Tcpip\Parameters\Interfaces\<interface>
Value Type: REG_MULTI_SZ—List of dotted-decimal IP ad-
dresses
Valid Range: Any set of valid IP addresses
Default: None
```

Description: This parameter specifies the IP addresses of the IP interfaces to be bound to the adapter. If the first address in the list is 0.0.0.0, the primary interface on the adapter is configured from DHCP. A system with more than one IP interface for an adapter is *logically multihomed*. There must be a valid subnet mask value in the *SubnetMask* parameter for each IP address that is specified in this parameter. To add parameters with Regedt32.exe, select this key and type the list of IP addresses, pressing ENTERS after each one. Then go to the *SubnetMask* parameter, and type a corresponding list of subnet masks.

NameServer
```
Key: Tcpip\Parameters\Interfaces\<interface>
Value Type: REG_SZ—A space delimited list of dotted dec-
imal IP addresses
Valid Range: Any set of valid IP address
Default: None (Blank)
```

Description: This parameter specifies the DNS name servers that Windows Sockets queries to resolve names. In Windows 2000, this and the *DomainName* are per-interface settings.

PPTPFiltering
```
Key: Tcpip\Parameters\Interfaces\<interface>
Value Type: REG_DWORD—Boolean
Valid Range: 0, 1 (False, True)
Default: 0 (False)
```

Description: This parameter controls whether PPTP filtering is enabled on a per-adapter basis. If this value is set to 1, the adapter accepts only PPTP connections. This reduces exposure to hack attempts if the adapter is connected to a public network, such as the Internet.

RawIpAllowedProtocols
```
Key: Tcpip\Parameters\Interfaces\<interface>
Value Type: REG_MULTI_SZ—List of IP protocol numbers
Valid Range: Any set of valid IP protocol numbers
Default: None
```

Description: This parameter specifies the list of IP protocol numbers for which incoming datagrams are accepted on an IP interface when security filtering is enabled (*EnableSecurityFilters* = 1). The parameter controls the acceptance of IP datagrams by the raw IP transport, which is used to provide raw sockets. It does not control IP datagrams that are passed to other transports (for example, TCP). An empty list indicates that no values are acceptable. A single value of 0 indicates that all values are acceptable. The behavior of a list containing the value 0 mixed with other, nonzero values is undefined. If this parameter is missing from an interface, all values are acceptable. This parameter applies to all IP interfaces that are configured on a specific adapter.

SearchList
```
Key: Tcpip\Parameters
Value Type: REG_SZ—Space delimited list of DNS domain-
  name suffixes
Valid Range: Any set of valid DNS domain-name suffixes
Default: None
```

Description: This parameter specifies a list of domain-name suffixes to append to a name to be resolved through DNS if resolution of the unadorned name fails. By default, only the value of the *Domain* parameter is appended. This parameter is used by the Windows Sockets interface. See also the *AllowUnqualifiedQuery* parameter.

SubnetMask
```
Key:   Tcpip\Parameters\Interfaces\<interface>
Value Type: REG_MULTI_SZ—List of dotted decimal IP ad-
dresses
Valid Range: Any set of valid IP addresses.
Default: None
```

Description: This parameter specifies the subnet masks to be used with the IP interfaces bound to the adapter. If the first mask in the list is 0.0.0.0, the primary interface on the adapter is configured using DHCP. There must be a valid subnet mask value in this parameter for each IP address specified in the *IPAddress* parameter.

TcpAllowedPorts
```
Key:   Tcpip\Parameters\Interfaces\<interface>
Value Type: REG_MULTI_SZ—List of TCP port numbers
Valid Range: Any set of valid TCP port numbers
Default: None
```

Description: This parameter specifies the list of TCP port numbers for which incoming SYNs are accepted on an IP interface when security filtering is enabled (*EnableSecurityFilters* = 1). An empty list indicates that no values are acceptable. A single value of 0 indicates that all values are acceptable. The behavior of a list containing the value 0 mixed with other, nonzero values is undefined. If this parameter is missing from an interface, all values are acceptable. This parameter applies to all IP interfaces configured on a specified adapter.

UdpAllowedPorts
```
Key: Tcpip\Parameters\Interfaces\<interface>
Value Type: REG_MULTI_SZ—List of UDP port numbers
Valid Range: Any set of valid UDP port numbers
Default: None
```

Description: This parameter specifies the list of UDP port numbers for which incoming datagrams are accepted on an IP interface when security filtering is enabled (*EnableSecurityFilters* = 1). An empty list indicates that no values are acceptable. A single value of 0 indicates that all values are acceptable. The behavior of a list containing the value 0 mixed with other, nonzero values is undefined. If this parameter is missing from an interface, all values are acceptable. This parameter applies to all IP interfaces configured on a specified adapter.

PARAMETERS CONFIGURABLE USING THE ROUTE COMMAND

The route command can store persistent IP routes as values under the Tcpip\Parameters\PersistentRoutes Registry key. Each route is stored in the value name string as a comma-delimited list of the form:

destination,subnet mask,gateway,metric
For example, the command:

```
route add 10.99.100.0 MASK 255.255.255.0 10.99.99.1 MET-
RIC 1 /p
produces the Registry  value:
10.99.100.0,255.255.255.0,10.99.99.1,1
```

The value type is a REG_SZ. There is no value data (empty string). Addition and deletion of these values can be accomplished using the route command. There should be no need to configure them directly.

NON-CONFIGURABLE PARAMETERS

The following parameters are created and used internally by the TCP/IP components. They should never be modified using the Registry Editor. They are listed here for reference only.

Dhc5pDefaultGateway
```
Key:  Tcpip\Parameters\Interfaces\<interface>
Value Type: REG_MULTI_SZ—List of dotted decimal IP ad-
dresses
Valid Range: Any set of valid IP addresses
Default: None
```

Description: This parameter specifies the list of default gateways to be used to route packets that are not destined for a subnet to which the computer is directly connected and for which a more specific route does not exist. This parameter is written by the DHCP client service, if enabled. This parameter is overridden by a valid *DefaultGateway* parameter value. Although this parameter is set on a per-interface basis, there is always only one default gateway active for the computer. Additional entries are treated as alternatives if the first one is down.

DhcpIPAddress
```
Key: Tcpip\Parameters\Interfaces\<interface>
Value Type: REG_SZ—Dotted decimal IP address
Valid Range: Any valid IP address
Default: None
```

Description: This parameter specifies the DHCP-configured IP address for the interface. If the *IPAddress* parameter contains a first value other than 0.0.0.0, that value overrides this parameter.

DhcpDomain
```
Key: Tcpip\Parameters\Interfaces\<interface>
Value Type: REG_SZ—Character string
Valid Range: Any valid DNS domain name
Default: None (provided by DHCP server)
```

Description: This parameter specifies the DNS domain name of the interface. In Windows 2000, this and *NameServer* are now per-interface parameters, rather than system-wide parameters. If the *Domain* key exists, it overrides the *DhcpDomain* value.

DhcpNameServer
```
Key: Tcpip\Parameters
Value Type: REG_SZ—A space delimited list of dotted dec-
  imal IP addresses
Valid Range: Any set of valid IP address
Default: None
```

Description: This parameter specifies the DNS name servers to be queried by Windows Sockets to resolve names. It is written by the DHCP client service, if enabled. If the *NameServer* parameter has a valid value, it overrides this parameter.

DhcpServer
```
Key: Tcpip\Parameters\Interfaces\<interface>
Value Type: REG_SZ—Dotted decimal IP address
Valid Range: Any valid IP address
Default: None
```

Description: This parameter specifies the IP address of the DHCP server that granted the lease on the IP address in the *DhcpIPAddress* parameter.

DhcpSubnetMask
```
Key: Tcpip\Parameters\Interfaces\<interface>
Value Type: REG_SZ—Dotted decimal IP subnet mask
Valid Range: Any subnet mask that is valid for the
  configured IP address
Default: None
```

Description: This parameter specifies the DHCP-configured subnet mask for the address specified in the *DhcpIPAddress* parameter.

DhcpSubnetMaskOpt
```
Key: Tcpip\Parameters\Interfaces\<interface>
Value Type: REG_SZ—Dotted decimal IP subnet mask
Valid Range: Any subnet mask that is valid for the
  configured IP address
Default: None
```

Description: This parameter is filled in by the DHCP client service and is used to build the *DhcpSubnetMask* parameter, which the stack actually uses. Validity checks are performed before the value is inserted into the *DhcpSubnetMask* parameter.

Lease
```
Key: Tcpip\Parameters\Interfaces\<interface>
Value Type: REG_DWORD—Time in seconds
Valid Range: 1-0xFFFFFFFF
Default: None
```

Description: The DHCP client service uses this parameter to store the time, in seconds, for which the lease on the IP address for this adapter is valid.

LeaseObtainedTime
```
Key: Tcpip\Parameters\Interfaces\<interface>
Value Type: REG_DWORD—Absolute time, in seconds, since
midnight of 1/1/70
Valid Range: 1-0xFFFFFFFF
Default: None
```

Description: The DHCP client service uses this parameter to store the time at which the lease on the IP address for this adapter was obtained.

LeaseTerminatesTime
```
Key: Tcpip\Parameters\Interfaces\<interface>
Value Type: REG_DWORD—Absolute time, in seconds, since
  midnight of 1/1/70
```

```
Valid Range: 1-0xFFFFFFFF
Default: None
```

Description: The DHCP client service uses this parameter to store the time at which the lease on the IP address for this adapter expires.

LLInterface
```
Key: Tcpip\Parameters\Adapters\<interface>
Value Type: REG_SZ—Windows NT device name
Valid Range: A legal Windows NT device name
Default: Empty string (Blank)
```

Description: This parameter is used to direct IP to bind to a different link-layer protocol than the built-in ARP module. The value of the parameter is the name of the Windows NT device to which IP should bind. This parameter is used in conjunction with the RAS component; for example. It is present only when ARP modules other than LAN bind to IP.

NTEContextList
```
Key:  Tcpip\Parameters\Interfaces\<interface>
Value Type: REG_MULTI_SZ—Number
Valid Range: 0-0xFFFF
Default: none
```

Description: This parameter identifies the context of the IP address associated with an interface. Each IP address associated with an interface has its own context number. The values are used internally to identify an IP address and should not be altered.

T1
```
Key: Tcpip\Parameters\Interfaces\<interface>
Value Type: REG_DWORD—Absolute time, in seconds, since
midnight of 1/1/70
Valid Range: 1-0xFFFFFFFF
Default: None
```

Description: The DHCP client service uses this parameter to store the time at which the service first tries to renew the lease on the IP address for the adapter by contacting the server that granted the lease.

T2
```
Key: Tcpip\Parameters\Interfaces\<interface>
Value Type: REG_DWORD—Absolute time, in seconds, since
midnight of 1/1/70
```

```
Valid Range: 1-0xFFFFFFFF
Default: None
```

Description: The DHCP client service uses this parameter to store the time at which the service tries to renew the lease on the IP address for the adapter by broadcasting a renewal request. Time *T2* should be reached only if the service is unable to renew the lease with the original server for some reason.

Within any administrator's daily routine there is the need for getting problems solved quickly. Having a library of software tools to draw from for quickly solving problems is invaluable for a system administrator. That's the purpose of the CD included with this book, it provides an entire series of tools for you to use in solving common problems that occur when supporting a workgroup, network, or an entire organization.

One of the foremost issues administrators deal with is security. Included on the CD are demonstration versions of applications that can secure desktops from unwanted access, whether they are connected to the Internet via DSL, cable modem, or T1 connections. These software firewalls have been tested and work seamlessly with Windows 2000 Professional. Taken together, the utilities on the CD provide you with quick access to valuable applications that have the potential of streamlining the tasks associated with keeping a network up and running securely.

IMPORTANT: *Please read the software licenses for the respective products for minimum system requirements and limitations on trial version periods.*

PDG Shopping Cart (trial demo)
PDG Software Inc.
http://www.pdgsoft.com

Included on this CD-ROM is a free 30-day trial version of PDG Shopping Cart for both Microsoft Windows NT and multiple UNIX platforms.

PDG Shopping Cart is PDG Software's core product and serves as a complete e-commerce solution. Some of its most notable features include real-time shipping cost calculation from UPS, easily customizable HTML templates, compatibility with multiple online payment processors, and the ability to sell softgoods (i.e. downloadable products such as computer software, electronic music files, and electronic documents).

Other Products From PDG Software

PDG Software offers a full suite of e-commerce solutions in addition to its PDG Shopping Cart application. The PDG product suite includes PDG Shopping Mall, PDG Distributor, PDG Auction and PDG Commerce. For more information, or trial downloads, please visit www.pdgsoft .com.

About PDG Software

PDG Software, Inc. provides Internet commerce solutions for small to mid-size companies. Based in the Atlanta area, the company was formed in 1997 and has a growing list of customers in the United States and abroad. Its core product, the PDG Shopping Cart, allows customers to create a customized Internet commerce web site. PDG Software, Inc. offers a complete suite of e-commerce products that includes PDG Shopping Cart, PDG Shopping Mall, PDG Distributor, PDG Auction, and PDG Commerce. The company also offers merchants a 30-day free trial period for its software and has been lauded for its customer-centric focus. PDG Software Inc. is a certified Microsoft [NASDAQ: MSFT] Solutions Provider and a certified Sun Catalyst Developer [NASDAQ: SUNW]. For more information about PDG Software Inc. and its suite of products, please visit the company's web site at www.pdgsoft.com.

Go!Zilla 3.92 Download Manager (trial version)
Radiate Corporation
http://www.radiate.com

Go!Zilla can download file clicks from your favorite browser, or you can drag and drop URLs. You can then download some or all of the files, schedule downloads for later when bandwidths have died down, and even resume broken transfers. Right-click a link for a host of options: download, schedule, edit, launch, delete, run SmartUpdate, or remove file from list. SmartUpdate searches the Web for the latest version of any file you download. Further features include Super Link Leech (to grab all the file links on a page) and Monster Update (which watches for updates to favorites), plus searching. Go!Zilla also offers subscriptions to channels with specific content. Configuration options abound: Set the program to auto-disconnect, or even auto-shutdown, after download. Also included is ZipZilla, an unarchiver/installer.

ZoneAlarm' (trial version)
Zone Labs, Inc.
http://www.zonelabs.com/

ZoneAlarm is an easy-to-use Internet security utility that sets up a personal firewall that's particularly well suited to DSL and cable modem users. Computers with such an always-on connection have a permanent IP address, making them especially vulnerable to information theft and other attacks. ZoneAlarm lets you select one of three security levels to separately apply to local and Internet traffic. In addition, you can designate which programs on your computer are allowed to access the Internet. Anytime an event occurs that ZoneAlarm blocks, you're notified with a pop-up window that details the offense and asks how to respond to future occurrences. The program also provides integrated protection against VB script email attachments. Its system-tray icon even provides a handy menu that enables you to disable all Internet activity with a mouse click. You can also set this "lock up" to activate automatically after a selected period of inactivity or whenever your screen saver activates. Detailed logging of all activity is maintained.

RegRun II v2.7 (trial version)
Greatis Software
http://www.greatis.com

RegRun II is a superb tool that enables you to keep a very close eye on the programs that automatically load with Windows. Its tabbed interface provides access to programs that start from the Registry, Win.ini, Config.sys, Autoexec.bat, Winstart.bat, NT Services, and your Startup folders. You can easily add and remove items from the Registry and Startup folders and edit your Config.sys and Autoexec.bat files. Beyond that, RegRun II offers a degree of security against Trojans (including all variants of the infamous Love Letter virus) that load when Windows starts. Each time it loads, any and all differences found are brought to your attention. Better yet, you can have the program load automatically with Windows and enable selective monitoring to make sure that no unauthorized programs are set to launch. RegRun II lets you save and restore startup profiles as needed, and does so from DOS if the need arises. Other features include Registry shortcut generation, system file editing, and a fine process viewer. Requirements: Windows 9x, NT, Windows 2000 Professional, or Windows Me.

Sandra v2001.0.7.10 (trial version)
SiSoft Software
http://www.sisoftware.demon.co.uk/

Sandra provides in-depth benchmarks and low-level information about your Windows PC. It's an excellent way to sum up your system performance, but keep in mind that not many benchmarks use the same scales and that the results must be obtained by the same testing program to be compared. Resembling a second incarnation of Control Panel with more than 50 separate modules, Sandra can graphically display system statistics such as drive benchmarks, CPU speed ratings, and DOS memory and physical memory bank information. You can easily retrieve detailed information about your mouse, keyboard, video, CPU and BIOS, drives, ports, sound card, and more. A report wizard collects user-specified portions of Sandra's arsenal of tests and writes the result to a file, your printer, or a fax, or saves it as a script. You can also run the performance wizard to conduct a battery of tests and generate a comprehensive list of performance improvement tips.

MemWasher v1.1 (trial version)
Intellisoft, Inc.
http://www.intellisoft-inc.com/MemWasher.html

Use MemWasher to improve the performance of your PC. It's convenient, easy to set up, and has a positive impact on performance by making more real memory available on your computer. MemWasher runs quietly from your system tray and automatically frees real memory depending on your settings. You can make memory available at a regular interval of your choosing, or base the process on need, by setting a low threshold value. Of course, you can also free memory manually as needed. Total memory load and available memory are shown as text and as real-time graphs.

DB to HTML Express v1.4 (trial version)
XlineSoft Corporation
http://www.xlinesoft.com/

DB to HTML Express makes it a breeze to get data out of your ODBC database and onto the Web. Just choose your data source, select a format, and generate the HTML. DB to HTML Express handles any ODBC database, so you can use it with anything from Access to Informix. It even features support for exporting graphics from your database and performing graphics format conversions. You can also create and save your own templates to gain more control over the output. Once you've selected a table, you can control the output by entering a SQL statement or loading a SQL script to be run against the data. When you publish your data, you can choose between index-card or datasheet-style formats, with control over fonts, colors, and alignment.

HyperText Studio v3.03 (trial version)
Olson Software, Ltd.
http://www.olsonsoft.com/

HyperText Studio is a professionally designed, visual hypertext authoring tool. With it, you can create Websites without having to know HTML. You can also create Windows Help files and printed documentation. The program supports HTML 3.2, Windows Help 3.1 and 4.0, and HTML Help 1.0, and you need to learn only one set of tools to be able to create any of those formats. One of the great strengths of HyperText Studio is that it allows you to develop in any format and then switch at a later time. It displays your site/help structure on the left and provides an editing window on the right. It makes great use of drag-and-drop editing and context-sensitive toolbars and comes with excellent documentation. Requirements: 32 MB RAM, Pentium 100, and Windows 95, 98, or Windows2000 Professional.

Dr. Hardware 2000 v1.5.0e (trial version)
Gebhard Software
http://www.drhardware.de/

Using this utility you can step through the test results one by one or access individual reports by category — Hardware, Devices, PC Setup, Operating System, and Benchmarks. Use the program to retrieve detailed information about serial and parallel ports, mouse, joystick, drives, sound board, video adapter, BIOS, CMOS, and memory. Benchmarking results are shown as attractive 3D graphs that provide valuable comparisons against similar computers. You can selectively print the test and benchmark results to a custom report. Excellent documentation is included. Requirements: Windows 9x, NT, or 2000.

CIW Foundations Demo Practice Test
i-Net+ Demo Practice Test
MCSE Windows 2000 Demo Practice Test

Self Test Software
http://www.selftestsoftware.com

See individual demos on the CD-ROM for system requirements and information.

Self Test Software, a Kaplan company, produced the first practice test in the industry and has maintained its stature as leader in the development of IT practice tests since 1992. From their first Big Red Self Test for Novell certification, to today's extensive line of practice tests including Microsoft, Novell, Cisco, Lotus, Oracle, CIW and CompTIA (A+ and Network+, I-Net+), they provide a wide range of certification assistance.

For more information about Self Test Software practice tests, visit the Self Test Software Web site at http://www.selftestsoftware.com or call 1-770-643-3600

Index

Applets in, 15, 52 83, 112
hyperlinks on, 50–51
CRC (Cyclic Redundancy
Checks), 58–59
cross-compatibility of network
operating systems,
377–378
Currency Settings, 78
cursors, setting options, 70–71
Cyclic Redundancy Checks
(CRC), 58–59

D
3D graphics and Open GL, 377
data exchange and Universal Disk
Format (UDF), 18
data link layer of OSI, 186,
329–330
Data Sources (ODBC) Administrative Tool, 345
data synchronization and Intellimirror management capa-
bilities, 5
databases
DHCP vs. DNS database content, 216–217
Internet Database Connector,
292
querying via PWS, 292
Date/Time applet, 57–58
debug mode, 393–395
defragmentation utilities, 18
"Dellizing," 384–386
Desktop properties, 99–107
Display properties, 99–104
Shortcuts, creating, 104–107
device drivers
Adapter device driver,
100–104
compatibility of, 30
Input Locals for languages, 67
installation and Plug-and-Play,
31
location of, 29
printer drivers, 144, 150,
175–176
Device Manager snap-in, 17
DHCP (Dynamic Host Configuration Protocol). See Dy-

namic Host Configuration
Protocol (DHCP)
diagnostic tools. See
Troubleshooting
dial-up connections, 246–247
see also Dial-Up Networking
Dial-Up Networking
AutoDial, 270–271
dial-up settings, 269–270
installation, 252–255, 266–267
logging on, 269
phonebook entries, 267–268,
272
reconnection, automatic, 271
Remote Access Services (RAS)
and, 255–256
troubleshooting with Networking Monitor, 272
user profiles, 270
Dialing Properties, 118–119
directories, 279
Active Directory Service Interface (ADSI), 378
architectures of (differing), 336
DEN and directory service
functionality, 381
directory browsing, 311
Directory Service Migration
Tool, 378
Secure Sockets Layers (SSL)
required, 315
security and SSL, 315
WWW directory access,
308–310
Directory-Enabled Networking
(DEN), 381
Directory Service Manager Net-
Ware (DSMN), 277
DirectX 7.0, 377
Disk Management snap-in, 13
disk space, 18
diskless workstations and DHCP,
203–204
DISKPERF switches, 357
disks, defragmenting, 18
Distributed File System (Dfs), 9
DNS (Domain Name Services).
See Domain Name System
(DNS) services

#DOM LMHOSTS keyword,
225 226
Domain Name System (DNS)
services, 206–207, 378
basic description of, 215–217
database communication
values, 345
database content, 216–217
Dynamic DNS, 10
vs. Dynamic Host Configuration Protocol (DHCP),
216–217
HOSTS files, 207
initialization sequence, 216
Internet addresses, 249–250
Nslookup for troubleshooting,
393–395
user interface parameter configuration for, 419, 420
domains
domain models and Active
Directory, 373
domain names, 249–250
Internet vs. Windows NT
domains, 215–216
DOS
application compatibility,
121–122
emulation in Windows 2000,
122–124
memory configuration to
support, 136–138
DPA, security, 12
DPMI Memory, configuring,
137
drivers
DirectX 7.0 multimedia driver,
377
Win32 Driver Model, 20
Dynamic DNS, 10
Dynamic Host Configuration
Protocol (DHCP)
addressing in TCP/IP, 182,
203–204
as competitive advantage in
e-commerce, 336–337
configuring TCP/IP to support, 212–214
database content, 216–217